CITIZEN JOURNALISM

Global Perspectives
VOLUME 2

EDITED BY
Einar Thorsen and Stuart Allan

PETER LANG
New York • Washington, D.C./Baltimore • Bern
Frankfurt am Main • Berlin • Brussels

The Library of Congress has catalogued the first volume as follows:

Citizen journalism: global perspectives / edited by Stuart Allan, Einar Thorsen.
p. cm. —(Global crises and the media; vol. 1)
Includes bibliographical references and index.
1. Citizen journalism. 2. Online journalism. I. Title.
PN4784.C615 C58 070.4'3—dc22 2009012804
ISBN 978-1-4331-0296-7 (volume 1, hardcover)
ISBN 978-1-4331-0295-0 (volume 1, paperback)
ISBN 978-1-4331-2283-5 (volume 2, hardcover)
ISBN 978-1-4331-2282-8 (volume 2, paperback)
ISBN 978-1-4539-1357-4 (volume 2, e-book)
ISSN 1947-2587

Bibliographic information published by **Die Deutsche Bibliothek.**
Die Deutsche Bibliothek lists this publication in the "Deutsche Nationalbibliografie"; detailed bibliographic data is available on the Internet at http://dnb.ddb.de/.

The paper in this book meets the guidelines for permanence and durability of the Committee on Production Guidelines for Book Longevity of the Council of Library Resources.

29 Broadway, 18th floor, New York, NY 10006
www.peterlang.com

Printed in the United States of America

Table OF Contents

Section Two: Capturing Crisis

Section Three: Globalising Cultures of Citizen Journalism

Section Four: New Crises, Alternative Agendas

Series Editor's Preface

GLOBAL CRISES AND THE MEDIA

We live in a global age. We inhabit a world that has become radically interconnected, interdependent, and communicated in the formations and flows of the media. This same world also spawns proliferating, often interpenetrating, "global crises."

From climate change to the war on terror, financial meltdowns to forced migrations, pandemics to world poverty, and humanitarian disasters to the denial of human rights, these and other crises represent the dark side of our globalized planet. Their origins and outcomes are not confined behind national borders and they are not best conceived through national prisms of understanding. The impacts of global crises often register across "sovereign" national territories, surrounding regions and beyond, and they can also become subject to systems of governance and forms of civil society response that are no less encompassing or transnational in scope. In today's interdependent world, global crises cannot be regarded as exceptional or aberrant events only, erupting without rhyme or reason or dislocated from the contemporary world (dis)order. They are endemic to the contemporary global world, deeply enmeshed within it. And so too are they highly dependent on the world's media and communication networks.

The series *Global Crises and the Media* sets out to examine not only the media's role in the *communication* of global threats and crises but also how they can variously enter into their *constitution,* enacting them on the public stage and helping to shape their future trajectory around the world. More specifically, the volumes in this series seek to: (1) contextualize the study of global crisis reporting in relation to wider debates about the changing flows and formations of world media communication; (2) address how global crises become variously communicated and contested in both so-called "old" and "new" media around the world; (3) consider the possible impacts of global crisis reporting on public awareness, political action, and policy responses; (4) showcase the very latest research findings and discussion from leading authorities in their respective fields of inquiry; and (5) contribute to the development of positions of theory and debate that deliberately move beyond national parochialisms and/or geographically disaggregated research agendas. In these ways the specially commissioned books in the *Global Crises and the Media* series aim to provide a sophisticated and empirically engaged understanding of the media's changing roles in global crises and thereby contribute to academic and public debate about some of the most significant global threats, conflicts, and contentions in the world today.

Since the publication of *Citizen Journalism: Global Perspectives*, edited by Stuart Allan and Einar Thorsen, in 2009, the presence of citizen journalism in major events and crises around the world has continued apace. As more people become connected through social media and other self-directed forms of communication, and as increasing numbers of them seek to participate in or document the unfolding events that surround and embroil them, so increased traffic of images and ideas circumnavigate the globe. Sometimes with politically disruptive effects. Whether based on the deliberately pursued and circulated images by activists or the contingent scenes witnessed as bystanders, today's different forms of citizen journalism increasingly enter into the mainstream news arena, helping to invigorate news formats and, sometimes, challenging elite definitions of events. Evidently there is considerably more to citizen journalism than the media industry's preferred term of "user-generated content" (UGC). Paradoxically the inadequacies of this concept help to better illuminate citizen journalism.

UGC speaks to the media's commoditized interest in the entertainment value of humorous and quirky family videos or even the drama and human interest of recorded unexpected events, but it doesn't do justice to the more radical, pluralizing or mobilizing forms of communications that coalesce under the umbrella term of "citizen journalism." Originated in the professional world of broadcasting, "user-generated content" offers a particularly stunted and proprietorial view of "its" user audience and the corporate utility of "their" generated content. With its flattened view of content, UGC also distances the different aspirations informing new ways of engaging the media and the deliberate or performative nature of communications as forms of civic or political enactment. The industry view of UGC simply loses sight of the civic richness, motivations and appeals relayed in and through citizen journalism and offered for wider media circulation and representation—whether politically or textually conceived. The expanding rise of citizen journalism within major crises and conflicts around the world in recent years, such as the Arab Spring, mass protests against austerity and authoritarian rulers, or major disasters and bloody civil wars, all attest to these different and sometimes potent forms of political and civic enactment. They implicitly challenge the essentially technicist and politically evacuated notion of UGC that has long passed its corporate sell-by date. By contrast, "citizen journalism" notwithstanding the continuing need for further refined conceptualization and theorization signifies a more active, participatory and responsible sense of "being in the world," one that variously enacts and brings alive a sense of civic duty or civil society in action, whether as moral witness or agent of change. In comparison to the flat corporate view of UGC, then, citizen journalism couldn't be more different or plentiful in its meanings and progressive possibilities.

Notwithstanding earlier cynical claims, citizen journalism hasn't become entirely co-opted and corporately trivialized or rendered historically redundant by the prevailing logics of media production and consumption. With its continuing, expansive and deep imbrication within wider processes of societal reflexivity and change, citizen journalism enters into the world and makes its mark on it. To what extent and how clearly demands close empirical analyses as well as refined typological distinctions and theoretical frameworks capable of situating its evolving forms, practices and progressive charge, especially as it enters into a world of conflicts and crises. Einar Thorsen and Stuart Allan's *Citizen Journalism: Global Perspectives Volume 2,* can therefore only be welcomed. With the help of their expert contributors from different countries and regions, they make considerable headway into charting the recent sea change in citizen communications. For this, notwithstanding earlier historical precedents, is what it is—a communicative sea change. Providing up-to-the-moment analyses of citizen journalism in the world today, many based on case studies and reflecting on the institutional parameters, powered fields and progressive hopes that surround it, *Citizen Journalism: Global Perspectives Volume 2* interrogates actual practices and possible impacts of citizen journalism across diverse fields of struggle and in the responses to unfolding calamitous events. As the contributors amply document, citizen journalism around the world today assumes diverse forms and gives expression to different, often interconnected, world realities. It is practiced and performed in different political contexts and through different traditions, cultures and configurations of civil society, but it is essentially alive—experientially, humanistically, often politically, and on the move. And this is so notwithstanding those authoritarian regimes that would seek to censor and crush it or those corporate interests who would shrink its significance to the corporate free lunch of incoming "UGC."

Animated by differing conceptions of both "citizenship" and "journalism," and practiced under very different political regimes around the world, "citizen journalism(s)" now assert, with increasing confidence, it seems, their communicative presence and demand to be heard *outside*, *through*, and *within* today's mainstream news media. Such is the academic interest in the fast-evolving world of citizen journalism and its close association or, better, close affinity with processes of global change and issues of social justice, this second volume provides many examples of the significance of citizen journalism in contemporary issues, events and crises. In the years that have passed since Allan and Thorsen published their first volume of *Citizen Journalism* perhaps we can also begin to detect a growing sense of accommodation by mainstream news media toward these new upstart technologies and their intrusive flows of images and ideas. No longer, it seems, are professional journalist preoccupations quite so focused on "incoming" source veracity and news material authentication, and we may even glimpse a less deferential stance toward elites than in the past based on today's unavoidable journalist encounter with an incoming cacophony of views. The deluge of views and voices from some of the world's crisis hotspots can conceivably broaden not narrow news agendas and source dependencies, alerting journalists to not only the events in motion but also the contending perspectives that seek to define them. This needs serious, sustained examination and theorization. As the editors declare at the outset, their volume sets out to "identify and critically assess pressing issues confronting global crisis reporting," and they do so principally "by focusing on ordinary people's reportorial involvement—typically improvised under daunting circumstances—in recasting its priorities within a citizen-centred ethos." This surely is one of the most pressing and intriguing issues of citizen journalism and news reporting in the world today, of how it not only circumvents mainstream mediums to communicate directly with others but also how it becomes re-mediated in and through the accelerating and overlapping media formations and flows of the contemporary world news ecology—and contributes to changing these in the processes of its accommodation. *Citizen Journalism: Global Perspectives—Volume 2* provides an excellent vantage point from which to observe, take stock and better evaluate and understand these and other trajectories of citizen journalism around the world.

Simon Cottle, Series Editor

Introduction

EINAR THORSEN AND STUART ALLAN

More than a decade has passed since an anonymous blogger posting under the pseudonym "Salam Pax" brought to the boil a simmering debate over the perceived impact of the internet on crisis reporting. This incisive form of citizen journalism—produced, it would be later revealed, by a 29-year-old architect living in middle-class suburban Baghdad—succeeded in documenting telling, frequently poignant aspects of life on the ground in the months before the US-led invasion as well as its aftermath. While sceptics questioned the authenticity of his posts, others praised the raw immediacy of his insights into the lived experiences of besieged Iraqis caught up in the grisly horrors of conflict. This proved to be "embedded" reportage of a very different order, effectively demonstrating the potential of online blogging as an alternative mode of war correspondence. As Salam would reflect afterward, "I was telling everybody who was reading the web log where the bombs fell, what happened [...] what the streets looked like." Acknowledging the risks involved made his efforts seem almost "foolish" in retrospect, he added: "it felt for me important. It is just somebody should be telling this because journalists weren't" (cited in transcript, CNN International, 3 October 2003).

Many news organisations were acutely aware of the value such first-hand perspectives would bring to their coverage, but struggled to accept the notion that it was professionally responsible to incorporate citizens' contributions they could not independently verify. In the case of the BBC, its first online report referring to Salam, headlined "Life in Baghdad via the web," appeared on 25 March 2003, one week after the invasion commenced. "The online diary of an Iraqi man living in Baghdad is proving hugely popular with net users," it began. "The weblog describes what it is like to live through bombing raids, the effect of the bombing on everyday life in the capital and the views of ordinary Iraqis" (BBC News, 2003). Daniel Bennett's (2013) enquiry into the BBC's treatment of the conflict explains how Salam's first-person accounts were gradually incorporated into itscoverage from May onwards (*The Guardian* having established his identity by then), his blog providing "a way of accessing a compelling voice which might never have been heard without the communication possibilities afforded by the World Wide Web" (2013: 1). In so doing, however, the BBC was rapidly re-writing its own editorial guidelines, particularly where the use of eyewitness testimony was concerned. By July of that year, the "Baghdad Blogger," as Salam became known, had begun appearing in a series of short documentaries broadcast on *Newsnight*, a programme at the forefront of BBC experiments with new media strategies. This "alliance of the amateur and the professional" broke new journalistic ground, Bennett maintains, acting as a catalyst to encourage the wider development of what soon would be called "user-generated content" (UGC) at the Corporation.

Flash-forward a decade, and the extraordinary nature of such tentative, experimental ventures have become ordinary, even routine features of daily reportage for news organisations around the globe. On 6 July 2013, for example, vital details regarding the crash landing of Asiana Flight 214 at San Francisco International Airport were being relayed by citizen witnesses almost instantly, long before major television networks assigned their own people on the ground. Similarly, while officials and aviation experts scrambled to determine the scale of the incident, ordinary individuals were already gathering fragments of this eyewitness material via social networking sites, curating it into visible evidence confirming that there was likely to be a significant number of survivors. As the story came into sharper focus, it became apparent that two of the 307 passengers—both 16 year-old Chinese students—had lost their lives, while over 180 others were left injured (several in critical condition, one of whom subsequently died a week later). The immediacy of these first-hand perspectives lent the news coverage intense news value, ensuring the story attracted international attention while still unfolding in real time.

One of the people onboard the Boeing 777 arriving from Seoul, South Korea, was Samsung executive David Eun (@Eunner), who posted a photograph to the social networking site Path he took just outside the airplane, showing passengers evacuating. "I just crash landed at SFO," Eun wrote. "Tail ripped off. Most everyone seems fine. I'm ok. Surreal…" He then continued via his Twitter account: "Fire and rescue people all over the place. They're evacuating the injured. Haven't felt this way since 9/11. Trying to help people stay calm. Deep breaths …" (his Twitter handle of the airport, @flySFO, serving as a location identifier). Evidently within thirty seconds of the airplane's impact on the ground, Google employee Krista Seiden had tweeted "Omg a plane just crashed at SFO on landing as I'm boarding my plane," using her mobile phone to capture the plume of smoke in the distance. Soon after she pointed out on her blog:

> *The incredible thing about social media is how instant it is.* Within seconds twitter was exploding with news and photos of the crash. People kept asking if this was real because none of the news channels were reporting anything yet. Yes, it was real. And yes, the news channels were reporting it, but today that actually means that *anyone with a social media account and internet access is a news reporter and can break a big story.* What an incredibly powerful tool we all have in our hands today! (Seiden, 2013; emphasis in original).

Recognition for Seiden's efforts followed swiftly, summed up in one tweet by "Cory S" as "You broke the news before the News broke the news!" as well as corrections to some of her assertions ("Hearing reports the #planecrash at #SFO is an Asiana 777 in from Taipei"). Danielle Wells, similarly tweeting from an airport terminal, had observed: "Literally just witnessed a plane crash start to finish. I cannot stop crying I can't believe this," which she also accompanied with a twit pic of what was visible from her vantage point. Minutes later Stefanie Laine (@stefanielaine) tweeted "guys I just watched a plane crash at SFO," then added "we were walking back from breakfast, stopped to take a picture of the runway, and a landing plane came in at a bad angle, flipped, exploded." Meanwhile Chinese passengers were sharing their experiences on related social networks, such as SinaWeibo, QC and Xiaonei. Businessperson XuDa's first-hand account on Weibo included vivid details, such as: "I immediately heard a loud bang in the back, and the oxygen masks dropped in the cabin. I smelled something burning and saw fire" (cited in Lu, 2013).

For television networks and online news sites hurriedly marshalling video imagery for special reports piecing the news story together, most of what could be secured by their own journalists arriving after the crash featured long-range shots of the wrecked passenger jet stranded on the runway. Appreciably more compelling in visual terms were the short clips of precipitous footage provided by citizen witnesses who happened to be near the scene at the time. Not surprisingly, then, their impromptu forms of reportage were swiftly appropriated as material from "actual non-journalistic sources," in the words of one CNN reporter, who then added "this really is the rise of citizen journalism" (Avlon, 2013). YouTube user, Alek Yoo (sfprepper415) had recorded black smoke billowing from the airplane resting on the tarmac, its rear fuselage torn away, from where he stood in the airport terminal. On the third floor of a nearby hotel, 18-year-old Jennifer Solis—described by ABC News as an "amateur videographer" in its report—recorded the moment the emergency chutes were deployed from the exit doors, as well as the efforts of firefighters trying to dowse the fire ignited in the cabin. Having heard the sound of the crash, she had quickly grabbed her digital camera, later recalling: "I immediately went outside; I saw the big cloud of dirt, and I started recording immediately." Solis said she could not believe the "surreal" scene occurring before her, remarking: "[It] was just the first time I ever experienced something like this. I usually see things like this on the news" (cited in Louie, 2013). Even more remarkable was the "amateur video" shot by "aircraft buff" Fred Hayes, who happened to be videoing airplanes landing as he walked along San Francisco Bay with his wife during a weekend visit. "When I caught the plane coming into view, everything looked fine at first until I kind of fixed my gaze on him, and I seen his nose up in the air," Hayes told CNN. "And then I just totally locked on him. I thought he was going to take off and go up, and then he just kept going down" (cited in Smith and Hall, 2013). Hayes' dramatic footage captured the airplane's descent leading to the violent moment of impact with the sea wall and the ensuing carnage, thereby helping to resolve divergent views—"a maelstrom of conflicting information," in the words of one news commentator—regarding what had actually transpired. The 40-second clip, obtained exclusively by CNN, also included the audio recording of the couple's anguished responses as they looked on, lending an emotive sense of the personal distress they were experiencing.

The significance of these and related forms of citizen witnessing being uploaded across fluidly ad hoc collaborative networks, where the resources of sites such as Twitter, Facebook, Path, Flickr, Instagram, Tumblr and YouTube were mobilised to considerable journalistic advantage, seldom received more than passing comment in mainstream press reports. In marked contrast with the media plaudits Salam Pax received as a "celebrity blogger" a decade earlier, this blurring of reportorial boundaries may be read as being indicative of the relative extent to which "citizen journalism" has been effectively normalised where breaking news of crisis events is concerned. Remarkably, the recent decade has seen the gradual unfolding of a profound shift in public perceptions, namely that contributions of citizens who happen to be first at the scene have become so commonplace as to be almost expected (indeed, explanations for the absence of such material may well be necessary in ensuing news accounts). Citizens—be they victims, bystanders, first-responders, officials, law enforcement, combatants, activists or the like—together are actively engaging in newsmaking by crafting for their own purposes a diverse array of tools, methods and strategies to relay first-person reports, increasingly in real time as crisis events progress.

Our challenge, then, is to try to de-normalise, to make strange, the familiar tenets of these dynamics at the very moment so many of them are slowly, albeit unevenly consolidating into journalistic values, rules and conventions. It is in so doing that we invite greater self-reflexivity about the wider implications being engendered, not least for our changing conceptions of journalism's civic responsibilities within wider participatory cultures. In keeping with the remit for the

"Global Crises and the Media" book series, then, this volume endeavours to identify and critically assess pressing issues confronting global crisis reporting, namely by focusing on ordinary people's reportorial involvement—typically improvised under daunting circumstances—in recastingts priorities within a citizen-centred ethos. Crucial in this regard is the interweaving of varied inflections of both citizenship and journalism, particularly as they are inscribed, re-mediated and contested in discourses of citizen journalism striving to claim their purchase in what series editor Simon Cottle aptly calls "the increasingly complex flows and formations of today's world news ecology."

NEW AGENDAS

This second volume of *Citizen Journalism: Global Perspectives* seeks to build upon the agenda set in motion by the first volume by: 1) offering an overview of key developments in citizen journalism since 2008, including the use of social media in crisis reporting; 2) providing a new set of case studies highlighting important instances of citizen reporting of crisis events in a complementary range of national contexts; 3) introducing new ideas, concepts and frameworks for the study of citizen journalism; and 4) evaluating current academic and journalistic debates regarding the growing significance of citizen journalism for globalising news cultures.

Section One, titled "Re-imagining Citizen Journalism," extends the trajectory mapped in the previous volume. Our opening chapter by Yasmin Ibrahim, "Social Media and the Mumbai Terror Attack: The Coming of Age of Twitter," documents one of the initial points of journalistic engagement with social media which, with the benefit of hindsight, proved formative. This terror attack, which saw ordinary citizens relaying eyewitness accounts and imagery of the horrors across the webscape, raised troubling questions, including with respect to the role of broadcasting protocols in live telecasts of violent incidents. It also highlighted, she argues, the risks that emerge in social media platforms through the "act of sharing" during such moments of crisis. Lindsay Palmer, in her chapter "CNN's Citizen Journalism Platform: The Ambivalent Labor of iReporting," examines the use of CNN's citizen journalism platform, iReport. Taking the platform's citizen coverage of the 2009 Iranian uprising as her case study, she contends that individuals prepared to think of themselves as iReporters typically become involved in complex, sometimes contentious relationships with news organisations intent on making use of their work for their own purposes. Palmer's analysis reveals that "citizen coverage of global conflict is a story of both exploitation and subversion," one beset with tensions associated with the increasingly disruptive informational milieu consistent with network cultures.

The agenda-setting power of citizen journalism is the focus of Chris Greer and Eugene McLaughlin's "Righting Wrongs: Citizen Journalism and Miscarriages of Justice," with specific reference to miscarriages of justice. Their empirical analysis elucidates the interaction of media, political and judicial forces following the death of newspaper vendor, Ian Tomlinson, shortly after being struck by a police officer at the G20 Protests in London in 2009. The police denial that an officer had been involved was flatly contradicted by evidence revealed by the *Guardian* six days later, namely a video clip documenting the assault, handed over to it by an American visitor to the city. The rise of citizen journalism is shown to bring to bear countervailing imperatives for those institutions that traditionally have been able to control the information environment. Disruptions to the flow of communicative power in the name of justice similarly reverberate in Lilie Chouliaraki's chapter, "'I Have a Voice": The Cosmopolitan Ambivalence of Convergent Journalism." A vital dimension of journalism's performative ethos, she argues, is characterised by a shift from the professional act of informing towards citizen-driven acts of deliberating and witnessing—what she calls the disposition of "I have a voice." Whilst this shift in the epistemology of the news has been welcomed as a democratisation of journalism, her contrasting case studies—the Haiti earthquake (2010) and the Egypt uprising (2011)—demonstrate that variation in the use of citizen-driven journalism reflects a concomitant variation in the power relations of Western mediation. Such structures, Chouliaraki believes, "selectively give voice to distant others, silencing the voice of the Haiti earthquake victims but amplifying the voice of Egyptian protesters, and, in so doing, encourage recognition of the plight of the later whilst depoliticising the condition of the former."

Kristina Riegert's "Before the Revolutionary Moment: The Significance of Lebanese and Egyptian Bloggers in the New Media Ecology" addresses similar themes. In contrast with celebratory claims crediting social media, such as Twitter and Facebook, with ushering in the Arab Spring in 2011, she delves into an array of factors that helped to engender the demonstrations that swept through the region. Of particular importance, she shows, is the emergence of bloggers and citizen journalists in the years beforehand, not as causal agents but rather as part of a wider media ecology confronting authoritarian power structures. She traces the online relationships between popular bloggers in Lebanon and Egypt in order to identify how, when and why citizen journalism makes a difference in a crisis media event like that of the Arab uprisings. Staying with the focus on blogging in nearby national contexts, Neil Thurman and James Rodgers' chapter "Citizen Journalism in Real Time? Live Blogging and Crisis Events" analyses data gathered on the consumption of live online coverage of crisis events. They proceed to examine live blogging's relevance to debates about citizen journalism with reference to recent examples from Syria and Algeria. In assessing the opportunities and challenges for live blogging or utilizing citizen contributions—and thereby for journalism more widely—they argue that this type of reportage, when used with care, can enhance our news provision to considerable advantage.

Section Two, "Capturing Crisis," brings into sharp relief the efforts of ordinary citizens compelled to engage in reportage, to bear witness under formidable—and at times heart-rending—conditions. It begins with Donald Matheson's chapter, "Tools in Their Pockets: How Personal Media Were Used During the Christchurch Earthquakes,"which explores the roles of personal, portable media in people's responses—suggestive of what he calls a "supportive intimacy" amongst strangers online—to the twin earthquakes that hit the city of Christchurch, New Zealand in 2010 and 2011. Matheson also reflects on the limits of citizen media, such as when the importance of place-based, face-to-face forms of community interaction came to the fore after electricity-based media collapsed in the wake of the second, devastating quake. In "Hurricane Sandy and the Adoption of Citizen Journalism Platforms," Trevor Knoblich extends this discussion of natural disasters. He points out that as new tools for capturing text, photographs and videos emerge, citizen journalists have increasing opportunities to participate in documenting, sharing, and providing nuance to breaking news events. In turn, each crisis event illustrates that public familiarity with a social media or other communications platform, as well as persistent access, largely determines how citizens choose to share information. His investigation shows how citizen journalists used a variety of tools to share information about Hurricane Sandy in the Eastern United States, as well as how they adapted when access to a given platform of choice became limited.

The pressures of breaking news similarly feature in Einar Thorsen's "Live Reporting Terror: Remediating Citizen Crisis Communication." For individuals caught up in the two attacks carried out by Anders Behring Breivik in Norway in July 2011, tragic forms of self-publishing emerged. In the case of the Oslo car bomb, eyewitness observations offered a sense of raw immediacy; victims of Breivik's shooting spree on Utøya were publishing cries for help, confirming they were alive and desperate for information. Despite the different modalities, ordinary citizens' contributions to documenting the unfolding crisis were vital. Through a comparative analysis of international news organisations' live blogging of the attacks, Thorsen further highlights the global remediation of citizen eyewitness accounts—from the initial assumption that the attacks were grounded in "international terrorism" to the confirmation of Breivik as a domestic right-wing extremist. Several related challenges where news imagery is concerned are brought to light by Mette Mortensen's "Eyewitness Images as a Genre of Crisis Reporting." She observes that amateur, non-professional image makers frequently become the initial link in the chain of breaking news because of the on-the-spot documentation they provide. News organisations draw upon eyewitness images for several reasons, not least due to their association with an exclusive insider perspective and proximity in time and space to events. Still, Mortensen argues, even if they are habitually considered authentic on account of their urgency, immediacy, and handheld aesthetics, this type of imagery recurrently puts the norms, editorial routines, and professional self-perception of journalism to the test.

Stuart Allan's "Reformulating Photojournalism: Interweaving Professional and Citizen Photo-reportage of the Boston Bombings" is the first of two chapters investigating another crisis event where journalism was severely tested, namely the bombing of the Boston marathon in April 2013 that killed three people and left 264 others injured. Allan's analysis sets the experiences of professional news photographers that day in relation to the improvised photo-reportage shared by ordinary bystanders who happened to be on the scene. In so doing, he elaborates the concept of "citizen witnessing,"as he has developed it elsewhere (Allan, 2013), as one possible way to recast the prospects for refashioning journalism. Specifically, he argues for the need to create innovative spaces for a collaborative, co-operative ethos of connectivity between professionals and ordinary citizens in order to reinvigorate, in this instance, photojournalism's public service commitments. Graham Meikle's complementary chapter, "Citizen Journalism, Sharing, and the Ethics of Visibility," examines a further dimension of citizen involvement in the news coverage of the Boston bombings. The focus of his discussion concerns the use of the social media platform Reddit by networks of individuals who attempted to crowdsource the identities of possible bombing suspects by sharing images and speculation. Situating these events within the frame of citizen journalism, Meikle considers the centrality of sharing to social media and its uses for nonprofessional journalism and related forms of collaborative information provision. He argues that such uses reveal the need for an ethics of visibility to be secured on the basis of the lessons learned from what happened in Boston.

Section Three, "Globalising Cultures of Citizen Journalism," renders problematic certain familiar—which is to say Western—presumptions underpinning current debates. The scene is set by Silvio Waisbord's "Citizen Journalism, Development and Social Change: Hype and Hope," which pursues questions that go to the heart of the relationship between citizen journalism and public dialogue and opinion formation. All too often, he argues, analyses fall into a celebration of citizen journalism "as the crystallization of individual expression," which tends to obscure, in turn, important issues concerning communication rights and opportunities for collective actors, as well as the institutional contexts for participation and decision making. Clemencia Rodríguez's "A Latin American Approach to Citizen Journalism" elaborates a similar thematic, in part by exploring the factors shaping this approach as an alternative to the Global North's priorities. In Latin America, she argues, citizen journalism is a practice of resistance, one that brings together social movements, activists, and other social justice collectives in a shared refusal to accept that only professional news organizations can practice journalism. In seeking to expand and enrich the public sphere with information key to democratic processes, its guiding tenets are informed by commitments to social responsibility and the public interest. In the course of her discussion, Rodríguez draws upon examples from her fieldwork with citizens' media producers in regions of armed conflict in Colombia.

In "Getting into the Mainstream: The Digital/Media Strategies of a Feminist Coalition in Puerto Rico," Firuzeh Shokooh Valle considers several related issues as they have been taken up by women working for gendere quality. Specifically, she investigates how the feminist movement in Puerto Rico has employed a listserv—the coalition's main virtual platform—to secure a safe space for news and information to circulate, as well as discussions, deliberation, networking, strategizing, consensus building, decision making, and task distribution. The use of citizen media in the struggle for social change similarly figures prominently in Yomna Kamel's "Reporting a Revolution and Its Aftermath: When Activists Drive the News Coverage." Ranging from "the image of the Tunisian Bouazizi to the image of the Egyptian blue bra-girl who was dragged, stripped and brutally beaten by security forces in Cairo's Tahrir Square," she writes, "huge amounts of visual and textual content have been posted and circulated on social media platforms by citizen journalists, who are also activists after a cause." Kamel proceeds to examine the nature of their relationships with professional journalists, showing how diverse forms of networking via social media have helped to amplify their voices. Many of those blurring the boundaries between citizen journalism and activism have succeeded in making their stories heard in international media contexts, which she assesses in relation to Al-Jazeera, the BBC, CNN, Russia Today and XINHUA.

"While the IT avalanche of the past decade has made it technically possible for anyone to report anything from anywhere," Kayt Davies writes, "the biting reality from the frontlines of simmering repressive contexts, such as Indonesia's disputed territories, is that caution is required to keep citizen journalists safe." In Davies' chapter, "Citizen Journalism in Indonesia's Disputed Territories: Life on the New Media Frontline," she details the evolution of online media in these places and analyses the work done by humanitarian and new media groups. This includes consideration of how well their work fits existing definitions of media practices and roles, and the importance of allowing the people affected by conflict and repressive regimes to speak their truths. Karina Alexanyan, in "Civic Responsibility and Empowerment: Citizen Journalism in Russia," examines the reasons why it is necessary to disentangle the role of the citizen from that of the journalist, and how both are shaped by socio-political contexts. In semi-authoritarian environments like Russia, she argues, online and mobile technologies have not only facilitated the emergence of participatory journalism, they have engendered "a novel sense of civic society, responsibility and consciousness" with significant implications for civic empowerment and activism. This theme echoes in a different way in Last Moyo's "Beyond the Newsroom Monopolies: Citizen Journalism as the Practice of Freedom in Zimbabwe." Focusing on the Kubatana blog as a case study, he evaluates the extent to which it serves as a forum for the articulation of personal freedoms otherwise difficult to express. Efforts to recast blogging as citizen journalism, he argues, have as their aim the establishment of an alternate-subaltern space where citizenship is imagined in radical terms, namely as discursive, deliberative, participative and transformative.

The fourth and final section, "New Crises, Alternative Agendas," looks to the future, striving to anticipate how citizen journalism will continue to develop in the years ahead. It commences with Lisa Lynch's "'Blade and Keyboard In Hand': Wikileaks and/as Citizen Journalism." She begins by pointing out that although some observers have labelled Wikileaks a "citizen media outlet" in order to distinguish it from the professional press, closer scrutiny reveals that the relationship between Wikileaks and the citizen-journalist is not so straightforward. She explores Wikileaks' roots in citizen media, its turn towards engagement with mainstream media, and its more recent incarnation as a "read-only" resource indexing materials of interest to activists and scholars concerned with the exercise of US soft power. Questions of power similarly underpin Nik Gowing's chapter, "Beyond Journalism: The New Public Information Space," which contends that the proliferation of social media is creating new levels of near-instant accountability while, at the same time, provoking acute vulnerabilities for business, governments and social systems. In tracing the features of this information space, Gowing proceeds to pinpoint a range of factors impacting upon institutions, forcing them to adapt to changing circumstances in a climate of uncertainty. Illustrating his thesis with a range of examples from different case studies, he shows why, in his view, certain guiding principles of political and corporate governance need to change.

Related themes regarding crisis reporting and information flow resound throughout "The Evolution of Citizen Journalism in Crises: From Reporting to Crisis Management," by Hayley Watson and Kush Wadhwa. In reassessing the role citizen journalists play in contributing to the construction of news, they consider its impact for crisis managers. Certain advantages as well as problems come to light in the course of their enquiry into how managers work to incorporate the real-time news and information afforded by citizen journalism into their response efforts. Casting a valuable light on these dynamics is Lei Guo's chapter, "Citizen Journalism in the Age of Weibo: the Shifang Environmental Protest." While citizen journalism is generally considered to be flourishing on China's Twitter-like micro-blogging service Weibo, its effectiveness came under intense pressure when a large-scale environmental protest against a government-approved copper plant sought to use it as a platform to spread news about the ensuing conflict. Surveillance of a different register is examined in Mary Angela Bock's "Little Brother Is Watching: Citizen Video Journalists and Witness Narratives." In accentuating the importance of attending to the politics of witnessing giving shape to citizen journalism, she compares and contrasts the narrative strategies used by citizen and professional video journalists (VJs), respectively, as they endeavour to establish their authority. Her analysis is based on observations of practitioners, with one group—a self-described "cop watching" organization in the US—serving as an illustrative case study. Several stories posted by group members were chosen for textual analysis, with Bock paying particular attention to their authority-building narrative strategies.

"With institutional journalism flailing and failing in the internet Age," Kevin M. DeLuca and Sean Lawson write, "there is much wailing about the future of news and democracy." In their chapter, "OWS and Social Media News Sharing After the Wake of Institutional Journalism," they actively resist the familiar conflation of institutional journalism with news and democracy. Instead, they argue that the internet, especially social media platforms, enables a proliferation of decentralized news practices to develop, which they term social media news sharing. Occupy Wall Street serves as a case study to explore the emerging news practices and politics that social media news sharing is creating. Rounding out this section, and the book overall, is Sue Robinson and Michael L. Schwartz's chapter, "The Activist as Citizen Journalist." Drawing on a case study of the educational community in Madison, Wisconsin, they investigate some of the ways citizen activists are "working their beats" to report information in online public realms. Particular attention focuses on two "super-contributor" activists on opposing sides of a controversial charter school, showing how citizens with agendas are attempting to fill "structural holes" in an ever-evolving media ecological network through Facebook posting, commenting and blogging. To what extent, they ask, is the relevance of professional journalism being called into question by activists tapping into a maturing citizen journalistic ethos to connect civically on issues of such importance for the community.

It is our hope that readers will discover a host of intriguing, worthwhile issues worthy of their critical engagement on the pages of this second volume of *Citizen Journalism: Global Perspectives*. Its contributors present a shared commitment to challenging familiar assumptions in their pursuit of fresh ideas, approaches and perspectives on topics of pressing significance. Precisely how best to develop citizen journalism so as to help make the most of its remarkable potential is an open question, one inviting answers from all of us dedicated to the reinvigoration of journalism in the public interest.

REFERENCES

Allan, S. (2013). *Citizen witnessing: Revisioning journalism in times of crisis*. Cambridge: Polity Press.

Avlon, J. (2013). cited in transcript, CNN Reliable Sources, broadcast 11:00 am ESE on 7 July.

BBC News (2003). 'Life in Baghdad via the web', BBC News, http://news.bbc.co.uk/1/hi/technology/2881491.stm, 25 March

Beaumont, C. (2008). Mumbai Attacks: Twitter and Flickr Used to Break News. *The Telegraph*, 27 Nov, http://www.telegraph.co.uk/news/worldnews/asia/india/3530640/Mumbai-attackstwitter

Bennett, D. (2013). *Digital media and reporting conflict: Blogging and the BBC's coverage of war and terrorism*. London and New York: Routledge.

Louie, D. (2013). Video shows SFO crash passengers' escape from plane, abc7news.com, 8 July.

Lu, R. (2013). Weibo user on flight 214 posts first hand account and photos, TeaLeafNation.com, 7 July.

Seidon, K. (2013). How social media broke the story of the SFO plane crash, BloggerChica: Marketing Insights, 7 July.

Smith, M. and Hall, L. (2013). "Oh, Lord have mercy": Witness captures fatal jet crash, CNN.com, 8 July.

SECTION ONE

Re-imagining Citizen Journalism

CHAPTER ONE

Social Media AND THE Mumbai Terror Attack: The Coming OF Age OF Twitter

YASMIN IBRAHIM

The terrorist attack on Mumbai in 2008 is a significant case study for many reasons; it highlighted the role of social media in the absence of adequate press coverage, the appropriation of citizen accounts by mainstream press globally, the issues raised of citizens reporting crises events and the threats these posed to national security in turning terrorism into a broadcast marathon. It equally raised the ethics of consuming and producing live updates of terrorist events for national and global consumption. The Mumbai attack of 2008 has been described as the "coming of age of Twitter" by observers mainly due to the fact that social media dominated the reporting of the event in terms of its immediacy and vantage point from the streets. The absence of foreign press in India partly due to the Thanksgiving holiday period in America provided Twitter an opportunity to oust the story globally through citizen accounts.

The attack on Mumbai on the 26th of November 2008 has been described as India's very own 9/11 whilst other press headlines have termed the 60-hour-long siege of the city as nothing short of "the longest running horror show" (Khullar 2008). The terror attack which struck at the heart of the financial and tourist centre claimed at least 172 lives whilst wounding 250 in a series of gun and grenade attacks. A group called Deccan Mujahedeen claimed responsibility for the attack, which targeted multiple sites simultaneously including hotels, a Jewish Chabad centre, a café which was popular with foreigners, hospitals and a railway station.[1]

Mumbai, after the terror attacks in December 2008, was narrated as a "bleeding city" where hundreds "lit candles to remember the dead and to help deal with the trauma the city suffered" (Dodd 2008).

The city in postmodern memory becomes a backdrop for terror where the unexpected and volatile can unfold before a global audience. The city since the turbulence of 9/11 in America represents an instable space which contradicts the order, stability and security it is supposed to impose through its form and structure. Iconic landscapes of a city often function as a symbol of that city. Drawing parallels with 9/11, the Rand study, "Lessons from Mumbai" (2009), reiterates that "the attacks on landmark properties amplified the psychological impact." Additionally, the selection of multiple targets—Americans, Britons and Jews, as well as Indians—suggests that the terrorists intended the attack to serve multiple objectives that extend beyond the terrorists' previous focus on Kashmir and India and to globalise their struggle and illuminate it through international media coverage (Lessons from Mumbai 2009). These religious, political and cultural values were chosen in order to make a statement. According to the Indian Subcontinent Practice of Risk Advisory, the well-planned operations were carried out with an anti-Western aim with the "deliberate selection" of foreign hostages (Mumbai Attack shows 2008).

During the 60-hour siege of Mumbai, many of the eyewitness accounts emerged from social media including social networks such as Facebook, Twitter and Flickr. The explosion of reports in new media networks and the narration of terror as it emerged prompted many media critics to describe it as a "social experimentation" for new media and a coming of age for these new forms of media (See Lewis, Kaufhold, & Lasorsa 2010; Heussner 2008; D'Amour 2008; Beaumont 2008; Lewis 2008). The event also became a crisis point for traditional broadcasting where there was a public controversy over the reporting of the event. The broadcasting media was charged with irresponsible and sensationalist reporting where each network had succumbed to "market tyranny" by trying to outdo each other with exclusives and inside reports without exercising a sense of self-restraint or respect for those killed in these attacks. The live reporting by broadcast media was also seen as compromising the national security of the nation and anti-Pakistan sentiments amongst the general public.

This chapter examines the ways in which the trauma of Mumbai was constructed and narrated through social media during the attack. In particular it profiles how eyewitness accounts and the civilian gaze constructed the city during the terrorist attacks. The chapter also explores how the mainstream broadcast media came under intense criticism and scrutiny after the incident. The public and political backlash against the coverage of terror in the city raised the need to install new protocols in the reporting of terrorism or events which can compromise the national security of the country and the rescue operations of the authorities.

THE EYEWITNESS ACCOUNTS OF TERROR IN MUMBAI

The city of Mumbai is a financial as well as a tourist centre but more importantly it is a dream factory where its film industry provides escapism from harsh reality for millions of ordinary citizens. In recent years Mumbai has been frequently targeted by terrorist attacks blamed on Islamic extremists and these have included a series of bombings in July 2007 that killed 187 people (Fox News 26/11/2008 2008). In the 1993 Mumbai attack 257 died in 13 bomb blasts across the city. The city of Mumbai is not unused to terror. When the city came under siege in December 2008, new media provided an insight into events as they unfolded in the city. India ranks as the third top country in Asia (after Japan and China respectively) and is fourth largest in the world in terms of internet penetration rates (at 7.1%) with 81 million users as of November 2008 (www.internetworldstats.com). In terms of the population of users in Asia Indians account for 12.5%. The majority of users in India tend to be between 19 and 40 and there is a digital divide between urban and rural areas as well as male and female users with the latter only comprising 15%. Significantly, the city of Mumbai has the largest number of internet users at 3.24 million in 2008 with chatting and social networking sites being very popular amongst these users (indianroadband.net, 2008). This high degree of internet penetration and media literacy was important in showcasing the role of social media when simultaneous attacks occurred in the city.

In the Mumbai attacks Twitter in fact became a stream of snippets from observers on the ground with details of casualties, sieges, gunfights and even the suspected name of terrorists (Lewis 2008). It is estimated that around 70 "tweets" or messages tagged under the label of "Mumbai" were posted every five seconds when the news of the tragedy first broke, according to some estimates (Beaumont 2008). These minute-by-minute details included an update of events, calls for help, as well as a dissemination of emergency numbers. In this sense these sites, which emerged hours within the attack, provided the public service of diverting users to vital contact information such as foreign offices and helpline numbers. Within hours Google documents were created containing lists of the injured and killed whilst others solicited blood donors for those injured and needing emergency care.

Beyond informing the immediate and global community of the events in the city, spaces such as Twitter and Facebook provided a means for people to convey their situation to friends and family. Mumbai Help, a blog which updated information on the Mumbai attack, offered to locate people in Bombay for friends outside the city (Bell 2008). Besides Facebook and Twitter, blogs and file-sharing sites provided accounts and images from the ground. These conversations in the new media landscape also provided a signposting function providing the best news reports appearing on the web (Lewis 2008). These reports compiled by new media users got to the news more quickly than the television reports and newspaper websites. In many instances these Twitter and blog reports actually questioned the veracity of mainstream media accounts.

In addition to discursive accounts on the internet and those disseminated through mobile telephony, images played an important role in conveying the terror of the attacks. The convergence of technologies and the embedding of recording equipment in mobile phones and the ability to upload images onto the internet from personal mobile recording devices meant that events could be captured as they unfolded. Photos on Flickr uploaded 90 minutes after the attacks revealed the bloodied streets of Mumbai and helped to communicate the gravity of the situation. Some of these were viewed at least 110,000 times in the next 48 hours (CBS 2008; Lewis 2008).

A few hours into the attack, a Google map was created to show the location of buildings and landmarks at the centre of the incidents with links to news stories and eyewitness accounts (Beaumont 2008). Similarly a Wikipedia page was created within minutes of the news breaking and updated thousands of times, providing a vast amount of background information often in real time about the attack and these were detailed and corrected as the event unfolded (Social Media 2008; Beaumont 2008; Lewis 2008). These different sources of information created connections to other stories by providing Url links and as such they provided means to navigate the information that emerged on the nebulous internet. Dedicated blogs were created to update the events in Mumbai and these included the *Metroblog* set up by a group of bloggers based in Mumbai. Blogs became a site for information as well as commemoration and as such functioned as therapeutic devices to deal with trauma and terror and to equally share nostalgic accounts of what the city meant to its citizens. The Mumbai Heroes blog, for example was created to honour the victims of the attack (CBS 2008).

In addition to informing the outside world of terror attacks in Mumbai, Twitter also provided people who were trapped inside the affected hotels during the siege to communicate with the outside world. The hotel guests trapped in the affected hotels logged onto Twitter to find out whether they were under attack and to get an idea of the chaos that had engulfed them. Twitter captured the drama through conversations and information exchanged between people affected by the situation and those who were observing and engaging with the trauma through traditional and new media platforms. Event creation in the new media age is a complex exercise where there is an intricate enmeshing of information between traditional and new media sources. Mainstream media (both local and foreign) such as print and broadcasting used information that occurred on the internet in their reports. News stations including CNN used video clips sent in from people on the ground in Mumbai to illustrate their reports and many traditional media outlets including broadcasting stations and newspapers monitored Twitter and blogs in compiling their reports (Beaumont 2008).

The Mumbai attacks come in a long line of events in which the new media accounts have helped build a mediated event for the public. The eyewitness accounts and images published on new media platforms have become a vital part in composing the media event. They provide new forms of visibility and civilian narratives which are often conjoined to mainstream media narratives both through the links in web pages and also due to the fact that mainstream media are increasingly using these reports and accounts in their event creation, often asking civilians to send in their reports and photos to construct the event. Twitter, Flickr and other file-sharing sites became places to both bear witness and engage with the event.

The accounting of both natural and man-made calamities has become an open-ended phenomenon enabled by the convergence of technologies and the embedding of a vast array of capturing and communication features in mobile telephony. This enables the civilian gaze, which maps events through its own vantage points, to become part of event creation and then to contribute to the media event. In contrast to the mainstream media it is difficult to underpin this gaze to questions of authenticity or truth. Just as the photograph has been subjected to long-running debates on the validity of representation, the civilian gaze whether enabled through image or words is a problematic device in representing and corroborating an event. The gathering and exchange of information on new media sites during crises or watershed political events from politically engaged or civic-minded citizens or those who want to partake through conversations is a much more intricate process that cannot completely be addressed by the term "citizen journalism." Social media such as Twitter enable the ability to engage with "imagined communities" during crises and materialise these imagined communities through conversations and the exchange of information. Twitter technology was seen as "coming of age" and described as a "social media experiment in action" (D'Amour 2008) mainly due to the fact that people were drawn to produce, gaze, and disseminate information whilst connected with the rest of humanity within the city and outside during a crisis. New media spaces, by enabling the creation of content by audiences, broaden event creation and in the process interactive media platforms double up as therapeutic sites for recovery and individual and communal meaning making.

Beyond the functionalist paradigms of new media, the insatiable trading of information presented various challenges. These blow-by-blow Twitter accounts traded at a rate of 50–100 posts a minute in a message were nevertheless fragmented and sometimes false (Social Media 2008). Valid and inaccurate accounts were thus strung into streams of conversation and information on the event mirroring both new forms of empowerment and vulnerabilities evident in the ways in which Mumbai was narrated during the attack. One report on the internet claimed that the Indian government was trying to shut down the Twitter streams people were using to spread news and information. This story was picked up and reported by the BBC without verifying the validity of the report. The BBC came under criticism for its use of live reports from Twitter in its coverage of the attacks, prompting its news editor Steve Herrmann to acknowledge that the corporation would have to take more care in how it uses "lightening fast, unsubstantiated citizen posts from Twitter in future" (Sweney, 2008). On the internet information can spread very quickly and this gives people the opportunity to spread sensitive or compromising information (Bell 2008) without pausing to understand the implications of such publicised information. Equally, with a plethora of information flowing through these new media platforms it is often difficult to discern what information is credible and often there is a trade-off of accuracy for immediacy (Leggio 2008).

Ironically, whilst technology helped communicate the events to a global audience in this instance, mobile telephony also played a crucial role in enabling the perpetrators to execute their trail of terror in Mumbai. According to the Rand Study (Lessons from Mumbai 2009) the attackers reportedly used cell phones and a satellite phone—both their own and others taken from their victims—to co-ordinate their activities. The report found that the perpetrators also carried Blackberries and communicated with each other during the siege to discuss their manoeuvres. They made contact with the news media via cell phones to make demands in return for the release of their hostages and this led to some confusion in the Indian authorities who believed that they were dealing with a hostage situation, posing further challenges for their tactical response (Rand Study 2009). The role of telephony during the 2008 Mumbai attack was inevitably double-edged, presenting new forms of empowerment and vulnerabilities in accelerated modernity. In the aftermath of the attacks in Mumbai, when the public perceived that the situation was mishandled by the government, much of the urban middle class vented their anger against the government in new media sites including blogs, social networking sites such as Orkut and Facebook, and text messages (Lakshmi 2008). These media also became the platforms in which to raise objections against the traditional media's coverage of the events and to protest against the government. Internet-savvy protestors also turned to social media platforms to organise protests against the government in the aftermath of the attack.

BROADCASTING "TERROR"

Dayan and Katz's (1992: 196–97) conception of the "media event" captured television's ability to mark an event or moment in history through the interruption of routine broadcasting. The suspension of usual routines to carve a moment in time where the nation convenes over the televisual space evokes both the power of mass media to create events but equally the conceptualisation of national spaces to mourn and commemorate. The medium's predominance in a public event, Blondheim and Liebes (2002: 274) argue, should be situated through electronic journalism's adoption of the live coverage format which positions it in an intermediary role as a storyteller, negotiator and movable stage on which the drama is enacted.

Live coverage is a significant aspect of narrating terror in our accelerated modernity where the "liveness" creates immediacy whilst impressing the interconnectivity with the wider world. Often when terror happens it is a ritual for the world to watch the terror as it unfolds. At one point during the siege TV news channels in South Mumbai were blacked out for forty-five minutes following an order by the Deputy Commissioner of Police and this created more panic and unease amongst the people (Khullar 2008), emphasising our insatiable need to watch events live and without disruption. In the case of the Mumabi attack, the live coverage in constructing the media event became an issue of contention. More significantly, the aftermath of the Mumbai attack proved that the televised "media event" is not an unproblematic device in forging a collective consciousness. The construction and narration of the event became a point of contention, adding to the public's loss of faith in the government.

The live telecast of the attacks was 60 hours long and the extended coverage of the events which occurred in different parts of Mumbai posed many challenges for broadcasters. The event became a watershed moment for live broadcasting in the country for various reasons but primarily it was deemed as compromising the security of the nation and inciting passion against its neighbour Pakistan with whom it has a tumultuous historical and political relationship. The Indian government liberalized the broadcasting market in the early 1990s by dismantling state controls, encouraging privatization and relaxing media regulations (Thussu 1999: 126). The liberalization of the market and new communication technologies saw a dramatic growth in the number of television channels and transformed the television landscape radically (See Thussu 2007; Butcher 2003). India is one of the world's largest television markets with an expanding Westernized, middle-class audience of 300 million and this is an attractive lure for transnational media corporations to gain a slice of the lucrative Indian market (Thussu 2007: 594).

Inevitably, the 60-hour long coverage of the events prompted an intense competition amongst channels to outdo each other's reporting. Television channels were criticised for capitalising on human trauma and turning it into a "reality show" (Gupta 2009; Khullar 2008). Additionally broadcasters were criticised for sensationalising their reports but more importantly sabotaging the rescue operations and the national security of the country by revealing sensitive information without restraint or reflection. A media study undertaken by Newswatch, a media watchdog based in New Delhi, shortly after the attack, found that the government-run TV channel DD was the least sensational and most restrained compared to commercial stations (cf. Thakuria 2008). It also found that many channels were overtly sensationalist in their coverage. The survey included 9,906 responses and 74% felt that the reporting was theatrical. The TV coverage of the incident, especially by 24-hour news channels, was criticised as "TV terror" (Pepper 2008). The media was also accused of over-simplifying and dichotomising a complex situation by portraying "Pakistan is the enemy" or "politicians are villains" (cf. Chandran 2008).

In the days following the attack there were robust public discussions on the extent to which the media should be regulated in the public interest and the interests of national security (Gupta 2009; Thakuria 2008; Divan 2008). The live reporting came under scrutiny in parliament as well and it was noted that "live feed of air raids on the rooftop of Nariman House (where the Israeli hostages had been held captive) had taken away the element of surprise which is critical and crucial in rescue operations." There was also fear that live coverage could have been used as free intelligence by the planners of the attacks located far away from the incidents and allegedly guiding the attackers by means of satellite and mobile phone communication to take appropriate emergent measures against security forces (cf. Thankuria 2008).

Criticism of mainstream media was evident in blogs and amongst viewers and became a point of national discussion with politicians questioning the role of the media in such situations. The media was also accused of pointing the finger towards Pakistan without understanding the full nature of the attacks. The stations portrayed the commandos sent to rescue the hostages as brave whilst repeatedly showing Indian flags. In the days following the attacks, the Indian flag was often used by broadcasters as a visual backdrop with viewers' text messages expressing anger at politicians or Pakistan scrolling at the bottom of the screen (Chandran 2008). Such cultural references were seen as irresponsible in influencing and inciting public opinion against Pakistan given the state of tension between the two countries.

Pakistan's media duly criticised Indian broadcasters of being in a "race for propaganda" and "providing unsubstantiated" charges about the origins of the attackers (D'Amour 2008). With the coverage of the media coming under intense public scrutiny, a parliamentary committee expressed concerns over the repeated display in the media of human corpses during natural calamities, accidents, bomb blasts, arson etc. (Thankuria 2008; Pepper 2008). A few weeks after the attack the Ministry of Information and Broadcasting mooted a proposal to amend the existing Programme Code under the Cable Television Rules of 1994 by introducing 19 new amendments. These included proposals to introduce restrictions amongst other things on live coverage of war or violent law and order situations, disclosures about security operations, live interviews with victims, security personnel or perpetrators of crime (cf. Divan 2008). Additionally, the South Asia Media Commission (SAMC) in its report "South Asia Media Monitor 2008" slammed the media both in India and Pakistan for promoting hysteria (Gupta 2009).

The broadcast media were not seen as abiding by the self-regulatory code of ethics and standards adopted by the New Broadcasters Association comprising of 14 networks. Whilst the attacks illuminated the role of social media in such a situation, the print and broadcast media were seen as immature and not exercising the kind of self-regulation which was felt to be warranted in a highly explosive environment where they could have incited further violence or communal riots. The media on its part felt that the authorities did not have proper protocols in place in reporting on emergency situations. The Indian Broadcasting Federation and the News Broadcasting Association which represents many of country's top news channels criticised the government for "failing to keep up with the developments in the media industry" and not being proactive in "creating a procedure for the coverage of national emergencies" (cf. D'Amour 2008).

The authorities were seen as failing to protect the public as they lacked a clear information and communication management strategy. The lack of orchestration in feeding information to the media was raised as another important factor that led to uncontrolled and chaotic reporting. In the post-event discussions about the role of the media it was pointed out that the authorities such as the Navy themselves fed details to the media without restraint. In addition to this, different authorities gave separate and contradictory accounts and versions to the media. The absence of a concerted media management by the authorities was seen as contributing to the chaotic nature of reporting. Besides irresponsible reporting, TV stations were also seen as elitist in their reporting where they concentrated on the hostages at the Taj Mahal Palace and Trident-Oberoi hotels, which are the domains of the country's wealthy and ruling elites, whilst largely ignoring Chhatrapati Shivaji train terminus which was the site of the largest number of casualties and where a total of 58 people were gunned down (Pepper 2008). TV stations were also accused of other forms of bias beyond class divides. For example, the British and American media focused largely on their own citizens.

In response to the criticisms from the general public and from the members of parliament, the National Broadcaster's Association, which represents many of the country's top news channels, announced a new set of rules for the industry in December 2008. These new guidelines ban broadcasting of footage that would reveal security operations and live contact with hostages or attackers (Pepper 2008). The guidelines also request broadcasters to avoid unnecessary repetition of archival footage which might agitate viewers (televisionpoint.com). Many of these guidelines still hinge on self-regulation as the guiding principle. Undoubtedly the effectiveness of these guidelines and their ability to fully resist the "tyranny of the market" may well be tested if another such incident were to occur in India.

The broadcasting of the disaster marathon in Mumbai prompted the public to question the construction of the "media event" by mainly 24-hour news channels.

Unlike the new media, which was seen as reaching a new maturation point with the terror attack, the broadcast media was perceived as regressing and failing to observe a role that was ethical or responsible in protecting the national interest of the country.

CONCLUSION

The city of Mumbai, with its famous film production facilities, is a place for myth- and dream-making. In November 2008, when terror gripped the celluloid city, the urban space of accelerated modernity was transformed into a site of chaos, carnage, suffering and media spectacle. The 60-hour siege of Mumbai in became a testing ground for new media technologies and their role in responding to long and sustained periods of crises. The event has been dubbed as the "coming of age of Twitter" as it highlighted the role of citizen journalism and the vantage points it provided from the streets particularly in the absence of foreign press coverage in reporting the terrorist attack to the world. The citizen accounts were used by foreign press to substantiate news on Mumbai.

The potential for both new forms of empowerment and risks were highlighted by the new media technologies where immediacy rather than authenticity became primary. On the other hand, the political economy of news making in India has raised serious concerns about how chasing exclusive deals and sensationalising terror attacks or rescue operations can be detrimental for the country. The broadcast media through its engagement with the Mumbai siege has been forced to review its role in a crisis and to abide by new protocols. More importantly, the siege of Mumbai showed that the "media event" is not an uncontested terrain where broadcast media can tell a story through freeze-framing of images and seamless narratives. In Mumbai audiences engaged with the mediated accounts of terror provided by the media and duly questioned the media event. The media's power to be the supreme story teller or to appropriate and subsume an event through its production codes came under intense scrutiny. In the process the power of the media was questioned and the will of the audience was reasserted in demanding a more responsible broadcasting space.

NOTES

1. Evidence suggests that Lashkare-Taiba (LeT), a terrorist group based in Pakistan, was responsible for the attack (Lessons from Mumbai 2009: 11). The Rand study elaborates that the Pakistan-based terrorists see India as part of the "Crusader-Zionist-Hindu" alliance, and therefore the enemy of Islam. "Muslim" Kashmir ruled by majority "Hindu" India, provides a specific

cause, but LeT has always considered the struggle in Kashmir as part of the global struggle, hence the specific selection of Americans and Britons as targets for murder, and the inclusion of the Jewish Chabad center as a principal target (2009: 11).

REFERENCES

Beaumont, C. (2008). Mumbai Attacks: Twitter and Flickr Used to Break News. *The Telegraph*, 27 Nov. http://www.telegraph.co.uk/news/worldnews/asia/india/3530640/Mumbai-attacks-twitter

Bell, M. (2008). Web a-twitter with Terror Attacks.livemint.com, 28 Nov. http://www.livemint.com/articles/printArticle.aspx?artid=E718CBE8-BCAA-11DD

Blondheim, Mehanheim, & Liebes, Tamar, (2002) Live Television's Disaster Marathon of September 11 and its subversive Potential. *Prometheus*, 20:3.

broadband.net (2008). 'Some Statistics about internet Users in India', *www.broadband.net*, http://www.indiabroadband.net/india-broadband-telecom-news/11169-some-statistics-about-internet-users-india.html, 31 March.

Butcher, M. (2003). *Transnational Television, Cultural Identity and Change: When STAR Came to India.* New Delhi/UK/USA: Sage.

Chandran, R. (2008). Indian Media Under Fire for Mumbai Attacks Coverage. Reuters, 5 Dec 5, http://www.newswatch.in/print/2303

Dayan, D., and Katz, E. (1992) *Media Events: The Live Broadcasting of History*. Cambridge, MA: Harvard University Press.

D'Amour, R. (2008). 'India: Mumbai attacks create media frenzy', *WAN-IFRA, Editors Weblog*, http://www.editorsweblog.org/2008/12/02/india-mumbai-attacks-create-media-frenzy, 2 December

Divan, M. (2008). Coverage of Mumbai Attacks: The Way Forward. Halsbury's Law. http://www.halsbury's.incoverage-of-mumbai-attacks.html

Dodd, V. (2008). Bleeding City Comes Out Again to Honour the Dead. *The Guardian*, 1 Dec. http://www.guardian.co.uk/world/2008/dec/01/mumabi-terror-attacks/print

Gupta, A. (2009). Yellow Journalism versus State Intervention. ipcs.org, 29 Jan. http://www.ipcs.org/print-article-details.php?recNo=2810

Heussner, K. (2008). Social Media a Lifeline, Also a Threat. ABC News, 28 Nov. http:abcnews.go.com/print?ID=6350014

Internetworldstats.com (2008). 'Internet Usage in Asia', *Internetworldstats.com*, http://www.internet-worldstats.com/stats3.htm#asia, November.

Khullar, M. (2008). "In India, English-Language TV Stations Face Criticism and Ire for their Coverage of Mumbai Attacks", *The WIP*, http://thewip.net/2008/12/11/in-india-english-language-tv-stations-face-criticism-and-ire-for-their-coverage-of-mumbai-attacks/, 11 December/

Lakshmi, R. (2009). In Anger over Mumbai Attacks, Indians Vilify Their Politicians. *The Washington Post*, 5 Dec. http://www.washingtonpost.com/wpdyn/content/article/2008/12/04/AR2008120403694.html

Leggio, J. (2008). Mumbai Attack Coverage Demonstrates (Good and Bad) Maturation Point of Social Media. CBS Interactive, 28 Nov, http://blogs.zdnet.com/feeds/?p=339

Rand Corporation (2009). 'Lessons From Mumbai', *Rand Corporation*, http://www.rand.org/pubs/occasional_papers/2009/RAND_OP249.pdf

Lewis, P. (2008). Twitter Comes of Age with Fast Reports from the Ground. *The Guardian*, 28 Nov. http://www.guardian.co.uk/world/2008/nov/28/mumbai-terror-india-internet-twitter

Lewis, S. C., Kaufhold, K., & Lasorsa, D. L. (2010). Thinking About Citizen Journalism: The Philosophical and Practical Challenges of User-Generated Content for Community Newspapers. *Journalism Practice*, *4*(2), 163-179.

The Economic Times, (2008). 'Mumbai Attack Shows New Sophisticated Face of Terror', *The Economic Times*, http://economictimes.indiatimes.com/articleshow/3767754.cms?prtpage=1, 28 November.

Televisionpoint.com (2008). 'NBA: NO Live Coverage of Terror Attacks', *Televisionpoint.com*, http://www.televisionpoint.com/news2008/print?id=1229686272, 19 December.

Pepper, D. (2008). Indians Condemn Media Coverage of Mumbai Attacks. *The Christian Science Monitor*, 23 Dec. http://www.mcmclatchydc.com/255/v-print/story/58383.html

Fox News (2008). 'Wave of Terror Attacks Strikes India's Mumbai, Killing at Least 82', *Fox News*, http://www.foxnews.com/printer_friendly_story/0,3566,457885,00.html, 26 November.

Some Statistics about internet Users in India. (2008, March 31). www.broadband.net. http://www.indiabroadband.net/india-broadband-telecom-news/11169-some-statistics-about-internet-users-india.html

Sweney, M. (2008). 'BBC admits it made mistakes using Mumbai Twitter coverage', *The Guardian*, http://www.theguardian.com/media/pda/2008/dec/05/bbc-twitter, 5 December.

Thankuria, N. (2008). India Brickbat for Commercial Channel on Terror. *American Chronicle*, 16 Dec. http://www.americanchronical.com/articles/view/85145

Thussu, D. (1999). Privatizing the Airwaves: The Impact of Globalization on Broadcasting in India. *Media, Culture & Society*, 21, 125–131.

Thussu, D. (2007). The Murdochization of News? The Case of Star TV in India. *Media, Culture & Society*, 29(4), 593–611.

Fox News (2008). 'Wave of Terror Attacks Strikes India's Mumbai, Killing at Least 82', *Fox News*, http://www.foxnews.com/printer_friendly_story/0,3566,457885,00.html, 26 November.

CHAPTER TWO

CNN's Citizen Journalism Platform: The Ambivalent Labor OF iReporting

LINDSAY PALMER

Time Warner's Cable News Network (CNN) has long defined itself as a "unifying global force" possessing the ability to "tell the world about the world" (Küng-Shankleman, 2000: 118–19). Ever since media mogul Ted Turner launched the U.S.'s first all-news network in 1980, CNN has endeavored to achieve a global reach. CNN International appeared five years later, followed soon after by several other versions of CNN broadcasting in languages such as Spanish, Italian, and Turkish. Media scholars trace CNN's influence back to the network's pioneering use of satellite reporting during the Gulf War, which engendered an explosion of discourse about newer, more immediate ways of covering global conflict (Zelizer, 1999). The sense of immediacy and proximity facilitated by satellite news reporting operated as a marketing tactic for the growing network, best exemplified in the early CNN motto, "The sun comes up somewhere all the time" (cited in Volkmer, 1999). This motto assured viewers that CNN could bring every corner of the world into their living rooms.

Yet, in the past few decades, media conglomeration and the intensified commercialization of news have led to a growing amount of newsroom lay-offs (Barkin, 2003; Compton & Benedetti, 2010). This has in turn led to a different set of strategies for mapping the world, with news networks increasingly seeking the involvement of citizen journalists working within diverse national and socio-cultural contexts. Accordingly, CNN launched its own citizen journalism platform in August of 2006, "in an effort to involve citizens in the newsgathering process" ("iReport Turns

One Year Old," 2007). As its own early rhetoric suggested, CNN's iReport site draws upon the contributions of unpaid news enthusiasts, who upload pictures, video, and print-based information to the platform. If CNN producers see an iReport story that might be useful to them, they "vet" the story for copyright issues or factual error, and then air the information with the "iReport" logo hovering in the frame.

Compellingly, the 2006 launch of the iReport website seems to have coincided with an explosion of scholarly research dedicated to understanding citizen journalism in the age of the "prod-user" (Bruns, 2008). Much has been said about the rise of the "blogosphere" in the past decade (Barlow, 2007; Bruns, 2008; Tremayne, 2007), a phenomenon that Stuart Allan (2006) notes has helped to illuminate public discontent with traditional news since the 2001 attacks on the World Trade Center. Yet, unlike many amateur bloggers whose messages can sometimes only reach a limited number of people, iReporters draw upon CNN's clout to disseminate their messages across a wide variety of multimedia platforms. Their unpaid labor simultaneously bolsters the power of the CNN brand while also illuminating the social hierarchies long associated with traditional journalism, thus serving as an example of the increasingly "symbiotic relationship" between mainstream media and citizen journalists (Friend and Singer, 2007).

The citizen journalists' disruption of such hierarchies cannot solely be attributed to the rise of digital technologies, though these technologies do indeed optimize the propagation of citizen messages (Allan and Thorsen, 2009). Such disruption is also not merely a product of the crisis that Robert McChesney and John Nichols (2011) identify in the credibility and accountability of corporate news reporting, though such a crisis exists. Graeme Turner (2010) aligns the rise of the citizen journalist with a crisis in the credibility of professional news itself, as well as with the "ordinary" person's effort at bridging the alienating gap between traditional journalism and its public. Though many scholars still assert that professional journalism is ideally the guardian of democracy (Papacharissi, 2009), this guardianship is increasingly perceived as a failure. This suggests, in turn, the need for what the more optimistic proponents of citizen journalism identify as media witnessing (Gillmor, 2004; Frosh and Pinchevski, 2009).

Even so, there is reason for caution, especially in the case of the corporatized citizen journalism facilitated by iReport. As Lisa Parks (2009) asserts, the CNN brand has "the power to shape knowledge about and impact interventions into world affairs," and such power must be reinvestigated as information technology changes (2009: 1, 10). This chapter attempts such an investigation, deploying two specific methodologies: First, I offer a discourse analysis of the industry chatter addressing the 2006 launch of iReport, with the purpose of illuminating the profound anxiety inspired by CNN's affiliation with citizen journalism. Second, I offer an analysis of the interviews I conducted with Global Challenge participants in 2010, as well as with iReporters who covered a very different story—the Iranian uprisings of 2009, spurred by the announcement of the re-election of presidential incumbent Mahmoud Ahmadinejad.

I turn especially to iReport coverage of those occurrences in an effort to examine the possibilities and the pitfalls of the high visibility promised by CNN. I will show that the deployment of the CNN brand operated very differently for iReporters covering the Iranian uprisings than it did for the Global Challenge participants, raising questions about the distinct definitions of the word "citizen" and "journalist" in Iran (Sreberny and Khiabany, 2010). One thing that both types of coverage had in common was the fact that unpaid volunteers were doing a great deal of work, pointing to the undeniable exploitation of citizen journalism at the hands of CNN employees. Yet, rather than stopping at a lamentation of the tactics deployed against iReporters covering the Iranian uprisings—and thus, muting the voices of the people who engaged in this dangerous, unpaid labor—I will also show how these social subjects manipulated the technologies available to them in order to share their political messages with a larger public.

As my interviews reveal, the iReporters' desire to voice their competing interpretations of the events in Iran inspired their unpaid labor during the protests. Instead of focusing on monetary compensation or even on the minimal recognition promised to "star" iReporters, these particular iReporters eschewed recognition—something potentially dangerous for them—and instead sought the visibility of their opposition. This visibility was made possible through the iReporters' communicative interconnectedness, the same interconnectedness that also facilitated their exploitation. Following this, my case study reveals that citizen coverage of global conflict is a story of *both* exploitation and subversion, since hegemonic journalistic representations of world events ultimately unfold within the increasingly disruptive informational milieu that is the product of network culture.

IREPORT ACCORDING TO THE PROFESSIONALS

In order to properly delineate the impact of network culture on the journalistic coverage of global conflict, it will first be necessary for me to examine the U.S. news industry's anxious effort at maintaining its monopoly on meaning—on who gets to render which stories relevant and why. Especially in the context of CNN, this effort largely centers on the twin notions of credibility and validity, reserving professional credibility for paid journalists, as well as questioning the technological and conceptual validity of the images uploaded by amateur iReporters. Tiziana Terranova (2004) argues that journalists' obsession with credibility and validity is indicative of their dependence on the representational logic of signification, where meaning itself depends on a sender relaying particular signs to a receiver who is familiar with the sign system; this significatory logic in turn depends on a conceptualization of space that is based on perspective, where one subject addresses a separate subject across a three-dimensional distance (2004: 15–16). Professional journalism has long drawn upon this notion of perspectival space, situating the supposedly "objective" perspective of paid anchors and reporters as the reference point against which all other perspectives are measured—especially when these professional journalists are called to render global events intelligible for their audiences (Hall et al., 1999; Morse, 2004).

Yet, such perspectival space has imploded, Terranova (2004) maintains, disrupting the transparency of signs (2004: 35, 37). Increasingly, space can be defined by its "excess of sensory data, a radical indeterminacy in our knowledge, and a non-linear temporality involving a multiplicity of mutating variables and different intersecting levels of observation and interaction" (2004: 37). While signs still circulate across the mediascape, and while meaning can still be constructed and contested within this space, Terranova argues that meaning cannot exist outside of the much more complicated informational milieu that "exceeds and undermines the domain of meaning from all sides" (2004: 9). This is relevant because of the growing way in which communicative processes are linked to cultural processes—where cultural processes, like the citizen coverage of the Iranian opposition, are "increasingly grasped and conceived in terms of their informational dynamics" (2004: 7).

Terranova's philosophy complicates professional journalism's obsession with objectivity, suggesting that experience itself exceeds all efforts at representation, including those "objective" efforts that citizen journalists supposedly cannot emulate. Yet, even in this network culture, corporations like CNN draw upon the very perspectives they also denigrate, pointing to the annihilation of perspectival distance and to the increased potential for alterity within entities that still attempt to situate themselves as the dominant "self" against which the degraded "other" is measured. While this notion of dominance depends on industrial-era hierarchies of production and consumption, participation in the production of content increasingly reflects the openness of network culture (Bruns, 2008). As my analysis will show, the 2006 launch of the iReport website facilitated the merging of a previously closed production system with the open system of the networked community of citizen journalists, in turn engendering an explosion of ambivalent discourse from industry gurus.

When iReport first launched, CNN's own executives attempted to distinguish between professional correspondents and iReporters by invoking the qualities of emotional authenticity and embodied experience in their description of the iReporters' labor. Indeed, industry commentators celebrated such qualities, even as they also raised questions about iReporters' credibility as objective witnesses to traumatic events. For example, executives told *MediaWeek* that they were enthusiastic about citizen journalism because it is "emotional and real" (Woerz cited in

Shields, 2008), while in the same article underscoring the "comfortable" nature of audience participation in newsgathering. This reference to emotional authenticity and comfort diffused the adventurousness typically associated with the professional war journalist, who was imagined as travelling across rough terrain and sacrificing domestic comfort for access to the objective facts.

PR Week featured interviews with CNN executives who celebrated the first-person, experiential nature of the citizen journalist's reports, as opposed to the professional journalist's third-person accounts, where the authoritative reporter or news anchor could maintain the proper objective distance from the event in order to best determine its meaning (Martin cited in Washkuch, 2008). Trade publications also repeatedly flagged the emotional authenticity of the iReporters' work, an authenticity that was carefully distinguished from that work's technological or conceptual validity; credibility and validity were two terms that were far more often employed to disassociate professional journalists from the iReporters whose work was portrayed as merely supplemental to that of professionals (Martin cited in Washkuch, 2008: 10). These discursive strategies signaled an undercurrent of anxiety in professional circles, anxiety that citizen journalism could potentially hijack the professional journalist's monopoly on meaning itself.

Accordingly, CNN executives asserted that the citizen journalists' enrichment of professional labor was largely due to their ability to capture footage of breaking news on their smart phones with a speed that could not always be matched by professional reporters (Martin cited in Waskuch, 2008). In this sense, professional journalists simultaneously celebrated and denigrated the embodied experience of the iReporters, implying that while such proximity could facilitate first-person accounts of certain events, it could also result in the citizen journalists' inability to objectively make sense of their own video. If anyone could be a citizen journalist, then objective reporting could also give way to a glut of information, a frenzy of competing meanings. This was anathema to the professionals who had been trained to conceptualize their own perspectives as the reference point against which all other perspectives were measured (Hall et al., 1999; Morse, 2004).

Such professional anxiety was temporarily resolved through the dubious assertion that iReporters were simply global audience members with whom CNN was trying to "create a relationship" (Grant cited in Shields, 2008). Jim Walton, president of CNN Worldwide, declared that the use of iReport was a "natural extension" for CNN (cited in Shields, 2008), while David Almacy of Waggener Edstrom called iReport "the perfect marriage" between "the concept of citizen journalism" and a "branded media outlet" (cited in Longpre, 2008: 10). This notion of the "perfect marriage" seemed to position iReporters as the proverbial wives, whose unpaid labor was both disavowed and fundamental to CNN's ability to cover global conflict. CNN did not need to pay as many professional reporters to travel the world, so long as they had iReporters who could upload the images instead. The trick was to take advantage of this free labor without forfeiting the network's professional credibility and without allowing the network's newscasts to be inundated with a glut of information.

Industry ambivalence gained particular momentum during the media coverage of the Iranian elections and protests of 2009. At this time, the declarations that iReporters tended to be fickle and capricious, failing to remain objective and propagating dangerously inaccurate facts, acquired a new sense of urgency (Epps cited in Mahmud, 2008). Trade publications and mainstream media outlets like *The New York Times* began revisiting the notion that iReport was initially marketed as a "post-moderated site" and that participants' reports were only to be verified if producers thought they could use them in CNN's official newscasts (Grant cited in Shields, 2008). *Daily Variety* complained that CNN had become increasingly reliant on footage from iReporters during the protests, and then followed this statement with a detailed description of the "vetting" process that involved a "team of Farsi speakers" combing through each image to determine "when, where, and of whom each shot was taken" (Thielman, 2009: 3). Meanwhile, *The New York Times* accused CNN of "publish[ing] first, ask[ing] questions later" (Stelter, 2009: 1), while CBS correspondent Elizabeth Palmer bemoaned iReporters' lack of access to facts, subsequently conflating Iranian citizen journalists with her perception of the nation itself as a "country that runs on rumor" (cited in Thielman, 2009: 3).

This nasty backlash from professional journalists highlighted the aforementioned guardianship of their credibility, a credibility further problematized by their constant censorship during the Iranian protests (Thielman, 2009). ABC correspondent Jim Sciutto and his crew even began to use mobile phones and FlipVideo cams instead of their own more obtrusive equipment, going "under cover" as concerned citizens in order to get footage of the events (cited in Thielman, 2009: 3). Oddly enough, Sciutto's use of the mobile media so often associated with amateurs did not appear to hurt his professional credibility, nor did it call the validity of his video into question. While this technology shed doubt upon the iReporter's credibility, it only seemed to solidify the resourcefulness of the professional. This was partly because the professional could "correctly" channel the mobile media's potential to propagate multiple meanings back into the linear informational flow that depends on the traditional journalistic hierarchy between sender and receiver (Morse, 2004).

Still, in many cases, professional journalists were imprisoned or sent back home during this uprising, while the citizen journalists who supposedly lacked credibility were able to capture the footage that CNN needed. Despite their tireless labor for CNN, many iReporters complained that CNN producers often censored their work. In an email, CNN producer Henry Hanks told me that CNN simply *mediated* that work, removing video until it could be traced to a specific owner and verified for validity using a professional set of standards (Hanks interview, 2010). If iReporters immediately provided this information to CNN, then the process of distributing the video—according to Hanks—would go smoothly.

Hanks's statement reveals CNN's vacillation between the erasure of the citizen labor on which it depends, and its invocation of the iReporters' personal responsibilities to verify the validity of their footage. Andrejevic discusses a similar notion of responsibility, which he terms the "duty of interaction" (2007: 144). In this instance, the user is taught to take responsibility for his or her consuming practices and pleasures, even as those practices produce capital for the corporations that encourage them. Andrejevic argues that this dutiful interactivity should be understood as a corporate strategy, where the labor of production gets offloaded onto the consumer. CNN disciplines its iReporters in just this decentralized way, at times addressing them as though they were obsessive and ultimately volatile "fans" of the news organization, while at other times addressing them as students of professional journalism.

Both modes of address situate iReporters as amateurs who do not deserve monetary compensation for their labor. Yet, iReporters overwhelmingly assert that their unpaid labor is willingly given. I will now turn to an analysis of the complex labor relations involved in iReporting, relying on Bruns' explication of the prod-user's labor, as well as upon Terranova's claims that such labor—fundamental to network culture—is often "pleasurably embraced" while at the same time "often shamelessly exploited" (Terranova 2004: 78). Drawing upon a series of interviews completed with iReporters in 2010, I will complicate the neoliberal notion of monetary compensation, instead suggesting that many iReporters were seeking a very different type of reward during the protests of 2009.

IREPORT ACCORDING TO THE IREPORTERS

A look at iReport's fine print will reveal what Farooq Kperogi (2010) has referred to as the "many hidden assumptions" of the iReport site (2010: 321). In the iReport service agreement—a set of rules that all iReporters must agree to observe before their accounts go active—these terms are listed in relation to the status of participants' contributions to the iReport site as well as to CNN more generally:

By submitting your material, for good and valuable consideration, the sufficiency:

> and receipt of which you hereby acknowledge, you hereby grant to CNN and its affiliates a non-exclusive, perpetual, worldwide license to edit, telecast, rerun, reproduce, use, create derivative works from, syndicate, license, print, sublicense, distribute and otherwise exhibit the materials you submit, or any portion thereof in any manner and in any medium or forum, whether now known or hereafter devised, *without payment* to you or any third party (Terms of Service 2009, emphasis added).

While the service agreement stipulates that participants' contributions must be made for free, the document goes on to attest that iReporters cannot copy or redistribute any of CNN's own copyrighted material without the written permission of CNN executives. CNN's rule points to a different sort of hierarchy than the discursive distinction between the credibility of the professional and the spuriousness of the amateur; in this case, the iReporters' labor itself is monetarily devalued, while the production of CNN employees is protected by copyright law. This bald assertion of CNN's exploitation of the iReporters' production differs drastically from the language in the assignment section of the iReport site, where participants are encouraged to "share" their stories and hone their skills according to the guidelines listed under the sympathetic "Need Help?" link. Such rhetoric is repeatedly bolstered by promises that iReporters' work may soon run on the network or on CNN's official website. In this sense, the assignment section promises a small amount of professional recognition for those iReporters who put in the right amount of hours while the service agreement stresses that there will be no financial compensation for the iReporters' labor.

In order to get a better sense of the iReporters' own feelings on the merits of gaining professional recognition instead of financial compensation, I contacted numerous iReporters through the website's inbox messaging system, asking them to recount their experiences reporting for the site. On the one hand, I contacted Global Challenge participants, leaving out the less active users who had contributed less than two iReports. I also contacted iReporters who had contributed two or more iReports covering the Iranian protests between June and August of 2009. When asked how CNN responded to their "thoughts, reports, and photography," the Global Challenge participants who responded to my message optimistically referenced the possibility of professional recognition. One Global Challenge participant stated that after seeing her photos air on CNN's official newscast, her interest in iReport grew, leading her to take even more photos for the website. Other iReporters for the Global Challenge assignment emphasized how encouraging the iReport producers were, making them feel that their work was much appreciated. This reference to professional recognition and appreciation was a recurring theme.

Some of the Global Challenge participants specified that iReport producers had found their personal blogs online and were impressed enough to seek them out; these iReporters identified themselves as semi-professional writers or aspiring journalists. The possibility of using iReport as a platform to gain monetary compensation down the line seemed to be a primary factor in motivating these writers, a possibility that is not altogether unrealistic according to Hanks. He informed me that the CNN iReport internship is typically filled by a former iReporter and that CNN has also hired former iReporters as employees. Still, Hanks continued the tradition of distinguishing between amateurs and professionals by stressing the word "former"—paid CNN employees can only be "former" iReporters, while current iReporters work for free.

This statement not only echoes the aforementioned industry discourse on citizen journalism; at first glance, Hanks' assertion also resonates for Andrejevic's argument that uncompensated online activity has been increasingly appropriated by corporations, with much corporate labor being offloaded onto consumers (2007). Indeed, Farooq Kperogi (2010) has declared iReport as merely one more example of "corporate media hegemons... co-opting what was once the subversive sphere of citizen journalism" (2010: 316). Still, while the iReporters' labor is undeniably uncompensated, Bruns' (2008) notion of prod-usage complicates the notion of consumer exploitation. If iReport can be viewed not merely as a corporate construction, but also as a very real site of networked collaboration, then the activity unfolding there must be aligned with prod-users rather than consumers—with the "users" who "are always already necessarily also producers of the shared knowledge base" (Bruns, 2008: 2). While CNN is indeed "harvesting" such prod-usage, as Bruns might put it, the knowledge base upon which CNN draws is irrevocably changed.

Correspondingly, Terranova (2004) asserts that the notion of a monolithic corporate structure appropriating the "authentic" labor of the online user should be scrutinized, because "incorporation is not about capital descending on authentic culture" but about "collective cultural labor" being "voluntarily *channeled* and controversially *structured* within capitalist business practices" (2004: 80; emphasis in original). In this sense, not all uncompensated labor is wholly exploited labor, since the conditions that "make free labor an important element of the digital economy are based on a difficult, experimental compromise" between the "desire for creative production...and the current capitalist emphasis on knowledge as the main source of added value" (2004: 77). This kind of work does not happen simply "because capital wants [it] to," but because of a "desire for affective and cultural production which [is] none the less real because [it] is socially shaped" (Terranova, 2004: 77).

Through the lenses provided by Bruns and Terranova, one could argue that the Global Challenge participants' labor is driven by the desire to produce, and that they view the pleasure of production, as well as the resulting recognition, as compensation enough. The desire to produce also informed the citizen coverage of the Iranian uprisings, though for very different reasons. In that context, iReporters were attempting to produce a variety of political messages that could then be quickly communicated across wide distances. My correspondence with iReporters involved in the protests of 2009 revealed that participants' hope of telling the world about the events in Iran was the driving force behind the hours spent working without pay. The recognition so often mentioned on the iReport site, as well as by the Global Challenge participants, was in this case recoded as visibility—not the visibility of the iReporters themselves, but the visibility of the political information being distributed. The possibility of flooding the global news media with protesters' own opinions of the events—opinions that often competed with those of professional journalists based in the U.S.—inspired the dedication of the iReporters' unpaid labor. Thus, rather than utilizing iReport as a sort of training camp for a future career in journalism, these iReporters appropriated the CNN technology *only* so that they could distribute their political messages in a way that ultimately disrupted the traditional representational system upon which professional journalism depends.

One iReporter stated that he chose to work with the website during this time because "it was connected to a prominent and well respected establishment, it had many viewers, and it was visible to people from all around the world, including Iran." He argued that such visibility helped protesters within Iranian borders to realize that "despite all the hardship they were not alone. This seemed to give them additional strength." Thus, CNN's corporate communications technology was made to work *for* the iReporters in this case, sending messages of support to those who were being misrepresented by Western media (Mirsepassi, 2010). Most of the iReporters with whom I spoke mentioned this ability to encourage and support the protests through the use of sites like iReport. They also noted that iReport was useful to them because it served as one of many channels along which they could relay multiple interpretations of the events in Iran.

Numerous participants said they used several other social media technologies during the protests, drawing upon sites like Twitter and Facebook as well as iReport. This practice suggests that during the uprisings, iReport operated as one node in a larger network of communication, pointing again to Bruns's notion of the openness of systems in the age of the prod-user. Even more compellingly, when iReporters covering the protests felt they had maximized CNN's potential for their purposes, many all but abandoned their iReport accounts, ceasing their labor for the website and only periodically checking in to see what other iReporters were saying. In this sense, these iReporters ensured that CNN's effort at harvesting their labor could only be successful as long as they were getting something in return. Rather than losing their political identities in their affiliation with the CNN brand, these iReporters manipulated that brand to make it work for them. This proves that though their unpaid labor was structured within CNN's complex business practices, these iReporters were still not entirely controlled by CNN's business system, pointing again to Terranova's assertion that this kind of labor is not merely exploited by capitalism but is also evidence of the desire to creatively produce (Terranova, 2004).

My case study highlights the numerous obstacles placed in front of the citizen journalists who increasingly draw upon corporate visibility to tell their stories. In turn, it suggests that corporations like CNN are not "monolithic structures, but allow for internal complexity," (Goode, 2009: 1289) especially as network culture fuels the increased interdependence of corporate news organizations and citizen journalism. The interdependence of CNN and citizen reporting complicates the static maps of the world that professional journalism attempts to construct. While interactive projects like iReport do not fully unleash their own transformative potential, they are still important sites at which to examine the disruptive capabilities of citizen journalists and the proliferation of possibilities they engender. Such disruption illuminates existing inequalities that need further investigation. At the very least, such disruption reveals that the project of staking out the political, social, and affective topographies that unfold within network culture—the project of "mapping" the world, in all its unevenness and multiplicity—*should* be a collaborative effort, and not just in name.

REFERENCES

Allan, S. (2006). *Online news*. New York: Open University Press.

Allan, S., and E. Thorsen, ed. (2009). *Citizen journalism: Global perspectives*. New York: Peter Lang.

Andrejevic, M. (2007). *Ispy: Surveillance and power in the interactive era*. Lawrence: University of Kansas Press.

Barkin, S. (2003). *American Television News: The Media Marketplace and the Public Interest*. New York: M.E. Sharpe.

Barlow, A. (2007). *The rise of the blogosphere*. Westport, CT: Praeger.

Bruns, A. (2008). *Blogs, Wikipedia, Second Life, and beyond : From production to produsage*. New York: Peter Lang.

Compton, J. R., and P. Benedetti. (2010). "Labor, new media and the institutional restructuring of journalism." *Journalism Studies* 11(4): 487–99.

Friend, C., and J. Singer. (2007). *Online journalism ethics: Traditions and transitions*. New York: M.E. Sharpe, Inc.

Frosh, P., and A. Pinchevski. (2009). *Media witnessing: Testimony in the age of mass communication*. New York: Palgrave Macmillan.

Gillmor, D. (2004). *We the media : Grassroots journalism by the people, for the people*. 1st ed. Beijing and Sebastopol, CA: O'Reilly.

Goode, L. (2009). "Social news, citizen journalism and democracy." *New Media & Society* 11(8): 1287–305.

Hall, S., C. Critcher, T. Jefferson, J. Clarke, and B. Roberts (1999). "Policing the crisis." In *News: A reader*, edited by H. Tumber. Oxford: Oxford UP.

Hanks, H. (2010). Online correspondence with the author, 11 November 2010. *iReport*. (2010). http://www.iReport.cnn.com.

"iReport Turns One Year Old." (2007). *CNN.com*. http://www.cnn.com/2007/US/08/01/ireport.first.year/.

Kperogi, F. (2010). "Cooperation with the corporation? CNN and the hegemonic cooptation of citizen journalism through iReport.com." *New Media & Society* 13(2): 314–29.

Küng-Shankleman, L. (2000). *Inside the BBC and CNN: Managing media organisations*. New York and London: Routledge.

Longpre, M. 2008. "CNN takes gamble with unfiltered site." *PR Week*, 18 February, pg. 10.

Mahmud, S. (2008). "Community TV," *Adweek*, 17 March, no page number.

McChesney, R., and J. Nichols. (2011). *The death and life of American journalism: The media revolution that will begin the world again*. New York: Nation Books.

Mirsepassi, A. (2010). *Democracy in Iran: Islam, culture, and political change*. New York and London: NYU Press.
Morse, M. (2004). "News as performance: The image as event." In *The television studies reader*, edited by R. C. Allen and A. Hill, 209–25. New York and London: Routledge.
Nawawy, M. (2010). "The 2009 Iranian presidential election in the coverage of CNN and Al-Jazeera websites." *Media, power, and politics in the digital age : The 2009 presidential election uprising in Iran*, edited by Y. Kamalipour, 3–14. Lanham, MD: Rowman & Littlefield Publishers.
Papacharissi, Z. (2009). *Journalism and citizenship: New agendas in communication*. New York: Routledge.
Parks, L. (2009). "Digging into Google earth: An analysis of 'Crisis in Darfur.'" *Geoforum* 40(4): 535–45.
Shields, M. (2009). "CNN ireport's Iran filings fly." *MediaWeek*, 22 June, pg. 6.
Shields, M. (2008). "CNN: Power to the people." *MediaWeek*, 11 February, pgs. 4–6.
Sreberny, A., and G. Khiabany. (2010). *Blogistan : The internet and politics in Iran*. London and New York: I.B. Tauris/Palgrave Macmillan.
Stelter, B. (2010). "CNN fires executive who led makeover." *The New York Times*, 25 September, pg. 4.
Stelter, B. (2009). "Journalism rules are bent in news coverage from Iran." *New York Times*, 29 June, sec. B1
Terms of service. http://www.ireport.cnn.com. Last updated 2009.
Terranova, T. (2004). *Network culture: Politics for the information age*. London and Ann Arbor, MI: Pluto Press.
Thielman, S. (2009). "As Iran blocks media, the press improvises." *Daily Variety*, 24 June, pg. 3.
Tremayne, M. (2007). *Blogging, citizenship, and the future of media*. New York: Routledge.
Turner, G. (2010). *Ordinary people and the media : The demotic turn*. London: Sage.
Volkmer, I. (1999). *News in the global sphere: A study of CNN and its impact on global communication*. Luton: University of Luton Press.
Washkuch, F. (2008). "News sites gain from citizen journalism." *PR Week*, 28 April, pg. 10.
Zelizer, B. (1999). "CNN, the gulf war, and journalistic practice." In *News: A reader*, edited by H. Tumber, 340–54. Oxford: Oxford UP.

CHAPTER THREE

Righting Wrongs: Citizen Journalism AND Miscarriages OF Justice

CHRIS GREER AND EUGENE MCLAUGHLIN

This chapter uses a case study approach to demonstrate the agenda-setting power of citizen journalism. More specifically, we explore the interaction of media, political and judicial forces following the death of newspaper vendor Ian Tomlinson shortly after being struck by a police officer at the G20 Protests in London 2009. First, we map out key transformations in the contemporary news environment, and discuss the implications of these for the control of information during the policing of public order events. Second, we discuss the rise of the citizen journalist as an important and developing feature of this transforming news media landscape. Third, we describe the 'events' of the G20 protests, and consider the initial 'inferential structure' used by the news media to make sense of them.[1] Fourth, we analyse the news media maelstrom around the death of Ian Tomlinson, and examine how the initial inferential structure and flows of communication power were disrupted by the intervention of citizen journalists. Fifth, we consider the news media outrage at the Crown Prosecution Service (CPS) failure to prosecute the police officer filmed striking Tomlinson shortly before he collapsed and died, and the subsequent official responses, inquiries and prosecutions. Finally, we return to our conceptual framework to consider the wider implications of this case study.

THE TRANSFORMING NEWS ENVIRONMENT AND THE RISE OF 'CITIZEN JOURNALISM'

The contemporary reporting of crime and public protests takes place within a radically transformed information-communications environment. The police are increasingly enmeshed in a complex web of internal and external stakeholders and 'publics' with different agendas and needs who are willing and able to use the news media and internet to represent their interests. Cottle (2008) has noted the extent to which protest groups and demonstrators have become 'reflexively conditioned' to get their message across and activate public support. The contemporary news media environment offers 'new political opportunities for protest organizations, activists and their supporters to communicate independently of mainstream news media' (2008: 853; DeLuca and Peeples 2002; Hutchins and Lester 2006; Maratea 2008; McCaughey and Ayers 2003). Protesters are aware that their activities have to compete proactively for space in the fast-moving, issue-based attention cycle that defines the 24–7 news mediasphere (Oliver and Maney 2000). In addition, as Milne (2005) argues, there has been a notable shift in political perspective amongst sections of the Fourth Estate as they attempt to prise open the political process. Market-driven newspapers in particular are much more willing to initiate and/or support anti-government/establishment campaigns and protests, and in certain respects have become 'ideologically footloose'. Consequently, there is the increased possibility of damaging images and representations of state institutions such as the police materialising and circulating in the offline and online news media. Of crucial importance here is the rise of the citizen journalist.

Allan and Thorsen (2009) define citizen journalism as 'the spontaneous actions of ordinary people, caught up in extraordinary events, who felt compelled to adopt the role of a news reporter'. Peat (2010) provides a vivid description: 'Armed with cellphones, BlackBerries or iPhones, the average Joe is now a walking eye on the world, a citizen journalist, able to take a photo, add a caption or a short story and upload it to the internet for all their friends, and usually everyone else, to see'. In recognition of this unprecedented news-gathering potential, news organisations have established formal links to encourage citizens to submit their mobile news material (Pavlik 2008; Wallace 2009). Citizen-generated content, in turn, can generate other information and images, fuelling 'endless remixes, mashups and continuous edits' (Deuze 2008: 861). Citizen journalism has been instrumental not only in providing newsworthy images, but also in defining the news itself—in shaping representations of key global events. The defining images of the 7/7 London bombings in 2005, probably the watershed in the emergence of a highly interactive and participatory contemporary news production process, were provided by citizen journalists (Sambrook 2005). The emergence of the citizen journalist carries significant implications for professional news gathering organisations and official institutions who would seek to control the news. Novel forms of selecting, gathering, processing, and disseminating 'news' are transforming communication circuits. On the one hand, there are real issues of simulation, manipulation, partisanship and lack of accountability. On the other, 'right here, right now' citizen journalism can bring authenticity, immediacy and realism to news stories through the production of dramatic and visually powerful 'evidence' of events 'as they happen'. The G20 demonstrations in the City of London on 1st April 2009 provide an important insight into the disruptive impact of citizen journalism upon routinised police-news media relations. They also illustrate the shifting nature of definitional power in the 24–7 news mediasphere.

IAN TOMLINSON AND CITIZEN JOURNALISM: FROM 'PROTESTER VIOLENCE' TO 'POLICE VIOLENCE'

The April 2009 G20 London Summit involved a meeting of the Group of Twenty Finance Ministers and Central Bank Governors—the G20 heads of government or state—to discuss the financial crisis of 2007–2008 and the world economy. Given the high levels of public anger at the way in which the financial crisis was being managed, the summit would become one of the most high profile security events to be staged in the UK. The London Metropolitan Police Service (MPS) identified a number of 'unique' factors that had the clear potential to generate problems for the securitisation of the G20 Summit, codenamed 'Operation Glencoe'. First, an unprecedented number of public order events were taking place simultaneously across London, including: the arrival of G20 delegations, a state visit by the President of Mexico, and an international football match at Wembley. Since any one of these events could present a target for a terrorist strike, the logistical pressures on police resources was enormous. Second, the potential for trouble would be increased significantly by widespread public anger at the handling of the financial crisis. And thirdly, a coalition of anarchist, anti-globalisation, anti-war and environmentalist 'direct action' groupings had declared their intention to 'take' the financial heart of the City of London. These groups were using a range of media to communicate their plans and exchange views on how the days of protest would develop, where the 'flashpoints' would be, and the likelihood that the police would over-react (Greer and Mclaughlin 2010, 2012).

In the countdown to the G20 protests, both the police and the press drew from a well-established or default news frame in order to interpret and explain the unfolding events. This default news frame was 'protester violence': that is, there was a clear sense that the demonstrations would be marred by violence, and that this violence would come from the protesters (Gorringe and Rosie 2009). An initial inferential structure developed around the news frame of 'protester violence', and it was this framework—reflecting and reinforcing the police perspective—that shaped newspaper coverage in terms of 'what the story was' and 'how it would develop over time'. When, as predicted, protesters clashed with police on 1st April, the inferential structure crystallised and now explicitly set the context for newspapers' interpretation of events at G20.

At 11.30pm on 1st April the MPS released a statement disclosing that a man had died in the area of the Bank of England (MPS statement, 1st April). Partly due to the timing and context of the statement, the press situated the death within the existing inferential structure, and reproduced the police narrative which claimed that the man had died in the midst of chaotic protester violence. Journalists' reports and protest group websites conflicted over whether or not the dead man, Ian Tomlinson, was a protestor, and where he had collapsed. On 2 April the Independent Police Complaints Commission (IPCC) confirmed that it had been asked by the police to review Tomlinson's death. An immediate post-mortem examination established that he had suffered a heart attack and died of natural causes. Whatever Tomlinson's G20 protest connections, the police position was that he had not come into contact with officers prior to collapsing in the street.

One of the most noticeable characteristics of the 1st April protests was the sheer density and variety of recording devices being used by professional and citizen journalists, private businesses, demonstrators, the police, and passers-by. Furthermore, because of police containment tactics, police-news-media-protester-public interactions took place in extremely close spatial proximity, which simultaneously created a captive audience to surrounding events. The result was a hyper-mediatised, high-surveillance context within which control of the information and communication environment would be difficult to maintain. As photographs of Ian Tomlinson appeared in the news media and online, witnesses began to emerge, claiming they had seen the man interacting with the police on several occasions. Their testimonies, significantly brought first to the news media rather than the IPCC, challenged the official line that bottles had been thrown at police while they were attending to Tomlinson after his collapse. It soon transpired that Tomlinson, in attempting to make his way home from work, had in fact come into contact with the police on several occasions prior to collapsing at 7.30pm. In a pivotal news media intervention, on 3rd April the *Guardian* informed City of London Police, who were responsible for conducting the IPCC investigation into the death, that it had obtained timed and dated photographs of Tomlinson lying on the pavement at the feet of riot police. On 5th April *The Guardian* published several of these photographs, along with the testimony of three named witnesses who claimed they had seen Tomlinson being hit with a baton and/or thrown to the ground by officers. The next day the IPCC confirmed that Tomlinson had come into contact with officers prior to his death, but continued to contest reports that he had been assaulted.

Serious concerns about the policing of G20 were aired across the weekend news media on 4th and 5th April, accompanied by the first calls for a public inquiry. Ian Tomlinson was becoming a *cause célèbre*. The decisive moment came on April 7th, when the *Guardian* website broadcast mobile phone footage that appeared to provide clear evidence of police violence against Tomlinson minutes before he collapsed. The footage had been handed to the newspaper by an American fund manager who said, 'The primary reason for me coming forward is that it was clear the family were not getting any answers' (*Guardian*, 7th April 2009). It shows Tomlinson walking, hands in pockets, seemingly oblivious to an adjacent group of officers, some dog handlers, and others in riot gear. He presents no discernible threat to public order. Without warning, an officer in helmet and balaclava pushes Tomlinson forcefully from behind, knocking him to the ground. When slowed-down, the footage captures the officer swiping at Tomlinson's legs with a baton, and then pushing him hard in the back. Police stand and watch as passers-by help Tomlinson to a sitting position, where he appears to remonstrate with the officers in question. He is then helped to his feet, again by passers-by, and is seen walking away. Soon afterwards he will collapse beyond the view of this camera.

The *Guardian* shared the footage with the news channels of the *BBC*, *Sky* and *Channel 4*. It was also added to various online news sites, and to *YouTube*. The footage was picked up globally and was by far the most read story on the *Guardian's* website, with about 400,000 views. It initiated intensive blogging and a letter-writing campaign to parliament. Authenticated, real-time footage of events surrounding Ian Tomlinson's death provided a focus for the growing body of complaints, led by the Tomlinson family who had now established a campaign website (http://www.iantomlinsonfamilycampaign.org.uk), about (a) the overall policing of G20, and (b) the actions of officers attached to specialist units. On 8th April new footage shot from a different angle, retrieved from a broken *Channel 4* camera, showed an officer striking at Mr. Tomlinson from behind with a baton and then pushing him to the ground. This combined footage set the agenda not only for other news agencies, but also for the response of the MPS and the IPCC. The MPS subsequently confirmed that four officers had come forward in relation to the investigation into the death of Mr. Tomlinson.

The initial inferential structure around 'protester violence'—so routinely and un-controversially established in the run up to the G20 protests—had disintegrated, and a new inferential structure—initiated and driven by the raw content of citizen journalism—had crystallised around the news frame of 'police violence'. The emergence of this *dominant* inferential structure was evident in the shifting focus of news media interest, and how the 'story' of G20 was re-ordered and re-interpreted within that context. But further, and crucially, this dominant inferential structure was evident in the extensive and highly public official response that asked probing questions about the MPS's public order policing strategy, and foregrounded the importance of two media-related phenomena: the need for the MPS to develop more positive police-press relations, and the implications of the rise of the citizen journalist for the policing of public events.

'NO REALISTIC PROSPECT OF A CONVICTION': THE CROWN PROSECUTION SERVICE (CPS) DECISION ON IAN TOMLINSON

On 23rd July 2010, Keir Starmer, the Director of Public Prosecutions (DPP), confirmed that Ian Tomlinson did not pose a threat to any police officer he had encountered on 1st April 2009. His innocence was officially confirmed. Starmer verified that the officer's use of force had been disproportionate and unjustified. However, there was an 'irreconcilable conflict' between pathologists about the cause of Tomlinson's death (Starmer, *Sky News*, 22 July 2010)—one ruled that Tomlinson died from a heart attack, whilst two subsequent post-mortems by other pathologists concluded that he had died as a result of internal bleeding after a blow to the abdomen. Consequently, there was 'no realistic prospect' of pursuing a conviction for manslaughter or assault occasioning actual bodily harm. The IPCC immediately released a statement that it would now conclude its final report and present it to the Coroner so that preparations could be made for an inquest. At a news conference the Tomlinson family and their lawyer branded as a 'cover-up' the CPS decision not to bring criminal charges against the TSG officer. The outrage of the Tomlinson family registered immediately across broadcast news bulletins and newspaper websites. Coverage was contextualised by re-running or re-posting video footage of the policing of G20 and, in particular, Tomlinson's encounter with the TSG officer. The news media inferential structure was crystallising around the news frame of systemic, multi-agency 'institutional failure' – a failure of 'justice'—and explicitly set the tone for press interpretations of the Tomlinson case the following day.

The CPS decision dominated the front pages of the *Guardian*, the *Times* and the *Daily Telegraph*, the *London Evening Standard* and the *Metro* (one of London's several 'freesheets'), and was covered on the inside pages of the *Independent*, the *Sun*, the *Daily Mirror* and the *Daily Express*. Headlines were remarkably consistent, communicating a clear consensus across tabloid and broadsheet, left and right. Moral indignation in the form of the Tomlinson family's 'fury' and 'outrage' was the dominant emotional register. The police officer had been 'let off' and allowed to escape justice. News items, feature articles and editorials reinforced and advanced an inferential structure that had been developing since the footage of Tomlinson's assault had been made public. Now, the dominant inferential structure extended beyond the Metropolitan Police Service (MPS) to include the CPS and the IPCC. The Tomlinson story continued to evolve as a rolling news story. But it was no longer about the Tomlinson case alone. It constituted collective press outrage at the impunity of police officers and the ineffectiveness of the structures of accountability designed to deliver public protection and justice. With each new development in the Tomlinson case, the inferential structure built around systemic institutional failure was consolidated and strengthened, and the journalistic distrust in those who possess and exercise institutional power simultaneously appeared to be validated and amplified across the criminal justice estate.

THE MEDIATISATION OF THE TOMLINSON INQUEST

In England and Wales, an inquest is a fact-finding legal inquiry to establish who has died, and how, when and where the death occurred. It is held in public—sometimes with a jury—by a coroner, in cases where the death was violent or unnatural, took place in prison or police custody, or when the cause of death is still uncertain after a post-mortem. An inquest does not establish liability or blame (Ministry of Justice 2012). Because the nature of the proceedings is inquisitorial, rather than adversarial, the inquest is the only independent forum in which questions can be asked, enabling families of the deceased to understand the circumstances of the death. In theory, the inquest is one of the clearest manifestations of the principle of 'open justice' in England and Wales. For critics and some victims, this potential is seldom achieved and inquests—due to delays, inconsistent verdicts and built-in reluctance to censure criminal justice agencies—are as likely to perpetuate 'miscarriages of justice' as they are to deliver 'open justice'. Those dissatisfied with the inquest process have no avenue of appeal. The inquest of Ian Tomlinson began on 28th March 2011, and heard evidence until 21st April. We would argue that the video footage of what happened to Tomlinson played a critical role in the jury's deliberations. The level of controversy and news media interest surrounding the Tomlinson case ensured that the inquest would be a high-profile event. Its importance was further signalled by the fact that it was conducted by the chief coroner, Judge Peter Thornton QC, who replaced the City of London coroner, Paul Matthews.

To our knowledge this was the first inquest in England and Wales to take full account of the new media environment. The *Guardian* and other newspapers were given permission by the coroner to tweet live from the inquest. The Tomlinson inquest was the first in British legal history to be reported as it happened, and was made accessible in real time via tweets and live blogs to millions of virtual onlookers. A dedicated Ian Tomlinson inquest website was created, providing updated information for anyone interested in the proceedings. Transcripts and links to key video and photographic evidence were uploaded daily. Anticipating significant public interest, members of the public and accredited journalists who could not access the inquiry room were allowed to watch the proceedings from a specially equipped court annexe. This annexe was serviced by a live audio and video link to the court room and a running display of the transcripts.

Visual evidence played an unprecedented role in the inquest. The Independent Police Complaints Commission (IPCC) was instructed to compile a montage of video footage from citizen, professional and official sources including CCTV cameras, police helicopters, police surveillance teams, news organisations, bystanders and websites. The footage was ordered chronologically into an evidential documentary that sought to provide a 360-degree account of events running up to Tomlinson's death. The first video montage covered Tomlinson's attempts to negotiate various police cordons. The second followed PC Simon Harwood as he was seen engaging with various protestors during the course of the day. The inquest opened with this evidential documentary being shown to the jury. An IPCC representative was asked to provide an account of the methods used to construct a visual data base. Over the four week period, the jury heard of Mr Tomlinson's actions on the day, and police, including PC Harwood, described their involvement. The medical reasons for Mr Tomlinson's death were discussed in detail, and three of the four pathologists who carried out post-mortem examinations, as well as other medical experts, were called to give evidence. The visual evidence, along with photographs, was used in the cross examination of witnesses to guide the discussion and assess the accuracy of the evidence being given. The Coroner also allowed the jury to re-watch the visual material when they retired to consider their verdict. No qualitative distinction was made between official, professional and citizen footage.

This was also the first inquest where the testimony of a citizen journalist was recognised as central to the deliberations. Christopher La Jaunie spoke publicly for the first time about his filming of the moment PC Harwood struck Ian Tomlinson. From reading the initial news coverage he realised that there were discrepancies between the official police account of what had happened and what he had witnessed and filmed:

> 'I basically contacted every reporter who had followed the story by email to say "Hey, I have something that may be of interest to you", because at the time, as you know, the story that had come out was that he had just died of natural causes, completely unrelated to this... In my opinion, that footage was contradicting the story' (Testimony, Ian Tomlinson Inquest, 31st March 2011).

As we have noted, only the *Guardian* appears to have recognised the significance of La Jaunie's footage. On 30th March 2011, all the main national newspapers reproduced the same iconic still—taken from this footage—of Ian Tomlinson being pushed to the ground by PC Harwood.

> 'Tears as family sees G20 victim's final moments' (*Daily Mail*, 30th March 2011: 22)
>
> 'Riot violence of officer in G20 death: Footage of Harwood with other protesters' (*Daily Telegraph*, 30th March 2011: 13)
>
> 'Family weep over CCTV film over man's death at G20' (*Times*, 30th March 2011: 9)
>
> 'Family in tears as Tomlinson's last moments shown' (*Guardian*, 30th March 2011: 7)
>
> 'Widow's tears as she sees G20 shove: Inquest shown new evidence' (*Daily Mirror*, 30th March 2011: 6)
>
> 'Tears for G20 dad' (*Sun*, 30th March 2011: 21)
>
> 'Tearful Tomlinson family see footage of G20 death' (*Independent*, 30th March 2011: 20)

On 3rd May the inquest jury returned the verdict that Ian Tomlinson had been unlawfully killed by PC Harwood, though use of 'excessive and unreasonable' force. The headlines the following day were robust in their near unanimous calls for Harwood to be formally prosecuted. Once more the iconic still of Ian Tomlinson illustrated the highly critical press coverage. The Director of Public Prosecutions (DPP) announced that evidence emerging during the inquest would be reviewed to ascertain if, despite the DPP and CPS previous decision not to prosecute though 'no realistic chance of a conviction', Harwood should now be charged with manslaughter. On 24th May 2011, the DPP announced that there was a 'realistic prospect' of convicting PC Harwood, who would now face prosecution on the charge of manslaughter. On 18th June 2012, PC Harwood stood trial at Southwark Crown Court. On 19th July 2012 he was found not guilty of the manslaughter of Ian Tomlinson. The press response to the 'unjust' verdict was unanimously scathing:

> 'Call This Justice?' (*Daily Mirror*, 20 July, Front Page)
>
> 'Cleared but G20 cop had stormy past' (*Sun*, 20 July, p.17)
>
> 'Freed: 'Thug in Police Uniform' (*Daily Mail,* 20 July, Front Page)
>
> 'PC Cleared of G20 Protest Killing, But Now faces the Axe' (*Daily Express*, 20 July, Front Page)
>
> 'Policeman Cleared over G20 Death, but Questions Remain' *(Guardian*, 20 July, Front Page)
>
> 'Not Guilty but Not Innocent' (*Independent,* 20 July, Front Page)
>
> 'Stains on the record of G20 Officer' (*Daily Telegraph*, 20 July, p.7)
>
> 'G20 Officer had Long History of Misconduct' (*Times*, 20 July, Front Page)

The press coverage juxtaposed a photograph of PC Harwood leaving court with the now instantly recognisable image of Ian Tomlinson. On 17th September 2012, PC Harwood was found guilty of gross misconduct by a Metropolitan Police disciplinary panel and sacked with immediate effect. Ian Tomlinson's family left the hearing before PC Harwood was dismissed from the force, saying the process was 'pointless' and left them with 'no answers'. They said they intended to pursue the case in civil court to try and establish who was responsible for Ian Tomlinson's death.

CONCLUSIONS

The nature and intensity of the Tomlinson news coverage, substantiated by real-time citizen-generated content of this and other incidents of police violence, and reinforced by the internet, made the MPS public order policing strategy a live political and policy issue that had to be addressed. Following the G20 protests, a raft of official inquiries into 'Operation Glencoe' raised wider questions about public order policing and the news media in the 21st Century (HMIC 2009; House of Commons Home Affairs Committee 2009; IPCC 2010; Joint Committee on Human Rights 2009; Metropolitan Police Authority 2010). The resulting reports all expressed concern that the high-profile exposure of police violence, however isolated, could seriously damage public confidence in the police. The changing media environment featured prominently in discussion of: the poor state of police-news-media relations, which generated tensions, frustrations and conflict between professional journalists and on the ground officers; the sophisticated use of multi-media technologies by protest groups, which by far surpassed the static communicative capabilities of the police; and the significance of the citizen journalist for intensifying public scrutiny of individual and collective police action, and in shaping public perceptions of the police. At the request of the Tomlinson family, an IPCC investigation was established specifically to consider the way the MPS and City of London Police handled the news media in the aftermath of Ian Tomlinson's death.

Were it not for the incendiary visual evidence handed to the news media by citizen journalists, the 'story' of Ian Tomlinson may never have taken off, the MPS may have succeeded in denying or defusing allegations of police violence, and the policing of G20 may have been record officially as a resounding success. Because of citizen journalism, the operational integrity and institutional authority of the MPS was first of all questioned, and then successfully challenged. The citizen-generated coverage of the events surrounding Ian Tomlinson's death was validated systematically, first through the national press, then through various official inquiries and, most significantly, as documentary evidence in the inquest. In the process, it became not only part of the official record of the policing of G20, but its defining element. Feeding directly into the inquest proceedings, it was core to the task of uncovering the *truth* of what happened on 1 April 2009—a vital means of ascertaining the accuracy of verbal evidence, and a primary source of influence on the jury's deliberations. The Inquest findings, combined with sustained news media outrage at the lack of police accountability—all underpinned by visual evidence from citizen, professional and official sources—were central in reversing the CPS decision that there was 'no realistic prospect' of securing a conviction against Tomlinson's police assailant. This in turn was a necessary stage in the processes that ultimately led to Harwood's acquittal for manslaughter, and dismissal for gross misconduct.

On 5 August 2013, the MPS issued a formal apology to Ian Tomlinson's family. They acknowledged for the first time that 'excessive and unlawful' force had been used. Ian Tomlinson's family said the apology marked the end of a long legal battle blighted by untruthful accounts and obstruction by PC Harwood. Harwood's dismissal and the MPS apology are clear evidence of the influence that citizen-generated content can have on the justice process. Tomlinson's widow, Julia Tomlinson, said 'The public admission of unlawful killing by the Metropolitan police is the final verdict, and it is as close as we are going to get to justice'.

NOTES

1. The concept of 'inferential structures' explains how the same news content can be shaped into multiple configurations, establishing selectively representative frameworks of understanding that shape how both newsmakers and news consumers interpret the story (see also Lang and Lang, 1955).

REFERENCES

Allan, S., and Thorsen, E. (eds) 2009 *Citizen Journalism: Global Perspectives*, London: Peter Lang.

Cottle, S. 2008 'Reporting Demonstrations: The Changing Media Politics of Dissent', *Media, Culture & Society* 30(6): 853–872.

DeLuca, K. M., and Peeples, J. 2002 'From Public Sphere to Public Screen: Democracy, Activism, and the "Violence" of Seattle', *Critical Studies in Media Communication* 19(2): 125–151.

Deuze, M. 2008 'The Changing Nature of News Work: Liquid Journalism and Monitorial Citizenship', *International Journal of Communication* 2(5): 848–865.

Gorringe, H., and Rosie, M. 2009 'What a Difference a Death Makes: Protest, Policing and the Press at the G20', *Sociological Research Online* 14(5).

Greer, C., and Mclaughlin, E. 2010 'We Predict a Riot? Public Order Policing, New Media Environments and the Rise of the Citizen Journalist', *British Journal of Criminology* 50(6): 1041–1059.

— 2012 '"This Is Not Justice': Ian Tomlinson, Institutional Failure and the Press Politics of Outrage', *British Journal of Criminology* 52(2): 274–293.

HMIC 2009 'Adapting to Protest—Nurturing the British Model of Policing', London: Her Majesty's Inspectorate of Constabulary.

House of Commons Home Affairs Committee 2009 'Eighth Report: Policing of the G20 Protests', available at http://www.publications.parliament.uk/pa/cm200809/cmselect/cmhaff/418/41802.htm#evidence

Hutchins, B., and Lester, L. 2006 'Environmental Protest and Tap-Dancing with the Media in the Information Age', *Media, Culture & Society* 28(3): 433–451.

IPCC 2010 'IPCC Completes Assessment of Complaint by Jody McIntyre and Update on Number of Complaints Received Following Student Protests', London: Independent Police Complaints Commission.

Joint Committee on Human Rights 2009 'Demonstrating Respect for Rights: Follow Up', Parliament: Stationary Office.

Lang, K., and Lang, G. 1955 'The Inferential Structure of Political Communications: A Study in Unwitting Bias', *Public Opinion Quarterly* 19(2): 168–183.

Maratea, R. 2008 'The e-Rise and Fall of Social Problems: The Blogosphere as a Public Arena', *Social Problems* 55(1): 139–160.

McCaughey, M., and Ayers, M. D. 2003 *Cyberactivism: Online activism in theory and practice*, New York: Routledge.

Metropolitan Police Authority 2010 'G20: Report of Civil Liberties Panel', available at: http//:www.mpa.gov.uk/committees/mpa

Milne, K. 2005 *Manufacturing Dissent: Single-Issue Protest, the Public and the Press*, London: Demos.

Ministry of Justice 2012 'Guide to Coroners and Inquests and Charter for Coroner Services', London: Ministry of Justice.

Oliver, P., and Maney, G. 2000 'Political Processes and Local Newspaper Coverage of Protest Events: From Selection Bias to Triadic Interactions', *American Journal of Sociology* 106(2): 463–505.

Pavlik, J. V. 2008 *Media in the Digital Age*, New York: Columbia University Press.

Peat, D. 2010 'Cellphone Cameras Making Everyone into a Walking Newsroom' *Toronto Sun*, 1st February Edition, Toronto.

Sambrook, R. 2005 'Citizen Journalism and the BBC' *Nieman Reports*: available online at: www.nieman.harvard.edu/reportsitem.aspx?id=100542.

Wallace, S. 2009 'Watchdog or Witness: The Emerging Forms and Practices of Video-Journalism', *Journalism* 10(5): 684–701.

CHAPTER FOUR

"I have a voice": The Cosmopolitan Ambivalence OF Convergent Journalism

LILIE CHOULIARAKI

This chapter engages with convergent, or networked, journalism in two of its most compelling manifestations, the Haiti earthquake (2010) and the Egyptian Arab Spring (2011).[1] It does so in order to illustrate how, in their different ways, these manifestations perform a new and important public disposition, the disposition of *'I have a voice'* – a disposition of symbolic recognition that creates community by valorising the opinion and testimony of ordinary people. Whilst many have celebrated this disposition as facilitating cosmopolitan solidarity that transcends borders and makes a difference in people's lives, I show that the disposition of *'I have a voice'* is an ambivalent one that can both catalyse change and reproduce power relations between West and global South. The potential for cosmopolitan solidarity inherent in convergent journalism, I conclude, lies with the insertion of ordinary voice in a broader structure of Western journalism that challenges existing hierarchies of place and human life and thus enables the disposition of *'I have a voice'* to go beyond communitarian recognition—the recognition of people like 'us'—towards recognising the voice of distant others, too, as a voice worth listening and responding to.

I begin with a theoretical account of news journalism as *performative* practice, that is to say practice that constitutes the communities it addresses at the moment that it claims to represent them (see also Butler 1997, 2009). What differentiates convergent journalism from the journalism of press and television, I contend, is a fundamental re-articulation of this performativity from the primacy of information to the primacy of deliberation and witnessing. Whilst these re-articulations,

I argue next, turn convergent journalism into a novel mechanism of symbolic recognition, the cosmopolitan efficacy of this mechanism changes depending on three distinct, albeit interrelated, dimensions of mediation: the *re-mediation* of ordinary voice in major news networks, such as the BBC, the *inter-mediation* of the social media, which organise this voice in distinct types of intervention (say, crisis communication or cyberactivism), through these networks and, finally, the possibility for *trans-mediation*, that is the transfer of this voice from the symbolic realm onto the realm of physical action. Using the Haiti earthquake and Egypt protests as contrasting case studies, I show how variations in their performative mechanism reflects concomitant variations in the mediation of ordinary voice and, hence, the scope of symbolic recognition that each event enacts. Such variations, I argue, reflect not only differences in the nature of the event but, importantly, in the hierarchical positioning of each country in the contemporary system of global governance. In my chapter's conclusion, then, I emphasise the radically ambivalent nature of the disposition of '*I have a voice*', both as a note of caution against celebratory accounts of convergent journalism and as a call for more empirically grounded accounts of the role of this journalism in the formation of cosmopolitan solidarity.

THE PERFORMATIVITY OF 'I HAVE A VOICE'

Journalism is about doing things with words, not simply about using words to report facts. What journalism does with words, and indeed with pictures, is that it brings into being the community of people it addresses as its audiences. Journalism is, therefore, performative in the sense that it evokes or 'performs' the very publics that it claims to inform.

Journalism and recognition: Butler (2009) draws on war and conflict reporting to argue that journalism is performative to the extent that its stories of Abu Graib do not only provide information about the war in Iraq but, in fact, organise the field of perceptible reality within which we are allowed to think, feel and act upon this war. It is, she argues, this field of the perceptible, rather than any specific story of Abu Graib, that constitutes us as moral and political communities, insofar as it is this field that ultimately regulates 'whether and how we respond to the suffering of others, how we formulate moral criticisms, how we articulate political analyses' (Butler 2009: 64; see also Taylor, 1995; Fraser, 2010).

Crucial to this performative constitution is the process of recognition, a symbolic process by which journalism ascribes a specific identity to an undefined body of viewers, through designating who belongs to this body and who does not. Recognition is, in this sense, not only a symbolic act emanating from the stories and images of news reporting but, by virtue of this, also a normative act that participates in the constitution of moral and political communities of belonging. Who watches and who suffers, to draw on the Abu Graib example, reflects and reproduces the power relations between a safe West and a dangerous non-West—a zone of suffering, war and destitution that often de-humanises distant others and thus leaves them beyond the realm of 'our' recognition (Chouliaraki 2006). The key question for journalism is, in this sense, not only to become fairer in including more people within its norms of recognition but, as Butler puts it, '*to consider how existing norms allocate recognition differentially*' across an already hierarchical axis between the West and non-West (2009: 6).

It is precisely this performative function of journalism to produce community in the course of reporting the news that Muhlmann (2008) has described as an ambivalent force, at once 'unifying' and 'de-centering'. Whilst 'unifying' journalism, she argues, works to establish community through the authority of the journalistic 'I' who tells the story, 'de-centering' journalism seeks to challenge community by replacing the journalistic 'I' with a 'dizzy multiplicity of interpretations of experience' (2008: 235). Recognition is central to both journalisms, the difference being that, unlike unifying journalism, de-centering journalism interrupts dominant norms of recognition and invites us to expand the boundaries of our taken-for granted affiliations in the name of solidarity. Even though Muhlmann's analysis of de-centering strategies focuses on the writing practices of individual journalists, nowhere is the claim to de-centering more evident than in the practices of convergent journalism—the networking of citizen voices through major institutional news platforms (Deuze 2008).

Journalism and cosmopolitanism: In its most celebrated form, convergent journalism shares with de-centering journalism the same normative imperative, that is to challenge the professional voice of the journalist, the institutional 'I' of the news, with the multiple voices of ordinary people who can now tell the news themselves. Ordinary voice is, in this context, used to emphasis the valorisation of the private, everyday voice of citizens as a legitimate form of public agency and, in so doing, to draw attention to the new continuities (and discontinuities) of this voice both in relation to the professional voice of journalists (challenging the traditional distinction between 'media' and 'ordinary' worlds, Couldry 2000) and in relation to the expert voice of political elites (challenging another established distinction between publics and their representatives, Allan 2009). In valorising ordinary voice, convergent journalism contributes, thus, to *'a new participatory folk culture'* that, according to Henry Jenkins' definition of convergence culture, *gives 'average people the tools to archive, annotate, appropriate and recirculate content'* (Jenkins 2001: 93) and allows the news to juxtapose institutional authority with the authority of the person in the street—the performativity of *'I have a voice'*. As Russell, quoting Boczkowski, speaks of the Egyptian protest movement, '"the news moves from being mostly journalist-centered, communicated as a monologue, and primarily local, to also being increasingly audience-centered", thereby enabling this audience to '… deeply affect the news, in which the margins grow in power to shape the center' (2011: 1239). It is, indeed, the connectivity of ordinary voice, powerless and marginal as it is, with the nodes of powerful journalism, that, according to Castells, can swiftly insert this ordinary voice within a 'global network structure and enter the battle over the minds by intervening in the global communication process' (Castells 2007: 244).

Convergent journalism has thus been celebrated as, at once, an empowering journalism that places people at the heart of politics, signalling what Mona Eltahawy (2011) speaking of Egyptian citizens, calls '*the triumph of the 'I'*", and as a cosmopolitanising journalism that uses these voices as a means of introducing the plight of distant others in the West. As CNN put it, in relation to the Haiti earthquake, its citizen-driven news hub intends 'not only to report but also to connect…informing the public but also providing them a possibility to connect with the victim and help' (Bunz, 2010). In a similar vein, Director of the BBC's Global News Division, Richard Sambrook (2005), speaks of the corporation's news as a 'collaborative product', arguing that the lesson drawn from major news on distant suffering, such as the Asian tsunami, is that 'when major events occur, the public can offer us as much new information as we are able to broadcast to them'.

Convergent journalism becomes, here, a practice of mutual benefit as it can place the disposition of *'I have a voice'* at the service of both cosmopolitan encounters and corporate benefits. Indeed, relevant literature has not only celebrated the use of ordinary voice in news reporting as enhancing the cosmopolitanising potential of the news (Reese and Dai 2009; Cottle 2009) but has also drawn critical attention to the use of this voice as a strategy for renewing the declining legitimacy of global news networks, in the name of the people (Scott 2005; Beckett 2008). It is this combination of cosmopolitan with instrumental visions that have come to define convergent journalism as a *'liquid journalism'* (Deuze 2008)—one that, despite acting from within news corporations, can propose the 'new humanism' of ordinary voice as an "*an antidote to narrow corporate-centric ways of representing interests in modern society*" (Balnaves, Mayrhofer and Shoesmith quoted in Deuze 2008: 859).

To summarise, convergent journalism is celebrated for democratising the space of traditional journalism, in that it introduces ordinary voice into technological platforms of instant and global connectivity and, in so doing, both empowers this voice and cosmopolitanises the West. Yet, I argue, given that the technological possibilities of convergent journalism are still deeply embedded within traditional structures of media power between West and non-West, the question remains as to whether these structures may, indeed, enable processes of recognition that challenges the communitarian boundaries of the West. It is precisely this question that concerns me here: under which conditions may convergent journalism facilitate the articulation of ordinary voice by non-Western others? and in which ways may such voice manage to de-centre our sense of belonging towards a solidarity with these others—people we do not know and will never meet? In order to address these questions, I move from a theoretical approach to the convergent performativity of *'I have a voice'* towards an analytical account of performativity as an object of critical research.

TWO CASE STUDIES: HAITI EARTHQUAKE AND EGYPT UPRISING

My empirical focus falls on two major cases of trans-national reporting that can be characterised as *'ecstatic news'*—news so extraordinary that it warrants live coverage beyond the normal news bulletin, bringing global audiences together around a 24/7 mode of reporting (Chouliaraki 2006). These are the cases of the Haiti earthquake, 2010, and the Egypt uprising, 2011. By virtue of their content as disaster and protest reporting respectively, these two ecstatic pieces simultaneously make dramatic claims to cosmopolitan solidarity, aiming to address their publics as a 'universal' community of common humanity. Both pieces further share a similar communicative structure of journalistic iterability that occupies the space of global mediation, by mixing broadcast with social networking technologies, electronic with online genres, professional with ordinary voice narratives.

For analytical purposes, I suggest that we approach this structure of journalistic iterability in terms of its three constitutive dimensions of its mediation: i) the ***re-mediation*** of ordinary voice in one of the biggest trans-national news networks of the West, the BBC; ii) the ***inter-mediation*** of social media sites (such as Twitter or Facebook) that already articulate this voice in distinct forms of communicative activism, through this network and, finally, iii) the possibility for ***trans-mediation***, that is the transfer of this voice from the symbolic realm of communication to the realm of physical action, within or outside the scene of the news.

Even if their nature (ecstatic), claim (solidarity) and mechanism (journalistic iterability) establish a series of similarities across cases, there is significant variation between them. Such variation, I show, arises not only from the nature of the news, disaster in Haiti and protest in Egypt, but also, importantly, from the position of each location in existing socio-economic world hierarchies: Haiti being one of the poorest countries in the world whilst Egypt, characterised by a dynamic and media-savvy youth elite, being a key link between the Arab world and the West. It is these variations that make it possible to compare and contrast the two cases as exemplary of two different processes of recognition and, ultimately, formations of community—a unifying and a de-centering one. My analysis begins with the

Haiti earthquake and moves to the Egypt protests, discussing each as a system of journalistic iterability through the categories of *re-mediation*, how speech acts are structured in the global news flows; *inter-mediation*, how their structures interact with one another and *trans-mediation*, whether they turn into action or not.

The Haiti Earthquake

Convergent journalism in the Haiti earthquake is characterised by i) the primacy of witnessing and deliberation acts in BBC's live blogs; ii) the function of these blogs as, simultaneously, crisis communication platforms and iii) the interface between online communication and offline humanitarian activism.[2]

Remediation: The Haiti earthquake introduced the practice of live blogging in disaster reporting. A cutting-edge form of convergent journalism, live blogging uses platforms of Western networks, such the BBC, as online news hubs that aggregate messages from multiple sources into a timeline of event updates (Beckett 2010). Due to its multi-media platform, live-blogging enables the online combination of complex written, visual and aural material, such as eyewitness links to mobile phone footage, 'audio accounts' of survivors, Twitter and email messages as well as telegraphic journalistic updates and interactive maps of the affected area. The remediation of the Haiti earthquake refers, in this sense, to the global re-dissemination of these various voices of disaster, in the chronological sequence by which they reached the BBC news hub.

Given damages in the Haitian communication infrastructures, the remediation of the disaster, at least initially, was conducted primarily by Western NGOs. Whilst having an obvious informational value, these messages were primarily testimonial, providing first impressions of devastating suffering. A large number of remediated messages also came from ordinary people. Even though a few originated in Haiti, the vast majority of messages, in BBC's live blog, came from Westerners who expressed compassion towards the victims or shared their concern about their own family members in Haiti (Chouliaraki 2010). Rather than witnessing, such voices were primarily deliberative in that they established a common space where solidarity towards distant others could be symbolically performed. Side by side to these voices, international elites also expressed compassion with the earthquake victims and stated their readiness to supply aid.

Combining witnessing with deliberative speech, then, the remediation of the Haiti earthquake construed an emotional public realm, where ordinary and expert voices co-existed without hierarchies in a trans-national space of solidarity for distant sufferers. Whilst this emotional space is necessary for catalysing de-centering processes of recognition, its quality of emotionality is not sufficient to the task. Insofar as it marginalises the voices of those who suffer in favour of primarily Western voices of witnessing and deliberative action, the emotionality of convergent journalism ends up as a mechanism of recognition for people like 'us', unifying the West as a community of therapeutic communication whilst leaving non-Western others outside our remit. The remediation of Haiti thus becomes, in Castell's terms, a form of 'mediated mass self-communication': self-generated in content, self-directed in emission, and self-selected in reception by many that communicate with many' (2007: 248).

Intermediation: By virtue of aggregating and re-disseminating messages, the remediation of this disaster also operated as a parallel network of crisis communication. This is so because the remediation of NGO and ordinary people messages were not simply sharing emotion but further intended to contribute to the co-ordination of humanitarian action: '*The earthquake set a number of precedents in terms of … the use of new technologies for humanitarian response*', the ALNAP (2010) report said, '*Many UN and international NGOs and some military actors embraced technologies such as Twitter, Facebook, YouTube and Skype in their work, to coordinate, collaborate and act upon information from the ground generated by people directly affected by the earthquake*' (2010: 27).

Intermediation, in this context, refers to the function of convergent journalism to tap into these social media networks so as to establish its own circuits of communication amongst interested parties, with a view to facilitating the informational management of the disaster. During the first day of the disaster, for instance, the UN forwarded regular updates of its missing, Port-au-Prince UNAID staff in the BBC live blogs. Whilst specialised networks of crisis communication obviously operated outside major news networks, such as Ushahidi, the intermediation of NGO and ordinary voices through these news networks points to the instrumental dimension that the acts of witnessing and deliberation may acquire in convergent journalism.[3] Insofar as it is used to facilitate practical action, this proliferation of voices reflects more than the moral and affective aspects of a Western community of users. It becomes, simultaneously, an integral aspect of humanitarian activism, which uses these voices as resources in its efforts to respond to the immediacy of suffering.

It is precisely in its function as a site for the symbolic performance of humanitarian activism that intermediation inscribes the earthquake into what Craig Calhoun (2010) terms, an 'emergency imaginary', that is a collective understanding of the disaster as a 'sudden, unpredictable event emerging against a background of ostensible normalcy, causing suffering or danger, and demanding urgent response' (2010: 2) Whilst there is no doubt that saving lives was the most urgent priority in the aftermath of the earthquake, Calhoun's term draws attention to the fact that this priority has a significant cost: it tends to detract attention from the causes of the disaster—'it calls' as he puts it,' for humanitarian response not political or economic analysis' (2010: 2).

Transmediation: Two types of transmediation link the speech acts of disaster reporting with the scene of the earthquake: donations as public response to NGOs' appeals for financial support to victims and relief aid as specialised assistance offered on the ground. Even though both constitute a transfer from speech acts to material action, donations are typically associated with action at a distance and so with the transmediation of ordinary voice, whilst relief aid is typically associated with professional action and, therefore, with the transmediation of the voice of NGOs or other international elite actors—governments, religious or for-profit institutions and, exceptionally, journalists. Indeed, just as the speech acts of live blogging blurred distinctions between ordinary and expert voices, so action on the ground saw journalists turn into aid workers, thereby not only blurring the boundary between journalism and humanitarianism but also throwing into dramatic relief the subtle interface between the speech acts of informing and witnessing in the news. The controversy around this unique form of journalistic transmediation granted, the scope of the Haiti transmediations was impressive overall. With $220 million collected in donations, in the first week of the disaster in the USA alone, and 2000 rescuers present from 43 different NGOs on the ground already in the first hours after the earthquake, ample support appeared to be offered.

Yet, criticism emerged regarding both the short- and long-term management of the disaster, with the former focusing on failures in the co-ordination of immediate aid and the latter on failures to secure recovery both at the level of hygiene (resulting in the 2011 cholera outbreak), but also at the level of absorbing funds: 'In a country sometimes dubbed a "republic of NGOs",' as Claire Provost (2012) put it in January 2012, 'only 6% of bilateral aid for reconstruction projects has gone through Haitian institutions, according to UN figures. Less than 1% of relief funding has gone through the government of Haiti'. What such criticism points to is a fundamental weakness in humanitarian relief as a form of global governance, namely that asymmetries in the macro-economic map of development cannot be compensated for by an emphasis on micro-aid. Despite a constant outpouring of aid funds throughout the past forty years, Haiti remains, indeed, at the bottom of world rankings across a number of wealth indexes. Instead, then, of a politics of continuing dependence, as Collier (2009) warned already in 2009, what Haiti needs is a strategic plan for sovereign economic development: 'what is missing,' he adds, 'is a practical and focused economic strategy that clearly specifies the actions needed by all the actors that collectively determine whether Haiti will achieve economic security' (2009: 4).

In summary, the Haiti earthquake marked a new era of convergent journalism in disaster reporting, which celebrated the disposition of '*I have a voice*', placing the acts of witnessing and deliberation on a par with the expert voices of journalists, NGOs and political elites. The structure of journalistic iterability at work, however, prioritised a particular articulation of speech acts that i) prioritised the ordinary voice of Western media users rather that of the Haitian victims; ii) construed the suffering in Haiti as an emergency, that is a sudden and unpredictable event devoid of its historical or political economic context; iii) facilitated a solidarity of salvation, intensely focusing on aid relief, over a solidarity of justice, which would also ask the pertinent question of why such devastation was possible in Haiti, but not, say, in San Francisco's equally powerful earthquake (2003).

The Egypt Protests

The convergent journalism of the Egypt protests is characterised by: i) a, similar to Haiti's, primacy of witnessing and deliberation in BBC's live blogging; ii) the function of blogging as, simultaneously, a platform of political advocacy and iii) the interface between online communication and an offline activism of mass demonstrations on the ground.[4]

Remediation: The BBC live blogs consistently placed the ordinary voice of the Tahrir square and broadly the Egypt street, under global visibility. However, it was particularly after Mubarak imposed a shut-down of the country's internet infrastructure for a week (starting January 28th 2011) that trans-national networks, notably Al Jazeera but also Western ones, became instrumental platforms for the re-mediation of ordinary voice. Whilst the BBC aggregated and re-disseminated a number of diverse voices, from Obama to Hezbollah and from the Muslim Brotherhood to its own Jeremy Bowen, it regularly also included messages from ordinary people who monitored the day-to-day situation and offered accounts of violent or, tellingly, peaceful encounters with the army.

Available through Twitter and YouTube videos, which were resourcefully accessed by bypassing Egypt's official ICT network, the majority of such messages had a witnessing value, in that they did not simply describe action in the street but also took a stance towards such action: 'The protesters are shouting "Down, down, Hosni Mubarak" and "The people want the regime to fall", reports say' (BBC News, 2011a). In so doing, they worked to sensitise not so much the majority of Egyptians, largely cut-off from internet access, but, rather, 'the international media and the world', in Khamis and Vaughn's words, 'thus ensuring that the regime would not be able to cut them off from the world' (2011: xx). Such acts of ordinary witnessing, they argue, did more than simply connect the West with the scene of events, however, in that the use of ordinary voice also worked to add authenticity to these events: 'young Egyptians', as Khamis and Vaughn put it, 'were in the thick of it, mobile phones at the ready, often live-tweeting as skirmishes broke out' (2011).

Unlike the Haiti earthquake, then, where local voices were rare, the coverage of the Egypt protests incorporates the voice of distant others both as a testimony of violence and as a marker of the authenticity of reporting. Similarly to Haiti, such acts of witnessing entail a strong emotional dimension, which invites Western publics to engage with the Egyptian protesters and, potentially, construes a space of trans-national recognition between 'them' and 'us' - as, for instance, when a protester tweets: 'Army trying to starve and box in the #tahrir protesters, so orders r everything but shoot, fine, ppl won't back down' (BBC News, 2011b). This is, however, a different emotionality. Far from reduced to an expression of compassion for the voiceless sufferers, it articulates the affective discourse of the Egyptians themselves not only as sufferers but, crucially, as an empowered public that speaks out against the regime: 'in regimes where expression is controlled and restricted', as Papacharissi & Oliveira argue, 'it takes a lot of courage to utter affective statements indicating dislike, hatred and anger at a dictator' (2011: xx). And what made this discourse particularly forceful in Western contexts, I argue, is the intermediation of the event.

Intermediation: By virtue of re-disseminating ordinary voice from within the Egyptian scene, the remediation of the protests also operated as a parallel network of cyberactivism. This is because the remediation of such voice was not only about sharing emotion but also about deliberating on a political agenda with a view to mobilising publics to action: 'the goal of such activism', as Howard puts it, 'is to ...tell stories of injustice, interpret history and advocate for particular political outcomes' (2011: 145). Even though Egypt's cyberactivist infrastructure was well into place when the protests began (Eltahawy 2011), the deliberative use of social media throughout the events ensured that people had 'a sense of ownership' over 'the tools for crafting the revolution's narrative' (Brisson 2011).

Intermediation, in this context, refers to the ways in which convergent journalism tapped onto these already existing networks of online activism with a view to amplifying the voice of the protests on a global scale and, in so doing, 'to empower activists to associate and share ideas with others globally, enabling collaboration between activists in Egypt and Tunisia, as well as between protesters and Arabs in the diaspora; democracy activists in other countries' (Khamis and Wangh 2011). Central to this intermediation of ordinary voice was the Egyptians' appeal to democracy, which attached moral and political significance to the protests: 'The people of Egypt want their freedom, and they want it now', as another tweeting protester put it, 'We have lost all trust in a system that makes promises and does not fulfill them'.

It is precisely this demand for freedom that, unlike the emergency of the Haiti suffering, politicised and historicised the Egypt protests and managed to engage the West in concrete acts of solidarity for justice.

Transmediation: Two types of transmediation link the speech acts of protest reporting with the scenes of action: the continuing demonstrations in physical space, Tahrir square and elsewhere in Egypt and the Arab world, throughout the initial eighteen-day period of protests (January 25th – February 11th), and the parallel demonstrations in many Western capitals, in support of the Arab struggles. Whilst both forms of transmediation consist of the physical co-presence of publics protesting in the name of solidarity, the difference between Arab and Western transmediations lies in that the former, unlike the latter, entail a strong element of risk: 232 people died in Cairo alone with the total number of deaths around the country reaching 846 in this eighteen-day period. Crucial, then, in this risky shift from online to offline activism was, as Eltahawy (2011) put it, 'a mass of youthful protesters' who managed to turn their ordinary voice into an extraordinary act of revolution: 'Every revolution has its square', she says, 'and Tahrir (liberation in Arabic) is earning its name. This is the square Egypt uses to remember the ending of the monarchy in 1952, as well as of British occupation' (Eltahawy, 2011).

It is, I argue, this persistent and dramatic doubling of the speech acts of convergent journalism, witnessing and deliberating, with the embodied performance of self-sacrifice that sparked off the widespread manifestations of solidarity with Egypt around the world. Yet, while the spread of activism in the Middle East at large testifies to the formation of an Arab public unified around a similar demand of regime change, protests in the West point to a distinct process of recognition. This is a recognition that, rather than unifying the West around a voice of its own, a la Haiti, endorses and amplifies the voice of distant others. In so doing, it also de-centres the boundaries of its own community in the name of a solidarity of justice, thereby further activating meaningful links between 'us' and 'them', under the 'universal' claim to a common humanity: the Occupy Wall Street movement describes itself precisely as an extension of the Arab uprisings, articulating ordinary voice against the injustices of the global financial system (see below).

Whilst the Egypt protests, then, offer good evidence of the power of convergent journalism to reconstitute community around a sense of responsibility beyond ourselves, it is important to draw attention to the fact that Egypt, unlike Haiti, even if not topping global economic indexes, is an economically advanced country of the Arab world with a young, well-educated, urban elite whose socio-political dispositions greatly converge with Western ideals of democratic governance: 'there is', as Pintak says of Egypt, 'an army of media-savvy activists who have seized on tools like blogs, Twitter, Facebook and other forms of instant messaging as weapons… in their battle with entrenched regimes' (in Ghannam 2011: xx).

In summary, the Egypt protests, similarly to Haiti, also celebrated the disposition of *'I have a voice'*, placing the witnessing and deliberation of ordinary voice on a par with the expert voices of journalists, NGOs and political elites. Key differences in its structure of journalistic iterability, however, lead to a different articulation of these speech acts that now managed to i) reverberate the ordinary voice of Egyptian protesters in Western media; ii) construe the Egypt protests as a historical and political event strongly resonating with Western socio-political sensibilities; and iii) prioritise a solidarity of justice that focuses on political transformation over a solidarity of salvation that focuses on saving lives.

THE AMBIVALENCE OF 'I HAVE A VOICE'

Drawing on a view of journalism as performative, consisting of speech acts that call a community into being as they claim to address it, I theorised convergent journalism as an institutional form of journalism that is characterised by a shift from the professional act of informing towards the citizen-driven acts of deliberating and witnessing—the disposition of '*I have a voice*'. Whilst this shift in the epistemology of the news from the truth of institutional expertise to the truth of ordinary voice has been welcome as a democratisation of journalism that breaks down the boundaries between forms of expertise (for instance, between media practitioners and ordinary people) and catalyses processes of recognition that may cosmopolitanise the West, I argued for a more cautious, empirically grounded approach that attends to variations in convergence reporting.

My analysis of two contrasting cases studies sought to exemplify that such variation reflects, in fact, a concomitant variation in the power relations of global mediation. The Haiti earthquake and the Egypt protests are both ecstatic events that make claims to cosmopolitan solidarity, insofar as they invite us to act upon the misfortune of non-Western others, yet their claims to solidarity radically differ. Whereas the nature of the events, a natural catastrophe and a political uprising, obviously justifies this difference, my argument is that the Western structures of journalistic iterability via which each event is mediated are also crucially responsible for their variation. This is, I argue, because these structures selectively give voice to distant others, silencing the voice of earthquake victims but amplifying the voice of political protesters, and, in so doing, encourage recognition of the plight of the later whilst depoliticising the condition of the former.

In particular, the journalistic iterability of the Haiti earthquake is structured in ways that re-mediate the ordinary voices of a compassionate West, inter-mediate networks of crisis communication that focus on the management of the disaster and trans-mediate an activism of Western donations as well as NGO relief aid. The solidarity of the Haiti earthquake is, as a consequence, a solidarity of salvation that communicates the immediacy of saving lives, at the expense of debating the systemic causes that led to the devastation of the island. Rather than downplaying the humanitarian dimension of the event, my point is, rather, to emphasise the importance of challenging the assumption that salvation is the only possible response to Haiti's disaster. We need to consider, instead, why convergent journalism reproduces the positioning of Haiti within an emergency imaginary, that is a dominant explanatory framework for the misfortunes of distant others, which depoliticises the nation's post-colonial trajectory and relieves global governance institutions of the responsibility for the nation's continuing aid-dependency.

In contrast, the journalistic iterability of the Egypt protests is structured in ways that re-mediate the ordinary voice of Egyptians, inter-mediate networks of cyberactivism and trans-mediate into militant forms of civic engagement both in Egypt (and beyond) and the West. The solidarity of this event is, as a consequence, a solidarity of revolution that communicates a human-rights appeal for political freedom and social justice. Instead of de-politicisation, as in Haiti, this is a story that did more than gathering support for the Egyptian cause in Western capitals. It further managed to establish equivalential claims to solidarity between Egypt and the West, rendering 'their' and 'our' protests a common terrain of activism against authoritative structures that limit democracy: 'I am occupying Wall Street', as one American protester put it, 'because it is my future, my generation's future, that is at stake. Inspired by the peaceful occupation of Tahrir Square in Cairo, tonight we are coming together in Times Square to show the world that the power of the people is an unstoppable force of global change. Today, we are fighting back against the dictators of our country—the Wall Street banks—and we are winning' (cited in Huffington Post, 2011).

What this juxtaposition of the performativities of convergent journalism shows is that, *pace* celebrations on the democratisation of voice, citizen-driven journalism remains subject to global power relations that control the mediation of voice along the West-global South axis. Whilst the Egyptian voice, in its heroic combination of cyberactivism with self-sacrifice, broke through and catalysed a de-centred community in the name of a common cause, the Haitian one, locked in a fate of utter devastation, did not manage to reverberate in global media. This inability to speak is, of course, associated with the digital divide—35% of the Haitian population owns a mobile phone whilst internet penetration amongst Egyptians is close to 'saturation point', i.e., over 80% (Daily News Egypt, 2011).

There is, however, yet another barrier to voice—a symbolic barrier that reflects the profound misrecognition of the global South in a predominantly Western mediascape. Captured in Spivak's (1988) pertinent question 'can the subaltern speak?', symbolic misrecognition refers to the systemic inability of non-Western others to speak out and be heard in the trans-national flows of mediation. Even though the West has always had the disposition of 'I have a voice', exercising its authority to define the boundaries of legitimate community around itself, the South does not automatically possess such disposition. Egypt may have managed to speak out, having long prepared the articulation of its ordinary voice in public: 'For at least five years now', as Eltahawy (2011) put it, 'they [Egyptian youth, LC] have been nimbly moving from the "real" to the "virtual" world where their blogs and Facebook updates and notes and, more recently, tweets offered a self-expression that may have at times been narcissistic but for many Arab youths signalled the triumph of "I". I count, they said again and again'. Haiti, however, did not. As a result, it has been unable to participate in a mediation process that would challenge the West to recognise its suffering not only as yet another God-sent disaster but also as a matter of political injustice and historical responsibility.[5]

Rather than a matter of a technology-driven democratisation of voice, then, the disposition of 'I have a voice', is, crucially, a political matter of injustice, which throws into relief the deep inequalities of power that selectively enable some voices to be heard and recognised as worthy of our solidarity and not others—what I earlier called the symbolic power of mediation. It is, therefore, a matter of crucial importance to draw attention to the ways in which the iterability of convergent journalism may now re-structure the dominant arrangements of this power and allow for a different distribution of voice in the flows of mediation: 'overcoming injustice', as Fraser says, 'means dismantling institutionalized obstacles that prevent some people from participating on a par with others, as full partners in social interaction' (2010: 16). Insofar as convergent journalism remains embedded in structures of iterability that reproduce the dominant visions and divisions of community, cosmopolitan solidarity will remain, at best, a rare moment of revolutionary resonance between West and its others, or, at worst, a narcissistic reverberation of our own pity for these other's suffering.

NOTES

1. An earlier version of this chapter was published in *Journalism Studies* 14(2): 267–283.
2. The 2010 Haiti earthquake, which occurred on January 12th 2010, was a catastrophic, magnitude 7.0 Mw, earthquake, with an epicenter approximately 25 km (16 miles) west of Haiti's capital, Port-au-Prince. The earthquake caused major damage in Port-au-Prince, Jacmel and other settlements in the region, causing damage to vital infrastructure, necessary to respond to the disaster; this included all hospitals in the capital; air, sea, and land transport facilities; and communication systems. The human loss is estimated to about 316,000 whilst the number of displaced up to 1.5 million (Haitian Government and International Organisation for Migration report May 2011).
3. For the emergence of novel, citizen-driven crisis communication networks, such as Ushahidi, following the Haiti earthquake: http://www.knightfoundation.org/blogs/knightblog/2011/1/11/new-media-and-humanitarian-relief-lessons-from-haiti/
4. The Egyptian protests, which started on 25 January, were a form of non-violent civil resistance, consisting of demonstrations and other acts of civil disobedience and mobilizing millions of Egyptians around the demand to overthrow the regime of then-President Hosni Mubarak. Whilst the aim was achieved on February 11th 2011, after the death of 846 people, political unrest, due to continuing political conflict, continues in Egypt until today.
5. See Al Jazeera 'live blog' on Haiti for a historicizing perspective along these lines: http://blogs.aljazeera.net/americas/2010/01/13/why-haiti-earthquake-was-so-devastating.

REFERENCES

Allan, S., (2009). The Problem of the Public: The Lippmann-Dewey Debate in S. Allan (ed) *The Routledge Companion to News and Journalism* London: Routledge pp. 60–70.

ALNAP (2010). Haiti Earthquake Response: Context Analysis, July. http://www.alnap.org/pool/files/haiti-context-analysis-final.pdf

Bach, K., (2006). Speech Acts and Pragmatics in *Blackwell Guide to the Philosophy of Language.* London: Blackwell Wiley pp. 147–67.

BBC News (2011a). As It Happened: Egypt Unrest on Friday, 29 January. http://news.bbc.co.uk/1/hi/uk_politics/9380441.stm

BBC News (2011b). Egypt Unrest: Day 14 as It Happened, 7 February. http://news.bbc.co.uk/1/hi/world/middle_east/9390387.stm

Beckett, C. (2008). *Supermedia* London: Blackwell.

Beckett, C. (2010). *The Value of Networked Journalism* Polis/BBC College of Journalism Report Available online: http://www2.lse.ac.uk/media@lse/POLIS/Files/networkedjournalism.pdf

Bennett, L. W., Pickard W.V., Iozzi D. P., Schroeder C. L., Lagos T. C., and Caswell E. (2004). Managing the Public Sphere: Journalistic Construction of the Great Globalization Debate *Journal of Communication* Vol. 54 Nr 3 pp. 437–55.

Bourdieu, P. (1989). Social Space and Symbolic Power in *Sociological Theory* Vol. 7 Nr 1 pp. 14–25.

Brisson, Z. (2011). Institutional Overview: Independent Media Blog in *The Reboot: Egypt. From Revolution to Institutions* Available online: http://thereboot.org/blog/2011/04/18/institutional-overview-independent-media/

Bunz, M. (2010). In Haiti Earthquake Coverage, Social Media Gives Victim a Voice *The Guardian*, 14 January.

Butler, J. (1997). *Excitable Speech: A Politics of the Performative* London: Routledge.

Butler, J. (2009). *Frames of War: When is Life Grievable* London: Verso.

Calhoun, C. (2010). The Idea of Emergency: Humanitarian Action and Global (Dis-)order in D. Fassin and M. Pandolfi (eds) *States of Emergency* Cambridge MA: Zone Books.

Carey, J. (1999). In Defence of Public Journalism in T. Glaser (ed) *The Idea of Public Journalism* New York: Guilford Press pp. 47–66.

Castells, M. (2007). Communication, Power and Counter-Power in the Network Society *International Journal of Communication* Vol. 1 Nr. 1 pp. 238–266.

Chouliaraki, L. (2006). *The Spectatorship of Suffering* London: Sage.

Chouliaraki, L. (2010). Ordinary Witnessing in Post-Television News: Towards a New Moral Imagination *Critical Discourse Studies* Vol. 7 Nr 4 pp. 7 (4). pp. 305–319.

Collier, P. (2009). *Haiti: From Natural Catastrophe to Economic Security. A Report for the Secretary-General of the UN* Available online: http://www.focal.ca/pdf/haiticollier.pdf

Cottle, S. (2009). Global Crises in the News: Staging New Wars, Disasters, and Climate Change *International Journal of Communication* Vol. 3 pp. 494–516.

Couldry, N. (2000). *The Place of Media Power: Pilgrims and Witnesses in the Media Age* London: Routledge.

Couldry, N., Hepp, A. & Krotz F. (2009). *Media Events in the Global Age* London: Routledge.

Daily News Egypt (2011). Egypt's Mobile Market Nears Saturation at 80 pct Penetration, 3 January.

Dayan, D., and Katz, E. (1993). *Media Events* Cambridge MA: Harvard University Press.

Deuze, M. (2008). Liquid Journalism *International Journal of Communication* Vol. 2 pp. 848–865.

Eltahawy, M. (2011). We've Waited for This Revolution for Years. *The Observer*, 29 January.

Entman, R. (2003). *Projections of Power: Framing News, Public Opinion and US Foreign Policy* Chicago: CUP.

Fraser, N. (2010). *Scales of Justice: Reimagining Political Space in a Globalized World* New York: Columbia University Press.

Frosh, P., and Pinchevski, A. (eds) (2009). *Media Witnessing* London: Palgrave.

Ghannam, J. (2011). *Social Media in the Arab World. A Report to the Center of International Media Assistance* Washington DC. Available online: http://www.humansecuritygateway.com/documents/CIMA_SocialMediaintheArabWorld_LeadinguptotheUprisingsof2011.pdf

Howard, P. N. (2011). *The Digital Origins of Dictatorship and Democracy: Information, Technology and Political Islam* Oxford: OUP.

Huffington Post (2011). Occupy Wall Street Considering Sending Protesters to Egypt for Elections, 14 November.

Jenkins, H. (2001). Convergence? I Diverge in *Technology Review* p. 93 Available online: http://www.phase1.nccr-trade.org/images/stories/jenkins_convergence_optional.pdf

Khamis S., & Vaughn K. (2011). 'Cyberactivism in the Egyptian revolution', *Arab Media and Society*, Issue 14, http://www.arabmediasociety.com/?article=769

Muhlmann, G. (2008). *A Political History of Journalism* Cambridge: Polity.

Papacharissi, Z., & Oliveira, M. F. (2011). The Rhythms of News Story-Telling: Coverage of the January 25th Egyptian Uprising on Twitter. Paper presented at the World Association for Public Opinion Research Conference, Amsterdam. Available online: http://tigger.uic.edu/~zizi/Site/Research_files/RhythmsNewsStorytellingTwitterWAPORZPMO.pdf

Peters, J. D. (2001). Witnessing in P. Frosh and A. Pinchevski (eds) *Media Witnessing* London: Palgrave.

Provost, C. (2012). Haiti Earthquake: Where Has the Aid Money Gone? *The Guardian*, 12 January.

Reese, S., and Dai, J. (2009). Citizen Journalism in the Global News Arena: China's New Media Critics, in S. Allan and E. Thorsen (eds) *Citizen Journalism: Global Perspectives* pp. 220–31 New York: Peter Lang.

Russell, A. (2011). 'Extra-National Information Flows, Social Media, and the 2011 Egyptian Uprising", *International Journal of Communication*, 5, pp. 1238–1247.

Sambrook, R. (2005). Citizen Journalism and the BBC, Nieman Reports, Winter. http://www.nieman.harvard.edu/reportsitem.aspx?id=100542

Scott, B. (2005). A Contemporary History of Digital Journalism *Television and New Media* Vol. 6 Nr 1 pp. 89–126.

Silverstone, R. (1994). *Television and Everyday Life* London: Routledge.

Spivak, G. (1988). Can the Subaltern Speak? in C. Nelson and L. Grossberg (eds.) *Marxism and the Interpretation of Culture* Urbana & Chicago: Univ. of Illinois Press pp. 271–313.

Taylor, C. (1995). The Politics of Recognition in A. Heble et al. (eds) *New Contexts of Canadian Criticism* Ontario: Westview Press pp. 98–131.

Thompson, J. (1990). *Language and Symbolic Power* Cambridge: Polity.

Tuchman, G. (1972). Objectivity as a Strategic Ritual: An Examination of Newsmen's Notions of Objectivity *American Journal of Sociology* Vol. 77 pp. 660–79.

Zelizer, B. (2007). On Having Been There: Eye-Witnessing as a Journalistic Word *Critical Studies in Media Communication* Vol. 24 Nr. 5 pp. 408–28.

Zelizer, B. (2010). *About to Die: How News Images Move the Public* Oxford: OUP.

CHAPTER FIVE

Before THE Revolutionary Moment: The Significance OF Lebanese AND Egyptian Bloggers IN THE New Media Ecology

KRISTINA RIEGERT

There were numerous deep-rooted causes for the spread of demonstrations demanding "dignity" and "freedom" that swept the Arab region in 2011, and they were no doubt fuelled by the simultaneous expansion of the social media. However, it would be misleading to credit the social media for the mass character of this unrest, due to low and fluctuating internet penetration, not to mention economic and class-related factors. Instead, the interaction between social media platforms, and pan-Arab and Western satellite television meant that the ideas and actions circulating on the internet reached both national and international audiences in a whole new way. This is hardly surprising since the mainstream media have been busy in the last decade positioning themselves in relation to social media, utilizing citizen journalists as sources and honing crowd-sourcing techniques.

The evolving relationship between traditional and new media has been characterized as a "hybrid media environment" where mainstream media are challenged by new media actors in power struggles to "control, police and redraw boundaries" of political communication (Chadwick, 2011: 10). In authoritarian societies with entrenched state affiliated media, the internet offers portals for alternative information and hubs of civic activism and resistance (Lagerkvist, 2010; Srebreny & Khiabany, 2010; Kulikova & Perlmutter, 2007; Taki, 2010). Naomi Sakr recently argued that this "satellite-internet divide" was incrementally bridged in the years prior to the revolutionary moment in Egypt. Not only due to the rise of bloggers, but because "Concerned Egyptian citizens, journalists and politicians made heavy use of the online space for political communication precisely because mainstream offline media were largely closed to them" (2012: 334). The rise of bloggers and citizen journalists from 2005 forms the backdrop to the Arab uprisings of 2011, but they can hardly be considered as causal agents; rather, they are part of the wider media ecology that challenges or accommodates the power structures in authoritarian and transitional societies depending on the context and chain of events.

This chapter traces the online relationships between some seminal social media users—popular bloggers in two Arab countries Lebanon and Egypt—in the years prior to the uprisings. It focuses on the most linked to and visited bloggers, and their relationships to each other and to the mainstream media within each country's specific political and cultural context.[1] The development and status of these bloggers in each context provides keys to understanding when citizen journalism makes a difference in a crisis media event like that of the Arab uprisings.

BLOGOSPHERES AS ALTERNATIVE PUBLICS

It is not a forgone conclusion to ask whether the most "popular" Arabic blogs are those of citizen journalists, or even if they stretch the political and cultural norms governing mainstream media. The most popular blogs may be business or entertainment oriented or diary-style blogs. As websites written in reverse chronological order, blogs also contain a rich blend of autobiographical narratives, political engagement, consumerist critique, poetry, satire, YouTube clips, or commentary on daily life. Especially in countries with internet filtering and heavy-handed involvement in gender relations, religion, and entertainment, blogs may be more apt to focus on technological "gadgets" and restaurant reviews than political commentary (i.e., as common in the Gulf states). This means that analysis of the blog content is required in the latter case, whereas an idea about whether the blogs form their own networks and what their linkages are to the mainstream media can be gleaned by interviews and more quantitative means.

We define "popular blogs" as a combination of the most often "linked to" and "most visited" blogs for each blogopshere.[2] These blogs, we reasoned, are those most likely to be picked up and linked to by the mainstream media. Relationships between the news media and bloggers can be difficult to trace in Arab countries, due to "idea theft," censorship, filtering, and technological anomalies, but we can get an overview of blogger networks through their links to each other, and through citations of them by local media, as well as from bloggers' own accounts. Included here are the top ten Lebanese and Egyptian Arabic and English language non-commercial, active, individual blogs.[3] We interviewed 17 of these

21 bloggers personally or via Skype. Our textual analyses cover at least every tenth blogging day of these blogs during the period between 1 April 2009 and 30 April 2010.

Middle East studies scholar Mark Lynch (2011) has argued that the most likely effect of the internet will be an incremental widening of Arab public spheres rather than fundamental change, because Arab governments may yet adapt to and stifle the internet challenge. Whether blogs can be thought of as extensions of mediated public spheres, or as alternative or counter-public spheres, they are accessible online spaces where people gain publicity and interaction for issues the blogger(s) have decided are of concern. Many bloggers emphasise their independence from established media organisations, marking their authenticity with subjective audience address, signaling first-hand experience and trustworthiness (Atton, 2008: 43). The bloggers we interviewed in Lebanon and Egypt are no exception. Only a few admitted to belonging to a social or civil society organization. That said, most participated in an *ad hoc* way in charity drives, workshops, protest demonstrations, or human rights campaigns run by NGOs.

The Lebanese and the Egyptian bloggers said that the aim of their blogs was to freely express their opinions about politics, society and culture, many saying that opinions such as theirs were not adequately reflected in the mainstream media. This does not mean that all bloggers were citizen journalists (aiming to influence the mainstream media), nor does it preclude the fact that some used their blogs to launch a media career (whether artistically or journalistically). What it does point to is the subjective feeling of not being represented in mainstream society. Thus, these bloggers are part of an "alternative media sphere" insofar as they are marginalized publics that use campaigns and alternative media to make their voices heard in the dominant mediated sphere (Wimmer, 2009; Warner, 2002).

Those not aiming to be citizen journalists use the online space to post about their everyday lives and concerns. Through interacting with readers and other bloggers online and offline, they form their own community, a parallel to the mainstream (Fraser, 1990). Those bloggers we interviewed demonstrated an awareness of their readers: they knew how many readers they had, which countries they came from and, through the "comments" function and Twitter, they interacted with many of them. Secondly, they knew most of the other bloggers included here, either through reading them regularly, via other online platforms, or through meeting them at "bloggers" meetings', conferences, "tweet-ups," protest demonstrations, and other common causes. Figures 1.1 and 1.2 demonstrate that the bloggers link to each other, forming a network. The lighter squares denote Arabic-language bloggers and the darker circles are English- or mixed-language blogs.

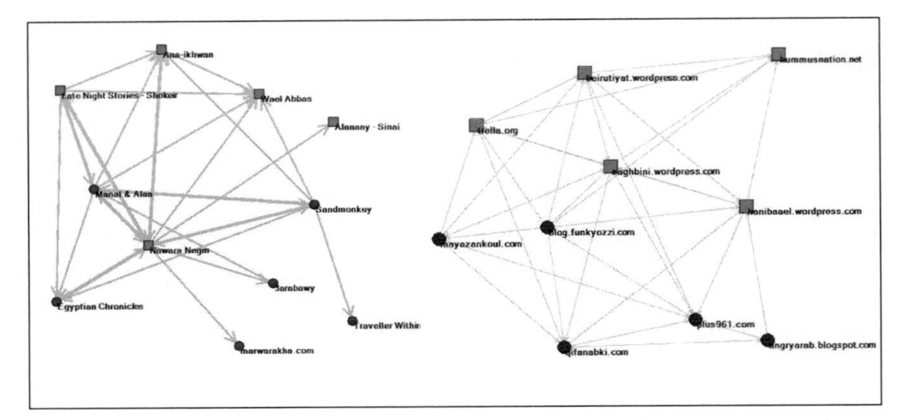

Figures 1.1 and 1.2 Egyptian (left) and Lebanese (right) Bloggers Interlinking Networks (with SocSciBot webcrawls; Thelwall, 2009), April 2010.

The Lebanese bloggers are more densely interlinked, whereas especially the Egyptian English-language bloggers are not. English is much more common in the smaller Lebanese blogosphere than it is in Egypt. Nonetheless, the link networks and our interviews demonstrate that bloggers, or "social media activists," as some call themselves, see themselves as belonging to a special type of community, whether or not they agree on substantive or ideological issues. Writing of Lebanese bloggers, Jurkiewicz (2012) describes intensive cooperation and conflict in both online discussion and offline meetings, as well as new friendships among bloggers.

EGYPT AND LEBANON: TWO VERY DIFFERENT CONTEXTS

To generalize, controversial subjects for public discussion in the Arab world have to do with sex (premarital, extramarital, homosexual), religion (sectarian issues, interpretations of one's own religion, criticizing religious authorities), and politics (criticizing the leadership, the army or governmental system). Taboos against public discussion of these issues vary, of course, widely across the Arab world as do the sanctions for breaking them (from censorship to threats, incarceration and death). The bloggers we interviewed felt that blogs were ideal spaces for airing controversial political and social issues, or criticizing cultural norms in their respective societies.

The conditions for free expression did however differ significantly between Lebanon and Egypt, for while the former had little formal government censorship, bloggers in the latter risked fines, harassment, physical harm, arrest or imprisonment for crimes such as insulting President Mubarak, the Supreme Council of

Armed Forces (SCAF) or Islam. For Egypt, it is a great paradox that despite its blogosphere being the largest and most lively in the Arab world, featuring early adopters of Facebook and Twitter, Egypt was cited in 2009 by Reporters without Borders as an "internet enemy."[4]

The rise of blogging in Egypt has been attributed to their documentation of incidences of election fraud and the ensuing Kefaya ("Enough") Movement protests in 2005 against the corrupt regime of President Mubarak. These pioneering bloggers were human rights activists and citizen journalists: they criticized government and business corruption, systematic torture, and labour exploitation, and defended women's and minority rights. Their online and offline mobilization became an inspiration for tech-savvy youth around the region as to what could be done in the face of heavy-handed government repression (Hamdy, 2009; Radsch, 2008). That same year in Lebanon, a first generation of bloggers flourished after the assassination of former Prime Minister Rafik al-Hariri. The ensuing mass demonstrations that came to be called the "Independence Intifada" forced the Syrian army to withdraw after 30 years in Lebanon.

Of our chosen Egyptian bloggers, most are first-generation veterans with an average age of 34—several were over 40 years of age when we interviewed them.[5] Most of these bloggers are well known internationally in the online activist world, and some could even be described as celebrities; they had good connections to foreign mainstream media and among the youth movement, both inside and outside Egypt. By 2011 several of the Arab-language bloggers had moved much of their activity to Twitter and Facebook, and were blogging infrequently; however, the English-language bloggers continued to be active and were linked to frequently by Western media.

In contrast, the top Lebanese bloggers were younger and most belonged to what some have called the third generation bloggers in Lebanon, which has diversified away from human rights and politics to arts and entertainment, business or lifestyle blogs (Jurkiewicz, 2011). While the average age was 29, most were closer to their mid-twenties. With a couple exceptions, most did not have the same status in relation to the news media as the Egyptian bloggers. By 2011 several bloggers had stopped blogging, opened new blogs, moved to Facebook or become more infrequent. Twitter and Facebook were not as often mentioned by Lebanese bloggers as alternatives or complementary tools as they were among the Egyptian interviewees. One can only speculate as to the reasons for the later adoption of Twitter among the Lebanese. One reason is that fewer of the Lebanese bloggers considered themselves to be citizen journalists.

The questionable staying power of even the most popular bloggers raises questions about the sustainability of blogospheres over time. After all, they are not paid, as journalists are, to produce a certain amount of text on a regular basis—only a few of them earned much money from ads. The great majority of bloggers tend to be active only during certain periods. On the other hand, the top bloggers in our sample that have slowed down have continued on other social media platforms. Khamis, Gold & Vaughn (2012) have noted how various social media platforms were used during the Egyptian Uprisings in 2011 for different purposes, so it is unlikely that successful bloggers simply disappear. How individuals utilize different social media platforms to create their online presence is one interesting and often overlooked avenue of study in analyses of social media. Traditionally, the focus has rather been on the blurred boundaries between offline and online activities, especially when it comes to activism.

THE TOP BLOGGERS AND LOCAL MEDIA

Western observers have noted that the adoption of citizen journalism and social media by the mainstream media is driven by conflict and crisis, when the latter are unable to be at the right place at the right time or when their watchdog role is compromised. This must be modified to have relevance to mainstream Arab media, which have by and large been appendages to power, either being state owned or controlled by powerful groups/business interests close to the state.

This is particularly true of Egypt, where state or privately owned (but government friendly) media have long been dominant. However, since the turn of millennium, this dominance has been challenged by the success of political talk shows on satellite television channels, the launch of online portals and or online versions of the new independent newspapers and from the rise of blogging (Sakr, 2012). Omnia Mehanna (2010) attributes the breakthrough of blogging in Egypt in 2005, in part, to the transnational Arab media. She says that the broadcast of Al Jazeera's programme *Taht Al-Mighar* (Under Examination) in May 2006, which mainly dealt with Egyptian bloggers' documentation of incidences of election fraud in the presidential and parliamentary elections of 2005 and human rights violations, was a turning point. This episode, featuring several of the bloggers discussed here, sparked an upswing in blogging among a whole generation of disgruntled youth and put bloggers on the Egyptian media radar. As she describes it, "…some newspapers started paying attention to the blogs, copying, sometimes without permission, stories and pictures from them." (p. 199)

Figures 1.3 and 1.4 depict the link citation relationships between our bloggers and some of the top news sites in Lebanon and Egypt.[6] The diagrams represent URL citations (of blogs or websites) found in the top blogs and the media sites of each other. These could, but do not have to be, hyperlinks (as were the previous set of diagrams); rather, they are a search engine's return of blog citations and their connectivity to local media sites.[7]

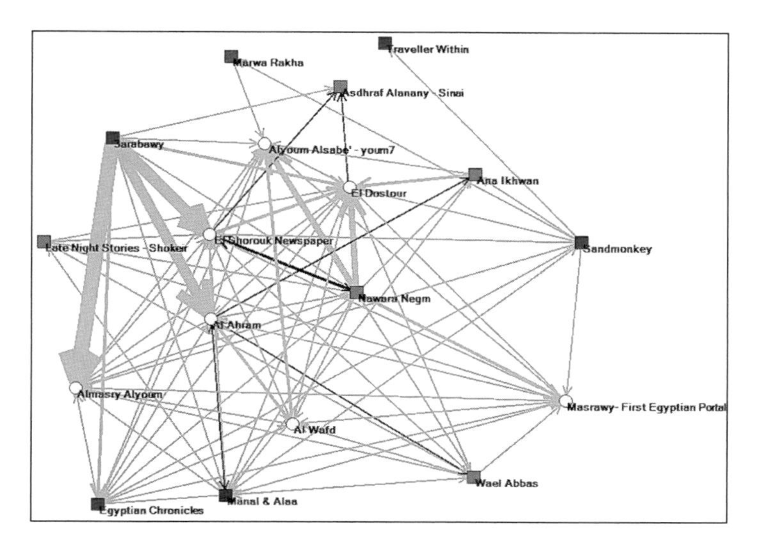

Figure 1.3 Network Diagram of Link Citations Between Top Egyptian Blogs and Local Media (thicker arrows = more links).

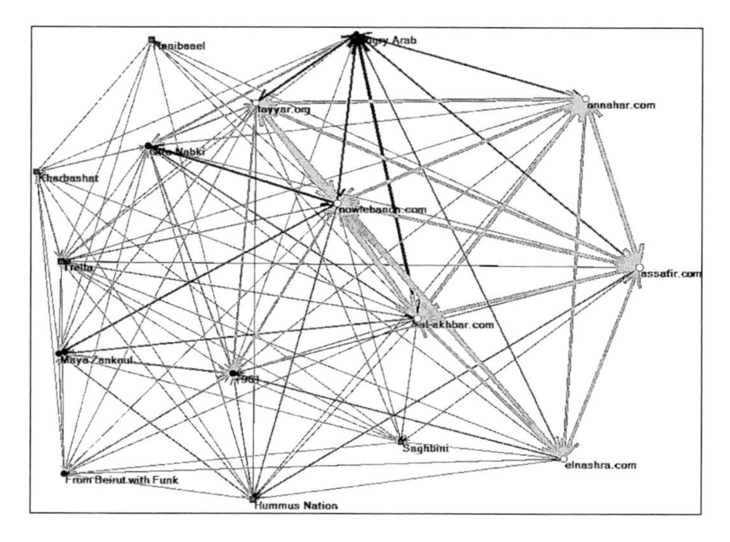

Figure 1.4 Network Diagram of Link Citations Between Top Lebanese Blogs and Local Media (thicker arrows = more links).

From both diagrams, it is clear that the Egyptian and Lebanese local news sites (nodes in light circles) tend to link to one another rather than to the blogs (English-language blogs are dark circles and Arabic-language in squares). The blogs cite the media and each other, but especially the Egyptian news sites cite the blogs more seldomly (denoted by black arrows). Some interesting exceptions should be noted, however. These are Nawara Negm (tahyyes.com) and Alanay, who are cited by *El Shorouk*,[8] and state-owned *Al Ahram*, which cites Wael Abbas, Manal and Alaa's Bit Bucket, and Ana-Ikhwan. Here it should be noted that five of the Egyptian bloggers were freelance journalists, columnists or had previously worked for news organisations. Nawara Negm (tahyyes) wrote a column for the independent newspaper *al-Dostor* (dostor.org). Abdel Monem Mahmoud (ana-Ikwan) and Hossam el-Hamalawy (3Arabawy) have worked for *Al Jazeera* and the *LA Times* respectively. Traveller within (Mohamed Dashan) freelanced with a column for *al-Masry al-Youm*. Wael Abbas (Misr digital), perhaps the most internationally famous of all the bloggers, had previously worked as a Middle East correspondent for the German news agency DPa. Obviously, writing for a media outlet has some explanatory power regarding citations, despite the paucity of independent news outlets in Egypt, since newspapers tend not to link to bloggers, but rather recruit them to write for the paper, whereas bloggers often link to their own articles in newspapers.

In a politically and socially polarized country such as Lebanon with 18 recognized religious confessions, it comes as no surprise that the Lebanese media system is divided along sectarian lines. This means that differences of opinion and various inflections on events are common in both the offline and online media system. Although confessionalism is part of the fabric of Lebanese society, two major political power blocs emerged after the retreat of the Syrian army in 2005: the March 14th movement (i.e., Future Movement Sunnis, various Christian and Armenian factions) and March 8th movement (i.e., Shia Muslim factions and Michel Aoun's Free Patriotic Movement). In this context, it is interesting to note the frequency of mutual citations between these online media, especially between the "old" press outlets, such as *an-Nahar* and *as-Safir*, and "new" online news outlets such as tayyar.org, elnashra.com, "Now Lebanon" news site and the leftist March 8th newspaper *Al-Akhbar*.

Lebanese media citations of blogs are more common than in the Egyptian case. Angry Arab is the most often cited by the media. Being a first generation blogger and a Middle East scholar, he writes a regular column for *Al-Akhbar*. Tayyar.org (supporting the Free Patriotic movement) and *As-Safir* cite +961, Saghbini, Hummus Nation and Trella. NowLebanon (Future Movement site) seems interested in linking to several of the English-language blogs. Here it should be noted that four of these Lebanese bloggers have freelanced for the leftist papers *al-Akhbar* and *As-Safir*.[9]

The Lebanese media mention Lebanese blogs more, and these are more interlinked than among the Egyptian bloggers. This, despite the greater fame and clout exerted by Egyptian bloggers in the local and international media. As noted above, this could be due to the Egyptian media simply "borrowing" ideas from the blogs without attributing them, recruiting bloggers as columnists, or to greater "caution" about blogs by the dominant government-affiliated online outlets. Various Lebanese mainstream media, on the other hand, link to our bloggers, despite the feeling expressed in interviews that the mainstream media marginalized bloggers. Here, I should caution that the existence of links says little about content or the context of the citation.

BLOGGING THEMES 2009–2010

Both the Egyptian and Lebanese top bloggers can be said to have pushed issues not prioritised in the mainstream media of each country during the period 2009–2010.

The Lebanese bloggers were critical of the politics of sectarianism and the media system it has spawned. Some criticized the Abrahamitic religions by giving examples of paganism in the region. Many engaged with the dysfunctions of daily life, the plight of the Palestinians and foreign domestic workers in Lebanon, environmental and infrastructural degradation, corruption and excessive consumerism. Four of the bloggers used satirical commentary, humour, irony or sarcasm to criticize governmental authorities, powerful groups in Lebanese society, ad agencies, or social norms. Though the English-language bloggers were less "socially activist" than their Arab-language cousins, almost all of them expressed deep frustration with the Lebanese confessional political system and how the mainstream media reinforced this. Even the less socially activist Lebanese bloggers "covered" the election in 2009, noting that it did not solve the polarized and deadlocked government. The bloggers were keen to distance themselves from the media by either ignoring "the partisan political" or dealing in issues that cut across the political boundaries, like anti-sectarianism, environmental and anti-consumerist stances. Similarly, gender discrimination was an issue in the (male) Arabic-language blogs and not exclusively the domain of the two female bloggers, Maya Zankoul and Independence 05 (now Funky Ozzy). While both of the latter from time to time pointed out daily gender discrimination and participated in women's rights manifestations, their blogs were much broader than this. These issues however only received limited support and attention from offline movements and the mass media.

The Egyptian bloggers included here exhibited greater ideological differences than those in Lebanon: from Marxist, to leftist Islamist, to Liberal, to Islamic intellectual, to socially conservative. Yet, in 2009–10 these were overshadowed by common themes related to the (il) legitimacy and corruption of the Mubarak government. Bloggers participated in or documented workers' strikes, protests against government corruption, arbitrary arrests and above all, the repression of free speech. They exposed widespread police brutality with video footage, the lack of accountability in the Egyptian administration and the business sector, the widespread poverty and exploitation of workers (lack of minimum wage), the discrimination of minorities, and criticism of the state-owned media for not giving a fair picture of events. The English-language bloggers all critically reviewed the much-hailed Cairo speech by US President Obama in June 2009. In contrast to the Lebanese bloggers' more diversified interests, the Egyptian top bloggers were all "social media activists" and citizen journalists. This is in a way unsurprising since these veterans had for the most part been part of the wider youth movement criticising the Mubarak government for six years when January 25, 2011, rolled around. In Egypt, where an authoritarian political system and censorship coexist with a vibrant intellectual culture, sectarian tensions and a revered religious orthodoxy, these bloggers, despite their differences, could coalesce around their desire to get rid of the hated symbol of all that was ill in society—the Mubarak regime. Despite this, it should be emphasized that these popular bloggers are in both cases disproportionately leftist and secular compared to the wider Egyptian and Lebanese blogospheres, as well as to mainstream opinion in their societies.

Based on the themes brought up by these top bloggers, one cannot talk about a common "Arabic blogosphere" but a multiplicity of national and even fragmented sub-national blogospheres (Jurkeiwicz, 2012; Etling, et. al., 2009). Each country's relationship to the outside world, its history, its mediascape, and types of cultural taboos have shaped the way the social media are being used by bloggers. This means while these bloggers certainly brought up pan-Arab political or cultural issues (i.e., the Palestinian issue, the Iranian election, sectarian conflict or the lack of basic human rights in the region), Hollywood film releases or popular music, most were mainly concerned with national and local issues, despite their otherwise cosmopolitan attitudes and lifestyles. One would expect that those blogging in English or geographically placed outside their home country would be more transnational in content, but neither of these determined whether a blog was more transnational or local. What language(s) a blogger uses is more an indicator of lack of available software, the bloggers' education, and cosmopolitan identity—and the frequent blending of English, Modern Standard Arabic (MSA), colloquial Arabic and French—demonstrates an address to like-minded and similarly linguistically gifted readers.

CONCLUSION

It is clear from the comparison between the Egyptian and Lebanese blogging content that the former clearly foreshadowed the events of 25 January that toppled

Mubarak. Even if they represented a tiny portion of larger networks of human rights activists, labour movements and others disgruntled sections of society. On the other hand, their clout as activists and citizen journalists with previous relationships to Western and local media made them disproportionately visible compared with the Muslim Brotherhood blogosphere. Since the Western media use mainly English-language blogs where secular and liberal views of events are dominant, there is a clear risk of misrepresentation using these citizen journalists as sources. Indeed, this warning should also apply to Lebanon, where the Arabic-language bloggers tended to be more leftist and more "social activist" than the English-language bloggers, and religious blogs are few.

Previously scholars have wondered whether Arab blogging is simply a pressure valve for the well-educated classes to "let off steam" or whether they can facilitate social change. While there is no doubt that online activism increases the visibility of certain issues or groups of advocates, we have also shown that Lebanese bloggers' issue agendas did not to have the same impact as their Egyptian colleagues, despite being "linked to" more frequently by the local online media. More links to various sectarian media outlets could not compensate for the deadlocked nature of the Lebanese sectarian system and the lack of popular mobilization for a secular political system. The Egyptian bloggers' anti-regime stance, on the other hand, resonated with many other offline and organized disaffected segments of society, all of which came together in January 2011.

So while social media moves quickly, social change is slow and there is no straight line between them. The blogospheres reflect the specificities of each country's mediascape, censorship and filtering, ownership structures, and to the extent that the most popular bloggers blog about the same themes, they also document a certain period in time. Over the long term however, blogs and social media platforms reflect the new media ecology including Arab satellite television and online news outlets. Bloggers are new voices that either find resonance in other media or become alternative platforms. It is interesting that neither of these alternatives in the year before the revolutionary moment appear to have predicted what would happen. Although it is clear in hindsight that the blogging network with a greater sense of themselves as citizen journalists, greater activist experience, wider networks and a common message with other disgruntled groups resonated with offline change. On the other hand, the real challenge for social media networks is how to build more lasting democratic institutions.

NOTES

1. The project on which this chapter is based is interested in the nature and impact of the most "popular" Arabic- and English-language bloggers in Lebanon, Egypt and Kuwait in the

pre-revolutionary period 2009–2010. It is a collaboration between Media and Arabic studies professors Kristina Riegert and Gail Ramsay and financed by the Swedish Research Council.

2. Alexa (www.alexa.com) is a web traffic analysis site that provides ranking and analysis of visitor statistics and averages for a three-month period. Alexa alone is not considered a reliable indicator, so after a link impact analysis (Thelwall, 2009) we generated a list of top 30 "most linked to blogs" that also had the most visitors (in June–August 2010).
3. The reason there are eleven blogs in the Egyptian case and only ten in the Lebanese case is because Manalaa.net, run by a couple, were not very active during our time period. Therefore not much could be said about the content, but since this blog was among the top most linked to and visited blogs, and also a seminal Egyptian blog whose founder started one of the most important Egyptian blog aggregators, we include it here.
4. At least three bloggers were imprisoned during 2010 and many more were detained, fined and had their computer equipment confiscated. Journalists also face fines, beatings, and imprisonment due to various laws and their inconsistent application. See http://www.freedomhouse.org/report/freedom-press/2011/egypt.
5. The onsite interviews took place in Beirut, November 2010 and in Cairo, March 2011.
6. The link analyses were made between April and June 2011.
7. URL citations are defined by Thelwall as "the inclusion of an URL (or URL without the http://) in a web page, with or without a hyperlink. See http://lexiurl.wlv.ac.uk/searcher/FAQ.html#URLCitation. Accessed: April 16, 2011. Having analysed hyperlink relationships for the Lebanese blogosphere, we found the URL citations to return a similar overall pattern of results.
8. The local media sites for Egypt are the independent online newspapers, almasryalyoum.com, dostor.org, shorouknews.com, and the state-owned ahram.org.eg, and the privately financed youm7.com and masrawy.com. *Al Youm al Sabe* is a liberal private newspaper rumoured to be owned by business interests close to Mubarak. Masrawy is one of the oldest online news sites in Egypt. It is owned by businessman Naguib Sawiris who also has shares in Al Masry al Youm and Orascom Telecom.
9. Angry Arab still writes a column. Hannibael (Hani Naim) freelanced for both *Al Akhbar* and *As-Safir* and Kharbashat (Assad Thubian) freedlanced for *As-Safir*. Saghbini had previously written for *Al Akhbar* but left it.

REFERENCES

Atton, C. (2008). "Bringing Alternative Media Practice to Theory: Media Power, Alternative Journalism and Production." In Pajnik, M. and Downing J. D.H. (eds.), *Alternative Media and the Politics of Resistance: Perspectives and Challenges*, Series: Politike Symposion. Ljubljana: Peace Institute, 31–48.

Chadwick, A. (2011). "The Hybrid Media System" Paper Prepared for delivery at the European Consortium for Political Research General Conference, Reykjavik, Iceland, August 25, 2011.

Etling, B., Kelly, J., Faris, R., and Palfrey, J. (2009). "Mapping the Arabic Blogosphere: Politics, Culture, and Dissent" Berkman Center at the Harvard. Research Publication No. 2009–06 June. http://cyber.law.harvard.edu/publications/2009/Mapping_the_Arabic_Blogosphere.

Fraser, N. (1990). "Rethinking the Public Sphere: A Contribution to the Critique of Actually Existing Democracy." *Social Text*, 25/26 (1990), 56–80.

Hamdy, N. (2009). "Arab Citizen Journalism in Action: Challenging Mainstream Media, Authorities and Media Laws," *Westminster Papers in Communication and Culture*, 6(1), 92–112.

Jurkiewicz, S. (2012). *Being a Blogger in Beirut: Production Practices and Modes of Publicness*. Dissertation Manuscript submitted to Faculty of Humanities University of Oslo. April.

Jurkiewicz, S. (2011). "Blogging as Counterpublic? The Lebanese and Egyptian Blogosphere in Comparison." In Schneidėr, N.C., & Gräf, B. (eds.), *Social Dynamics 2.0: Researching Change in Times of Media Convergence*. Berlin: Frank & Timme, 27–47.

Khamis, S., Gold, P., & Vaughn, K. (2012). "Beyond Egypt's 'Facebook Revolution' and Syria's 'YouTube Uprising:' Comparing Political Contexts, Actors and Communication Strategies." *Arab Media & Society*, 15 (Spring). http://www.arabmediasociety.com/?article=791.

Kulikova, S.V., & Perlmutter, D. (2007). "Blogging Down the Dictator? The Kyrgyz Revolution and Samizdat Websites" *International Communication Gazette*, 69(1), 29–50.

Lagerkvist, J. (2010). *After the Internet Before Democracy: Competing Norms in Chinese Media and Society*. Bern: Peter Lang AG.

Lynch, M. (2011). "After Egypt: The Limits and Promise of Online Challenges to the Authoritarian Arab State" *Perspectives on Politics*, 9(2), pp. 301–310.

Mehanna, O. (2010). "Internet and the Egyptian Public Sphere" *Africa Development*, 35(4), 195–209.

Radsch, C. (2008). "Core to Commonplace: The Evolution of Egypt's Blogosphere." *Arab Media & Society* 6 (September).

Sakr, N. (2013). "Social Media, Television Talk Shows and Political Change in Egypt." Special Issue Media and the Middle East. *Television and New Media*, 14(4) 322–337 DOI: 10.1177/1527476412463446.

Sreberny, A., & Khiabany. G. (2010). *Blogistan: The Internet and Politics in Iran*. New York: I.B. Tauris.

Taki, M. (2010). "Bloggers and the Blogosphere in Lebanon & Syria: Meanings and Activitie." PhD diss., University of Westminster.

Thelwall, M. (2009). *Introduction to Webometrics: Quantitative Web Research for the Social Sciences*. Synthesis Lectures of Information Concepts, Retrieval, and Services, #4, Morgan & Claypool. E-book. www.morganclaypool.com.

Warner, M. (2002). "Publics and Counterpublics." *Public Culture*, 14(1), 49–90.

Wimmer, J. (2009). "Revitalization of the Public Sphere? A Meta-Analysis of the Empirical Research about Counter-Public Spheres and Media Activism" In Garcia-Blanco, I., Van Bauwel, S. and Cammaerts, B. (eds.), *Media Agoras: Democracy, Diversity and Communication*. Newcastle upon Tyne: Cambridge.

CHAPTER SIX

Citizen Journalism in Real Time? Live Blogging and Crisis Events

NEIL THURMAN AND JAMES RODGERS

At 14:46 local time on 11 March 2011 an undersea megathrust earthquake hit the Pacific plate boundary 69 kilometres east of Tōhoku, Japan. The magnitude 9.03 quake was the fifth-largest ever recorded anywhere in the world, and Japan's most powerful. Such was its strength that the Earth shifted an estimated 25cm on its axis; but Tōhoku's most devastating effects were the result of the seawater it displaced 30 kilometres above its hypocentre. As the seafloor deformed, it raised the Pacific Ocean by up to eight metres over an area 180 kilometres wide. Although the resulting waves were relatively low in the open ocean, as they reached land their height increased, with terrible consequences. Along nearly 1,500 kilometres of the Japanese eastern seaboard—from Chōshi, Chiba in the centre, to Nemuro, Hokkaidō in the north—waves were recorded in excess of two metres and in places were much, much higher. A total of 562 square kilometres of land were inundated leaving over 18,500 people dead or unaccounted for. Some of the more than 1 million buildings damaged or destroyed were part of the Fukushima Daiichi nuclear power plant. Its loss of power ultimately caused a level 7 meltdown, releasing radiation into the atmosphere, and prompting widespread evacuations.

Media coverage of the quake, aftershocks, and tsunami was, predictably, extensive, with the ongoing search for survivors and the crises at Fukushima and elsewhere sustaining the story for weeks. News organisations across the world used live updating news pages—or live blogs—to cover the disaster. Live blogs have been defined as "a single blog post on a specific topic to which time-stamped content is progressively added for a finite period—anywhere between half an hour and 24 hours" (Thurman and Walters, 2013). Updates are regular—typically every 10 minutes—short—averaging about 100 words—and usually presented in reverse chronological order. It is conventional for live blogs to be relatively transparent in their correction practices; to quote extensively from, and link to, external sources; and to be authored by more than one journalist, often in different locations. Examining a range of live blogs that covered the Japanese earthquake and tsunami suggests that those news organisations more practised in the live blogging of breaking news had access to platforms that allowed journalists covering the disaster to relatively easily integrate videos, still images, maps, and social media. Examples include the live pages published by NYTimes.com (Goodman, 2011), Telegraph.co.uk, (Hough, Chivers, and Bloxham, 2011), and Guardian.co.uk (Batty, 2011). Other news organisations requisitioned established blogs from their sites, turning them into vehicles for the live coverage of the crisis. Using platforms not optimised for live coverage was limiting, and these live blogs were relatively text-heavy with minimal inclusion of still or moving imagery and social media. Examples can be found at FT.com (Bond, 2011), and at the websites of *The Wall Street Journal* (WSJ, 2011), and NBC News (NBC, 2011).

Looking at the content streamed by these live blogs reveals the contributions made by citizens on the ground. In some cases a contributor's social media activity led to a traditional interaction with a journalist—for example, the blog account by an academic visiting Tokyo that gave rise to an interview with *The New York Times* journalist, Maria Newman (Goodman, 2011). In other cases citizens' social media posts were quoted directly, for example the photos on Twitter from @K_TN72 and @Odyssey (Hough et al., 2011), the video of the tsunami arriving at Sendai Airport by 'Jack19661221' (Batty, 2011), and the direct accounts, via Twitter, of the aftermath of the earthquake from Obata Hiroshi and @_mego. @_mego's testimony was particularly dramatic. She tweeted that she was stranded in her house and was being forced to take refuge on the second floor. She appealed for rescue, giving her address and posting a photograph of her devastated, flooded neighbourhood (WSJ, 2011). As this chapter will show, although it is not unusual for such "citizen" sources to feature in the coverage of crisis events, some live blogging platforms are featuring unprecedented quantities of citizens' testimony.

Live blogs are not a new news format. Britain's second-most-popular newspaper website, Guardian.co.uk, was using them as early as 1999 to cover live sports matches on a "ball-by-ball" or "minute-by-minute" basis (Thurman and Walters, 2013). Such coverage found an audience amongst readers who wanted a convenient way to follow soccer, and, in particular, cricket matches—some of which last five days—whilst getting on with other things at home or work. Surprisingly, perhaps, it took a few years for live pages to break out—to any significant extent—of their niche in sports and other scheduled events and be adopted by journalists covering hard news.[1] The trend towards using live blogs to cover breaking news in general, and crisis events in particular, has been subject to scant research. This chapter aims to build on Thurman and Walters' (2013) research by:

- Analysing new data on the consumption of live online coverage of crisis events, and on readers' attitudes to live blogs.
- Discussing live blogging's relevance to debates around "citizen journalism" with reference to new data on readers' contributions—via live blogs—to the reporting of crisis events.
- Considering the influence live pages may have on the reporting of future crisis events.

THE CHANGING CONSUMPTION OF CRISIS NEWS ONLINE

For this chapter we mined data from three sources in an attempt to build up a picture of how live blogs covering crisis events are consumed by readers. Firstly, we compared the relative popularity of coverage of the Tōhoku earthquake and tsunami across a live blog, a picture gallery, and a video story, all published by Guardian.co.uk on Saturday 12 March 2011. As figure 1 shows, the live blog received 3.5 times more visits and nearly four times more page views than the video story; and almost 16 times more visits and close to five times more page views than the photo gallery. In terms of engagement (as measured by time-spent) the live blog was even more successful, attracting 5.7 times the attention of the video story, and over 16 times more than the picture gallery (see figure 2).[2]

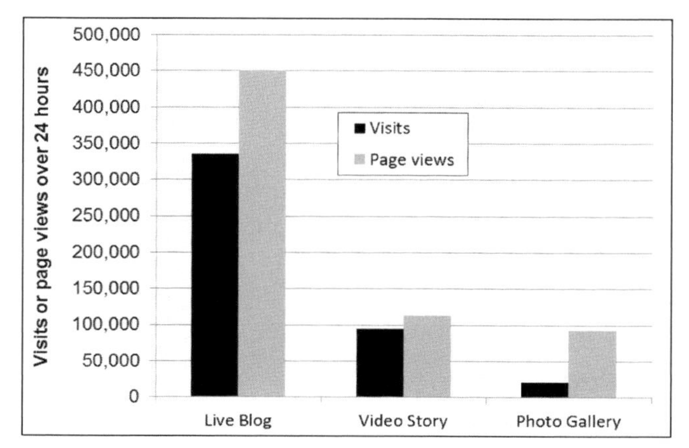

Note: Visit and page view counts are for the 24-hour period starting 00:00 on 12 March 2011. The live blog (Batty, 2011) began at 09:46 on 12 March, the video story (Guardian.co.uk, 2011a) was published on 11 March (time not specified), and the photo gallery (Guardian.co.uk, 2011b) at 10:56 on 12 March.

Figure 1: Relative popularity—by unique visits and page views—of a Guardian.co.uk live blog, a picture gallery, and a video story covering the 2011 Japanese earthquake and tsunami.

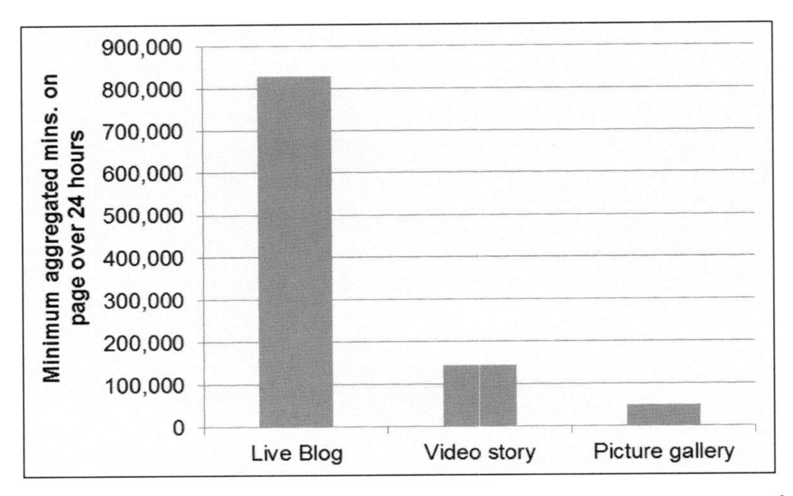

Note: The values here are minimums because, firstly, visits of 15 seconds or less were not counted and, secondly, the lowest value in each time band was used. For example, all visits in the 20–30 minutes band were counted as 20 minutes. Values are for the 24-hour period starting 00:00 on 12 March 2011. The live blog (Batty, 2011) began at 09:46 on 12 March, the video story (Guardian.co.uk, 2011a) was published on 11 March (time not specified), and the photo gallery (Guardian.co.uk, 2011b) at 10:56 on 12 March.

Figure 2: Aggregated attention (in time spent over a 24-hour period) received by a Guardian.co.uk live blog, a picture gallery, and a video story covering the 2011 Japanese earthquake and tsunami.

We also analysed data from ScribbleLive, a Canadian company whose live blogging platform is used by a range of news organisations. This showed that, across 20 live blogs covering 2012's Hurricane Sandy, the average length of engagement per visitor was 18 minutes. Similar levels of engagement were achieved by live blogs covering the ongoing Syrian civil war—19 minutes per visitor (Ekaterina Torgovnikov, personal communication, 19 March 2012). These figures compare favourably against levels of engagement across newspaper websites as a whole. For example, in 2012, the average duration of visits to US newspaper websites was 3.75 minutes (NAA, 2012).

Although some live blogs appear to be able to engage some readers relatively deeply, how broad is their popularity? Our third data source, the 2013 Reuters Institute/YouGov digital news survey,[3] offers some insights. It shows that 11 per cent of UK news consumers had followed a live news page in the previous week. By the same measure live blogs were even more popular in the US, Brazil, Italy, Spain, and, especially, in France (19 per cent) and Japan (35 per cent). Their notable popularity in Japan is likely to be the result of the influence of Yahoo! News Japan—the country's flagship news website. It carries a "breaking news" tab leading to a live page. This prominent placement, combined with the rather conventional approach to editorial presentation it displays on the rest of its homepage, is, according to

Yasuomi Sawa, Deputy Editor of the New York Bureau of Kyodo News, the likely explanation for the popularity of live pages in Japan (personal communication, 10 March 2013). The Reuters Institute/YouGov data show the consumption of live blogs is not confined solely to a hard-core of news junkies. However, live blogs are used to cover more than just breaking news. Sports events, ongoing issues—often related to politics—and scheduled events such as the Oscars ceremony are also commonly reported using the format. Given the evidence we have of the popularity of sports and celebrity stories with online news consumers (Boczkowski, 2010: 146), could it be that the popularity of live blogs with the Reuters Institute/YouGov sample of news consumers is, in large part, a result of their consumption of "sports" and "scheduled event" live blogs? Interestingly, the survey indicates this is not the case, showing that live blogs covering breaking news like natural disasters are the most popular category with both US and UK news consumers (see figure 3).

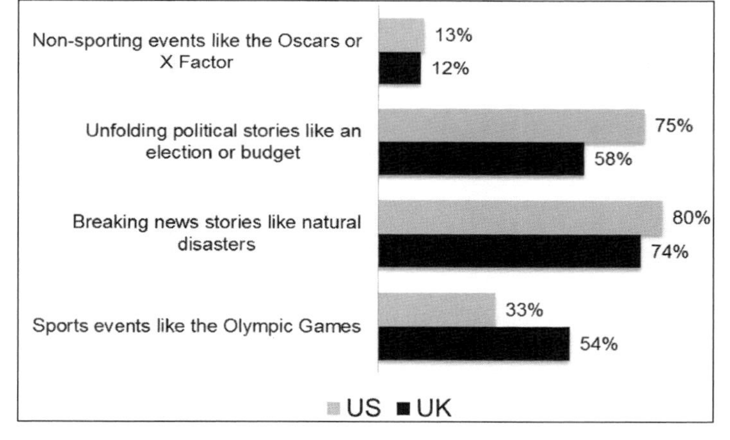

Note: Based on an online survey conducted by YouGov Plc for the Reuters Institute for the Study of Journalism. Sample size 229 (UK) and 208 (US). Fieldwork was undertaken between 22–31 January 2013 (UK) and 29 January–6 February 2013 (US). Respondents who had used a live news page in the preceding week were asked which of the above types of live news pages they use. Figures were weighted and are representative of all adults (aged 18+) who had consumed news in the previous month and have access to the internet.

Figure 3: Popularity of different types of live news pages (live blogs) with US and UK news consumers, January 2013.

LIVE BLOGGING AND CITIZEN JOURNALISM

So far this chapter has demonstrated that live blogs are an increasingly common feature of news websites and that they can, on occasion, attract and engage visitors to a greater extent than some traditional online news formats. We have also shown how their uptake within the general population of online news users is—using Rogers'

(2003) classic definition—moving beyond the "innovators" and "early adopters" and into the "early majority,"[4] and how they are—for the moment at least—popular for their coverage of breaking news and unfolding political and economic stories.

All well and good, but how do live pages relate to the theme of this volume, "citizen journalism"? The rise of professional live blogs can be seen as a response to the way social media and blog sites introduced the idea of bite-sized chunks of content arranged in a chronological order. This new format proved useful for audiences in keeping up-to-date with friends and family but soon started to become useful in telling certain types of news stories. Twitter in particular, through the introduction of hashtags, effectively enabled its community of users to run something like their own live blog on major events, from the Iranian elections to the World Cup. And the development of software like Storify has made such so-called micro blogging more accessible, and increased its visibility, by allowing anybody to quickly and easily curate a news story by aggregating social media items in a coherent way on a single page.

Although social media coverage of breaking news has become relatively accessible via aggregating platforms such as Storify, most consumption of live breaking news online still takes place on, or via, mainstream media sites.[5] Despite the low cost of live blogging platforms such as ScribbleLive, CoveritLive, LiveBlogPro, and Blyve, the format, unlike regular blogging, requires considerable time and resources: live blogs have been shown, on average, to run for six hours, include 40 separate updates, weigh in at over 4,000 words, and utilise multiple authors (Thurman and Walters, 2013). They are, therefore, perhaps not a format that "citizen journalists"—in the solitary, casual, and spontaneous mould discussed by Liu and colleagues (2009), Nip (2009), and others—can easily produce independently of the mainstream media. So, although professional live blogs have been, and will continue to be, influenced by social media, the main relevance of live blogging to the broad concept of citizen journalism relates to its potential to elicit, collate, and distribute citizens' contributions, enlarging the mediated public sphere and diversifying the voices heard in the media. This potential exists because:

- Readers have said they are more than twice as likely to participate in live blogs as in other articles types (Thurman and Walters, 2013).
- Live blogging platforms often support the seamless integration of social media content (such as tweets) and comments; and the conventions of the genre are to quote heavily from, and link to, secondary sources, some of which are non-official or citizen. For example, in the BBC's description of the mix of content in its new local live blogs, the Corporation gave "tweets [and] emails from members of the public" a similar level of prominence as "quotes from BBC reporters" (BBC, 2013a).

Previous research about the use of citizens' contributions in the coverage of crisis events has examined the extent to which reporters use official and non-official sources, and how technologies such as email or social media mediate their source interactions. Evidence on how much journalists rely on non-official sources during crises is mixed. Case studies on the reporting of the 9/11 attacks (Li and Izard, 2003), environmental incidents (Sibbison, 1988), and business crises (Powell and Self, 2003) showed a reliance on official, mainly government, sources. On the other hand, studies of anti-American attacks (Nacos, 1996); the Virginia Tech shootings (Wigley and Fontenot, 2009); shootings in Tucson, Arizona, that injured US Representative Gabby Giffords (Wigley and Fontenot, 2011); and Hurricanes Andrew (Salwen, 1995), and Katrina and Rita (Fontenont, Boyle, and Gallagher, in press) showed that non-official or "citizen" sources can dominate.

Less research has been published on the extent to which technologies, such as email or social media, mediate between journalists and their citizen sources in crisis reporting. In their study of the Virginia Tech shootings, Wigley and Fontenot (2009) found that 6.5 per cent of quoted sources were both non-official *and* new-technology mediated (what Wigley and Fontenot also refer to as user-generated content). In a follow-up study of the Tucson, Arizona shootings the authors found a similar proportion—9.5 per cent (Wigley and Fontenot, 2011). Both studies used a sample of stories from national newspapers and "articles" from four television news websites. In the 2011 study there appeared to be "little difference in the amount of citizen generated content used by [...] news websites and newspapers," whereas the 2009 study showed that "news websites were much more likely to use citizen generated content than newspapers."

Table 1: Proportion of content contributed by journalists and readers in 30 live blogs covering four crisis events hosted on the ScribbleLive platform

	Journalist contributions published	**Reader contributions submitted**	**Reader contributions published**	**% Reader contributions**
Japanese earthquake(n=4)	3,913	33,000	11,988	75%
UK riots of 2011 (n=4)	2,298	6,759	896	28%
Hurricane Sandy (n=20)	27,550	80,911	5,199	16%*
Syrian civil war (n=2)				11%**

Note:

* For the five live blogs covering Hurricane Sandy that received the most traffic, reader contributions averaged 39 per cent.

** Detailed breakdown unavailable.

Source: ScribbleLive (Ekaterina Torgovnikov, personal communication, 19 March 2012).

For this chapter we analysed 30 live blogs covering four crisis events to find out the proportion of citizen contributions. The live blogs were published using the ScribbleLive platform.[6] The results, presented in table 1, show much higher levels—ranging from 11 to 75 per cent—of non-official new-technology mediated content than observed in these previous studies. Furthermore, because some of the journalists' own updates (not used to calculate the proportions reported in table 1) contained non-official new-technology mediated content these figures are a conservative estimation.[7] These results contrast with the levels of citizens' contributions to live blogs hosted by one mainstream news provider's proprietary platform. At a sample of 20 Guardian.co.uk live blogs, just 7.5% of updates were made up of readers' tweets and comments "above the fold" (Thurman and Walters, 2013: 91). One reason for such differences is likely to be the fact that all the ScribbleLive live blogs analysed covered crisis events where (as Nacos, 1996; Wigley and Fontenont, 2009, 2011; Salwen, 1995; and Fontenont, Boyle and Gallagher, in press, have shown) citizen sources can dominate. Of the 20 Guardian.co.uk live blogs analysed by Thurman and Walters (2013) only a quarter covered breaking news, and not all of those breaking news stories were crisis events. Another reason is technical: the ScribbleLive platform publishes readers' contributions with the same level of prominence as contributions from journalists.

Allowing citizens' contributions to come to the fore in such a way has been editorially unthinkable and technically impossible for much of the history of journalism. While vox pops, radio phone-ins, and letters to the editor have been enduring examples of the way editors have sought to involve citizens, live blogging, at least in some manifestations of the format, seems to be increasing the proportion of user-generated content that is appearing in the mainstream media's coverage of crisis events. This has implications for the way in which reporters work too. Consider the reporting of the capture of Saddam Hussein, in December 2003. Alongside the more conventional breaking news coverage on TV and radio, the BBC tried to make original use of its news website, then still only a few years old. While they did not use a live blog in a form we would recognise today—the title, indeed, referred instead to a *Reporters' Log* (BBC, 2003)—it can nevertheless be seen as something of a pioneering example of what has come to be so widespread now. True, these are not "citizen" contributions, because they draw on the words of various BBC journalists, but they do bring them together on a single web page, and thus conform in a significant way to the definition of live blog which we have offered above. As Rodgers (2011) has previously suggested, one of the main shortcomings of the way in which this event was covered was the lack of contributions from people who did not share the occupying forces' jubilation at the capture of the deposed Iraqi leader. Live blogs offer a possibility that such shortcomings might be addressed—provided, of course, that there is the editorial will to do so.

Increasingly, that seems to be the case. The events of the Arab uprisings since the spring of 2011 have coincided with the continuing growth of internet access across the Middle East, North Africa, and beyond. More and more people in those regions have used mobile technology not only to follow news events, but to contribute to them. These citizen contributions have increasingly been used by established news organisations, most notably in their live blogs. The British-based *Guardian*, with its "Middle East Live" pages (The Guardian, 2013), and *Al Jazeera English*, with its "Syria Live Blog" (Al Jazeera English, 2013), are just two such examples. During the war in Syria, access for established news organisations has been extremely difficult, and extremely dangerous. The death in Homs in February 2012 of the *Sunday Times*' vastly experienced correspondent Marie Colvin served to remind journalists, and the companies for which they work, of those hazards. For this reason, citizen, and in many cases activist, contributions, have been especially valuable to mainstream media. This is particularly true of video material. A glance at the content that *Al Jazeera English* posted on its live blog on 13 April 2013 (Al Jazeera English, 2013) serves as an illustration. While there is a preponderance of material from official news agencies—the page carried some lines from a Reuters story about the Israeli military having fired artillery after its forces were shot at on the occupied Golan heights, and an AFP story on rebels having apparently tortured a Kurdish man in Aleppo—the video material Al Jazeera used that day could all be considered citizen contributions. This included pictures of an opposition "Friday of the Free" rally which had apparently taken place in Aleppo; and other video material showing "razed buildings and battered belongings" apparently in the "Al-Qosoor neighbourhood on the outskirts of Homs."

Naturally, there are endless editorial considerations here concerning the authenticity or otherwise of this material. Aware of this, established news organisations have devised detailed systems for checking content in order to authenticate it before publication. There is even a page on the BBC news website in which a BBC journalist, Alex Murray, explains the lengths to which the Corporation goes in order to verify "eyewitness/citizen journalist/user-generated content," which, Murray notes in the article, "has become increasingly complicated as the material has become more sophisticated" (Murray, 2011). Such detailed checking may be beyond the means of those many news organisations that do not enjoy the same resources as the BBC, but it has its purpose. For all the potential benefits offered by the mass of material coming out of Syria and elsewhere, some high-profile fakery found in citizen contributions has served to sully the reputation of some user-generated content. Perhaps the most notorious example was the "Gay Girl in Damascus" blog. That turned out to have been written neither in Damascus nor by a "gay girl," but not before it had been widely taken to be just that. As Bennett noted, the "hoax highlighted the pitfalls of operating as a journalist in the digital era" (2011: 190). Few developments in the history of journalism have been without their potential drawbacks, and live blogging, reliant as it is on such material, is no exception. In 2012, Britain's *Channel 4 News*, perhaps reflecting the frustrations that might come with dealing with these endless editorial dilemmas, ran a report in which they exposed the fact that some Syrian activists were embellishing their video material in order to increase its impact. The story, "Syria's video journalists battle to tell the 'truth,'" broadcast on 27 March 2012, paid tribute to the undoubted courage and determination of the citizen journalists of Homs—while also showing them setting fire to a tyre in order to provide a column of black smoke in the back of the shot for a piece to camera (Channel 4 News, 2012).

Despite these misgivings (and, after all, journalistic hoaxes are probably as old as journalism itself—there is no reason why the advent of digital journalism in all its various forms should suddenly have marked their passing), the live blog at its best does, from an editorial point of view, offer great potential. In covering the "Algeria Hostage Crisis" of January 2013 (where an armed group, described by the BBC as "Islamists," took over a gas plant in the Algerian desert, and subsequently killed some of their captives), the BBC used the medium to good effect. Their page, "As it happened: Algeria hostage crisis" (BBC, 2013b), combined official sources such as the UK Foreign Secretary, William Hague; excerpts from despatches from AFP, Reuters, and the Algerian State News agency APS; and user contributions. The user contributions helped to bring important context to the unfolding story. One issue much discussed was the Algerian security forces' approach to retaking control of the plant. There were suggestions that more could have been done to avoid hostage deaths. Citizen contributions to the live blog on this subject brought a degree of insider knowledge—and also Algerian views of the crisis—to a global audience.

One comment came by email from Guido in Skikda, Algeria

> The only people who has to be blamed for this disaster is the Algerian army! I worked on that site for three and a half years. That area is completely flat, and the Algerian army who has a base there is supposed protect and create a security area for the gas plant and living compound (BBC, 2013b).

A contributor named as Meridja from Algiers offered a different assessment

> Dozens of Americans die quite often during hostages crisis or shootings in schools and random public spaces and yet the claim that the death of ONE American hostage is a catastrophe and that Algerian army need to explain themselves? That's just unacceptable (BBC, 2013b).

In both cases, these comments—one from a contributor with firsthand knowledge and experience of the area, and one offering an Algerian view of this big international story—really helped to explain the situation. They added context that a reporter on the ground could not have contributed so easily (and in this case, the remoteness of the region where the crisis was unfolding meant that there were no journalists at the scene anyway), and shared insights that journalists without expert knowledge of Algeria (i.e., the majority of non-Algerian journalists) could not have offered.

DISCUSSION

Citizen contributions can challenge and complement both official sources and information provided by established news organisations—keeping both on their toes, and giving a more detailed picture of a developing story, whether in Algeria, Syria, or elsewhere. As long ago as July 2005, when a group of suicide bombers attacked the London transport system in the morning rush hour, killing 52 people, citizens were making important contributions. As Beckett subsequently wrote of the coverage of what became known as the 7/7 attacks, "citizen journalism made an impact with people *sending in* phone images … to relate the day's events" (2008: 69, emphasis added). In 2005, the year before the public launch of Twitter and Facebook, *sending in* images (or textual comments and accounts) to established news organisations, in the hope they would publish them, was the primary distribution mechanism for user-generated content. Social media networks have radically changed this model and, in doing so, have created a key challenge for mainstream journalism: how to keep up with social media which now, for some breaking news stories, including crisis events, is the "first" place to look, behind "continuous TV news" (Mark Thompson, president and chief executive of *The New York Times*, quoted in Perry, 2013, speaking about coverage of the Boston Marathon bombings). The increasing deployment of live blogs has been one way news organisations believe they can keep up, increasing their "ability to be fast and accurate with the latest stories" (BBC, 2013c). However, their editorial processes—including fact and source checking, and negotiations with rights holders—mean they can rarely compete with social media platforms on pure speed. Although these editorial processes slow the news streams emanating from established news sites, like weirs in a river they also serve a purpose: to reassure audiences; what ITV News' web Editor, Jason Mills, calls the "'we filter so you don't need to' principle" (personal communication, 23 July 2013). As well as filtering social media updates, mainstream media also add analysis and insight as events are unfolding, a skill that Steve Herrmann, Editor of the BBC News website, says is "becoming even more important" (personal communication, 20 July 2013).

In this chapter we have shown how live blogs hosted on the ScribbleLive platform are incorporating unprecedented quantities of citizens' testimony on crisis events. Whether this model, which is giving that testimony the same level of prominence as contributions from journalists, will endure, or even spread, remains to be seen. Given that mainstream news organisations need to be able to differentiate themselves from social media channels, it is likely that they will continue to see the selective aggregation that, in part, characterises their live blogs as a unique selling point. However, the live blog, even in its relatively filtered form, *has* made it easier for news organisations to include an increasingly diverse range of content in their output, including from non-official or citizen sources. Jason Mills believes that as live video is incorporated more and more into online storytelling, mainstream news companies will "start merging their on-air and online content to provide different solutions for different devices, all with the 'live' principle at their heart" (personal communication, 23 July 2013). That 'live' principle has embraced citizens' contributions to a degree that at least equals, and probably surpasses, practice found in most other forms of journalism. The consequences of that embrace will be fascinating to follow as the future unfolds.

ACKNOWLEDGEMENTS

The authors are grateful to the Reuters Institute for the Study of Journalism and YouGov for permission to use data from the 2013 Digital News survey, to ScribbleLive for analytics data on the live blogs hosted on their live blogging platform, and to Nic Newman for his help sourcing some of the quotes used in this article.

NOTES

1. One of the first breaking news stories to be reported via a live blog was the London bombings of 7 July 2005 (McIntosh, 2005).
2. It should be noted that the popularity of live blogs at Guardian.co.uk is not necessarily mirrored at other mainstream news sites. For example, when a live blog is run side-by-side with an article at the BBC News website the two are "often equally visited" (Steve Herrmann, personal communication, 20 July 2013).
3. A fuller analysis of the survey's findings as they relate to live blogs has been published by Thurman (2013).
4. According to Rogers (2003), "innovators" and "early adopters" make up 16 per cent of the population of consumers who will ultimately adopt a new technology. The Reuters Institute/YouGov data show that, across the nine countries surveyed, the average number of news consumers (weighted by country population) that had accessed a live blog in the previous week was just over 16 per cent.
5. Compare, for example, the average number of views for stories on Storify in 2012: 662 (Storify, 2012). with the median number of pages views for a sample of live blogs at Guardian.co.uk: over 150,000 (Thurman and Walters, 2013: 86).
6. ScribbleLive supplied the data in anonymised form. Given the number of reader contributions, it is likely the live blogs were originally published by some of their mainstream media clients,

who include: Al Jazeera, CBS, the New York *Daily News*, Globeandmail.com, CBC, MSN, News International, CNN, the *Toronto Star*, the *Wall Street Journal*, and the Telegraph Media Group (ScribbleLive, n.d.). Tweets from readers were included. On the ScribbleLive platform reader contributions appear in the main part of the live blog amongst journalists' updates.

7. For example, the Reuters' "Japan Earthquake" live blog powered by ScribbleLive contains a number of updates by Reuters' journalists that contain only non-official new-technology mediated content. For example, Reuters' Ross Chainey posted "some amateur footage [from YouTube] reportedly captured in Sendai when the earthquake struck" (Reuters, 2011).

REFERENCES

Al Jazeera English (2013). "Syria Live Blog." Aljazeera.com. http://blogs.aljazeera.com/liveblog?page=1&f[0]=field_ns_topic%3A153, accessed 15 April 2013.

Batty, David (2011). "Nuclear Meltdown Fears After Japan Earthquake—Live Coverage." Guardian.co.uk, 12 March. http://www.guardian.co.uk/world/2011/mar/12/japan-earthquake-tsunami-aftermath-live, accessed 15 March 2013.

BBC (2003). "Reporters' Log: Saddam's Capture." BBC News Online, 15 December . http://news.bbc.co.uk/1/hi/world/middle_east/3317945.stm, accessed 12 April 2013.

BBC (2013a) "Opinion Survey." http://ecustomeropinions.com/survey/survey.php?sid=190809629, accessed 22 July 2013.

BBC (2013b) "As It Happened: Algeria Hostage Crisis." BBC News Online. http://www.bbc.co.uk/news/world-africa-21073655, accessed 15 April 2013.

BBC (2013c) "Local Live—Giving Users Access to BBC Newsrooms." BBC News website, April 22. http://www.bbc.co.uk/news/uk-england-21045859

Beckett, Charlie (2008). *Supermedia*. Oxford: Blackwell.

Bennett, Daniel (2011). "A 'Gay Girl in Damascus,' the Mirage of the 'Authentic Voice' and the Future of Journalism." In John Mair and Richard Keeble (Eds.), *Mirage in the Desert? Reporting the Arab Spring*. Bury St Edmonds: Abramis, pp. 187–195.

Boczkowski, Pablo J. (2010). *News at Work: Imitation in an Age of Information Abundance*. Chicago: University of Chicago Press.

Bond, Shannon (2011). "Japan's Earthquake." The World Blog, FT.com, 11 March. http://blogs.ft.com/the-world/2011/03/japan-earthquake/, accessed 18 March 2013.

Channel 4 News (2012). "Syria's Video Journalists Battle to Tell the 'Truth.'" Channel4.com, 27 March. http://www.channel4.com/news/syrias-video-journalists-battle-to-tell-the-truth, accessed 15 April 2013.

Fontenot, Maria, Boyle, Kris, and Gallagher, Amanda H. (in press) "Civic Respondents: A Content Analysis of Sources Quoted in Newspaper Coverage of Hurricanes Katrina and Rita." *Newspaper Research Journal*.

Goodman, J. David (2011). "Updates on the Earthquake and Tsunami in Japan." The Lede, NYT.com, 11 March. http://thelede.blogs.nytimes.com/2011/03/11/video-of-the-earthquake-and-tsunami-in-japan/, accessed 18 March 2013.

Guardian.co.uk (2011a) "Japan's 8.9 Magnitude Earthquake Triggers Tsunami—Video." Guardian.co.uk, 11 March. http://www.guardian.co.uk/world/video/2011/mar/11/japan-earthquake-tsunami-video, accessed 15 March 2013.

Guardian.co.uk (2011b) "Japan: The Day After the Earthquake and Tsunami—in Pictures." Guardian.co.uk, 12 March. http://www.guardian.co.uk/world/gallery/2011/mar/12/japan-earthquake-tsunami-in-pictures, accessed 15 March 2013.

Guardian.co.uk (2013). "MiddleEastLive." Guardian.co.uk, http://www.guardian.co.uk/world/middle-east-live?INTCMP=SRCH, accessed 15 April 2013.

Hough, Andrew, Chivers, Tom, and Bloxham, Andy (2011). "Japan Earthquake and Tsunami: As It Happened March 11." Telegraph.co.uk, 11 March. http://www.telegraph.co.uk/news/world-news/asia/japan/8377742/Japan-earthquake-and-tsunami-as-it-happened-March-11.html, accessed 18 March 2013.

Levy, David, and Newman, Nic (Eds.) (2013). *Reuters Institute Digital News Report 2013*. Oxford: Reuters Institute for the Study of Journalism, Oxford University.

Li, Xigen, and Izard, Ralph (2003). "9/11 Attack Coverage Reveals Similarities, Differences." *Newspaper Research Journal*, 24(1), 204–219.

Liu, Sophia B., Palen, Leysia, Sutton, Jeannette, Hughes, Amanda L., and Vieweg, Sarah (2009). "Citizen Photojournalism during Crisis Events." In Stuart Allan and Einar Thorsen (Eds.), *Citizen Journalism: Global Perspectives*. New York: Peter Lang, pp. 43–63.

McIntosh, Neil (2005). "Bomb Blasts Plunge London into Chaos." Guardian.co.uk, 7 July. http://www.guardian.co.uk/news/blog/2005/jul/07/explosionsplun, accessed 22 March 2013.

Murray, Alex (2011). "BBC Processes for Verifying Social Media Content." BBC Academy College of Journalism, 18 May.http://www.bbc.co.uk/blogs/blogcollegeofjournalism/posts/bbcsms_bbc_procedures_for_veri, accessed 15 April 2013.

NAA (2012). "Newspaper Web Audience." Newspaper Association of America, 26 December. http://www.naa.org/Trends-and-Numbers/Newspaper-Websites/Newspaper-Web-Audience.aspx, accessed 21 March 2013.

Nacos, Brigitte L. (1996). *Terrorism and the Media: From the Iran Hostage Crisis to the Oklahoma City Bombing*. New York: Columbia University Press.

NBC (2011). "Live Blog: Huge Tsunami Hits Japan After 8.9 quake." World Blog, NBCnews.com, 11 March. http://worldblog.nbcnews.com/_news/2011/03/11/6243734-live-blog-huge-tsunami-hits-japan-after-89-quake?lite, accessed 13 March 2013.

Nip, Joyce Y.M. (2009). "Citizen Journalism in China: The Case of the Wenchuan Earthquake." In Stuart Allan and Einar Thorsen (Eds.), *Citizen Journalism: Global Perspectives*. New York: Peter Lang, pp. 95–105.

Perry, Phillip M. (2013). 'Thriving in a "Twitter First" World', *Editorial Calendar*, http://editorialcalendar.net/thriving-twitter-first-world/, 29 May.

Powell, L., and Self, W.R. (2003). "Government Sources Dominate Business Crisis Reporting." *Newspaper Research Journal*, 24(2), 97–106.

Reuters (2011). "Japan Earthquake." Reuters.com, 11 March. http://live.reuters.com/uk/Event/Japan_earthquake2?Page=0, accessed 22 March 2013.

Rodgers, James (2011). "Capturing Saddam Hussein: How the Full Story Got Away, and What Conflict Journalism Can Learn from It." *Journal of War and Culture Studies*, 4(2), 179–191.

Rogers, Everett M. (2003). *Diffusion of Innovations, 5th Edition*. New York: Free Press.

Salwen, Michael B. (1995). "News of Hurricane Andrew: The Agenda of Sources and the Sources' Agendas." *Journalism and Mass Communication Quarterly*, 72(4), 826–840.

ScribbleLive (n.d.) "Customers." ScribbleLive.com, http://www.scribblelive.com/About_Customers.aspx, accessed 22 March 2013.

Sibbison, Jim (1988). "Dead Fish and Red Herrings: How the EPA Pollutes the News." *Columbia Journalism Review* 27(4), 25–28.

Storify (2012). "Storify Year in Review—2012." Storify.com, http://storify.com/storify/storify-year-in-review-2012, accessed 22 July 2013.

Thurman, Neil (2013). "How Live Blogs are Reconfiguring Breaking News." In David Levy and Nic Newman (Eds.), *Reuters Institute Digital News Report 2013*. Oxford: Reuters Institute for the Study of Journalism, pp. 85–88.

Thurman, Neil, and Walters, Anna (2013). "Live Blogging—Digital Journalism's Pivotal Platform? A Case Study of the Production, Consumption, and Form of Live Blogs at Guardian.co.uk." *Digital Journalism*, 1(1), 82–101.

Wigley, Shelley, and Fontenot, Maria (2009). "Where Media Turn During Crises: A Look at Information Subsidies and the Virginia Tech Shootings." *Electronic News*, 3(2), 94–108.

Wigley, Shelley, and Fontenot, Maria (2011). "The Giffords Shootings in Tucson: Exploring Citizen-generated versus News Media Content in Crisis Management." *Public Relations Review*, 37(4), 337–344.

WSJ (2011). "Live Blog: Japan Earthquake." Japan Real Time, WSJ.com, March 11. http://blogs.wsj.com/japanrealtime/2011/03/11/live-blog-japan-earthquake/, accessed 18 March 2013.

SECTION TWO

Capturing Crisis

CHAPTER SEVEN

Tools IN Their Pockets: How Personal Media Were Used During THE Christchurch Earthquakes

DONALD MATHESON

Many accounts from those affected by two large earthquakes to hit the New Zealand city of Christchurch in 2010 and 2011 begin with their mobile phones. These devices, along with the internet and in particular the social media they connect to, are now so much part of how many people interact every day that they are now a significant part of how disaster is mediated. One Christchurch woman, who had just moved north to Wellington, remembered the second quake as follows:

> I was in the middle of my first introductory lecture when I got a tweet from a Christchurch friend. It was a comment on a 'big' aftershock.
>
> Everyone gets on Facebook and Twitter to talk about the aftershocks and debate how big they were, so I ignored it. I had more important things to think about.
>
> I was just about to turn off my phone when I got a text from my father.
>
> 'Big quake. Lots of deaths. I was in town and buildings collapsed around me'. (Sophia 2011)

In societies where media are now ambient—always on and always to hand for many individuals—the question arises of the role they play in people's response to the breakdown of the urban fabric. In particular, the potential of these networked media to empower, even to democratise, raises significant hopes of ameliorating the worst of the chaos and anguish of such moments.

This chapter explores citizen participation in mediating the news of the two large earthquakes which struck Christchurch, a city of 300,000 people, in September 2010 and then in February 2011. In the first quake, measuring 7.1 on the Richter scale, damage was localised, affecting most through the large amounts of mud and silt that liquefaction of the soil pushed to the surface. After the second, magnitude 6.3, quake five months later, 185 people died as central city buildings collapsed, in what insurers later called one of the most expensive natural disasters in history (Booker 2012).

In assessing the roles of personal and portable media in Christchurch people's response to the twin disasters, I want to step back a little from framing their participation as citizen journalism. That is, I want to follow Rosen (2008) only part of the way in looking for civic participation foremost in tools of public communication or in imagining that participation as a desire of people to inform one another. Instead, Christchurch appears to me a case study of the limits of that frame in understanding the role of networked media in coping with an urban disaster. Firstly, I suggest that a dominant experience of those in the city on the day of the second, devastating, quake was of the absence of these media. Secondly, I discuss the story of the Student Volunteer Army, an enormous mobile media crowdsourcing initiative, as partly about young people performing primarily social interactions that stopped short of being political. Thirdly, I emphasise the supportive intimacy amongst strangers which was enabled by networks such as Facebook. Lastly, I contrast the level of use of networked media for community re-engagement with the overwhelming resort to long-established placed-based and often face-to-face forms of community interaction. Overall, the chapter follows scholars such as Aldrich (2012) by observing that, after the disaster, how people were connected individually to each other and to their own communities was highly important to them. Connecting across more public uses of media was secondary. The consequences of an analysis such as this are not only to 'decentre' media, as Couldry (2006) puts it, but also to position the citizen participation as often quite weakly political and perhaps less transformative of major social institutions than expectations of citizen journalism might suggest.

BROKEN MEDIA

One irony for a communication scholar is that digital networks were part of what was broken by the second, larger quake. For perhaps half of the city, electricity was not restored for days and in some cases weeks afterwards. The rupture of infrastructure meant that images of the enormous destruction in the city centre were more easily available to people in the northern hemisphere than for many residents a kilometre away, whose televisions and computers were unusable. Landlines in many places were down and the overloading of the mobile phone system in the city meant that voice and data usage were initially very limited; even text messages (SMS) took sometimes hours to get through. In one vivid example, the husband of Dr Tamara Cvetanova tried to phone her 25 times over a 12-hour period as she lay dying in a collapsed building, getting through only intermittently (Radio NZ 2012). The overloading was exacerbated as emergency batteries in mobile phone towers in parts of the city without electricity began to fail. Then, in the weeks that followed, recharging mobile phones became a major challenge for those still without power in the worst-hit areas, and a key service at community hubs was the provision of charging stations.

By contrast, the local newspaper, *The Press*, published a 24-page edition the morning after the earthquake, despite its building collapsing and killing a staff member. It printed the paper instead at a nearby town's presses. For many, radio remained a key resource, including both national radio networks and small community broadcasters such as the student station RDU (which bought and converted a horse truck so it could resume live service). Yet a review by the disaster agency, Civil Defence, found that even radio was an uncertain communication form for those without electricity, as relatively few people possessed transistor radios with working batteries. Since the quakes, the Red Cross has launched a programme to give away 35,000 cheap solar-powered/wind-up radio-torches through the region's schools.

Civil Defence researchers reported a sudden silence for many people: 'Because of these communication failures many people felt isolated and threatened not knowing what was happening around them' (McLean et al, 2012: 76). Disaster researcher Jeanette Sutton reached a similar conclusion:

> In this day and age, when emergency managers are frequently developing their communication strategies by building online information portals and delivering information via web-based and social media communications, disaster affected individuals are at risk of being in a vulnerable vacuum of information just when they need it most. (Sutton 2012: 1)

Certain individuals were, in dramatic ways, placed at the forefront of initial reporting of the news through their mobile media tools. One user, @dyedredlaura, posted twitpics within minutes, including one of the badly damaged Anglican cathedral which received more than 14,000 views within an hour. Another user, ypud, quickly uploaded a video onto YouTube of a house crushed by rocks from the cliffs above, which has at time of writing attracted 800,000 views (see Mabbett 2011). More illuminating about the general experience, though, is the story of the media set up in the badly damaged suburb of New Brighton, the site of no deaths but with little working infrastructure throughout February and March 2011. There an independent-minded police officer teamed up with the local MP to set up their own basic media form: 'an A4 sized information sheet was produced and distributed to local residents to keep them informed and regularly updated, as the lack of electricity meant many were unable to access television, radio or telephones' (Review of Civil Defence, op.cit.).

A key point, then, is that citizens' experiences were highly varied, at a moment when the urban fabric became badly damaged—electricity down, workplaces closed, roads impassable—and the shared life of the city fragmented. So too were their experiences and uses of communication media, to help each other, regroup and begin to rebuild. My personal memory, as a Christchurch resident after the quakes, is of becoming narrowly focused on my family, not wanting to connect much with the wider world, and certainly not to listen to too much of the horror I could hear on the radio. Some studies suggested that for many others too life became suddenly much simpler—focused on food, clean water and working toilets and on family and neighbours. The media that serve the complexities of urban life became less relevant. Others talked of the mobile phone and the laptop as allowing them to continue their working lives with little disruption, other than being stuck at home. Others still—at least among those who had electricity—experienced a strong need to connect with both friends and strangers through social media. One wrote a year later, as the aftershocks continued to wear residents down: 'Facebook is my friend in the dead of the shakey night' (Sarah 2011).

EXPECTATIONS OF TRANSFORMATIVE CITIZEN MEDIA

Those experiences do not accord in any simple way with an academic literature that sets up expectations that the media produced in networks of citizens will be transformative in natural disaster situations. After the 2004 Indian Ocean tsunami, mobile phones allowed Scandinavian authorities to quickly track down and help their citizens who had been holidaying in the worst-hit Sri Lanka and Thailand resorts (Kivikuru 2006). After the Haiti earthquake in 2010, there was a 'firestorm of social media use' among non-profit organisations, raising large amounts of funds and coordinating longer-term aid (Muralidharan et al. 2011). During forest fires in southern California in 2007, significant numbers of residents turned to social media to augment what they saw as inadequate mass media reporting and slow official provision of information, as well as to deal with the distress they were experiencing (Sutton et al. 2008). Some of these information searchers then became information providers, passing details and advice on, with some producing sophisticated Google map overlays that matched snippets of information with location or teaming up with firefighters or news organisations. One resident told researchers:

> I was plugged in to everything I could find and I knew that a lot of my less tech savvy friends were having problems getting real information from the news, so I just soaked up

> as much as I could from the internet and regurgitated it through text messages, instant messengering, twitter, and my blog. (ibid.)

Sutton et al.'s survey found that a recurring theme in residents' recollections was a sense of frustration with established sources of information. Local news was slow and hard to access when people were evacuated, while, one wrote, 'national news websites were completely worthless as they ignored everything except the comparatively minor Malibu fire which burned near some celebrity homes' (ibid).

Disaster management as a field has slowly begun to value this kind of free and horizontal flow of information. Traditionally, communication has been regarded as something to control so as to mitigate risks of rumour, panic and poor risk decisions. In Christchurch, there was some evidence to reinforce that perspective. One of the NZ Blood Service's first strategic responses to the second, large earthquake, was to send out messages that blood donations were not needed, as its stocks of most blood products were excellent. That job became harder when it became apparent that a platelets donor, who had been called in to donate, sent out a general call on Facebook that was then picked up by professional media (Flanagan 2011). In parallel with renewed attention to collective and grassroots action in disaster management theory, networked media examples such as those above were being given prominence. Crowdsourcing of information is seen as a valuable supplement, sometimes even a replacement, for top-down, one-way and regulated forms of information provision during calamitous events. Attention often rests on the technology itself. Palen et al. (2010: 1) write: 'by viewing citizenry as a powerful, self-organising and collectively intelligent force, ICT has the potential to play a remarkable and transformative role in the way society responds to mass emergencies and disaster.'

THE STUDENT VOLUNTEER ARMY: 'USING THE TOOLS IN OUR POCKETS'

There were parts of Christchurch that were largely unaffected in the shaking, including the area around Canterbury University. There, technology did enable a remarkable instance of self-organising individuals. After the first earthquake in 2010, political science student Sam Johnson rang Civil Defence to offer his services as a volunteer in the clean-up of the sometimes metre-deep piles of silt in parts of the city. The clean-up task was enormous. Yet, Johnson later wrote: 'A Civil Defence official gave me a lengthy phone interview, established I had no "skills" to offer, declared this to be a situation for 'experts,' and advised me to go home and check on my neighbours' (Johnson 2012). He and his friends, frustrated at sitting at home waiting for the university to reopen, set up a Facebook event, sent it to 200 of their friends and over the next two weeks organised the placement, transport and feeding of 2500 people who responded. Between them they shovelled an estimated 65,000 tonnes of silt. After the much larger 2011 earthquake, the 'Student Volunteer Army' Facebook page became a focus of a huge volunteer effort, attracting 13,000 volunteers in its first week and 27,000 likes within three weeks. It coordinated hundreds of volunteers each day for two months, becoming involved also in distributing leaflets and checking on residents who were isolated.

In an article titled, 'Students vs the machine', Johnson credited the group's success to two factors. The first was the high degree of responsiveness of the organising team to the changing tasks. Systems evolved as the volunteer effort grew and as requests for help arose, in what could be characterised as distributed and emergent structures rather than traditional forms of organisation. Johnson wrote:

> Every evening, we posted plans for the next day. Our volunteers knew to check Facebook daily at 8pm, giving details on when and where to meet, what to bring, and what to wear. They could ask questions and, as the page was constantly managed, get answers very quickly. Beyond this, as anyone could view and post on the page, people from all over New Zealand and the world could post messages of support and encouragement. (Johnson 2012: 19)

Johnson and his colleagues trained team leaders to take responsibility and use common sense, rather than follow rules or structures. The organisation, he wrote, was instead coordinated through mobile phones connected by an off-the-shelf dispatching app: 'The tools in our pockets—cell phones, Google maps, Facebook, Twitter and everything in between—were the key to our success' (ibid.: 20). The second factor he cited was the readiness of the volunteers, used to organising their social lives over mobile phone and social media networks, to respond to calls to action in this fashion. Indeed, many students later commented that taking part in the volunteer army was a great social event in a city that had ground to a halt.

Johnson points to a frustration, similar to the feelings expressed after the Californian forest fires cited above, with official networks and a contrasting readiness to work in other ways. '[T]he official processes and manuals were stagnant, outdated and irrelevant to our generation's spontaneous, modern and impatient volunteers' (ibid.). Other commentators in the city point to an enthusiasm among young people (labelled by marketers as Generation Y) for solution-oriented action over protest and for social entrepreneurship that combines a faith in technological solutions with the power of ideas. 'Gen Y has grown up in an era of ever-widening horizons—an explosion of ideas and possibilities due to the information-rich internet and the globalisation of world culture' (McCrone 2011: C5). Strikingly absent in the descriptions of ideas and assumptions shared among these young people was any politics. Johnson (a politics student) describes a failure of Civil Defence to allow young people to take part, but it is social not political participation that he emphasises. The mobilisation of students here as willing workers, brought together across Facebook 'friend' networks, expressed a faith in networking as a way of getting things done. Students gained social capital in their networks—to not be part of the 'Army' connoted social isolation for some—and Johnson gained much political capital, to the extent that Hillary Clinton presented him with an award when she visited the city. But very little rhetoric of any public or civic movement attached itself to volunteer activity.

MEDIA OF REASSURANCE

In some ways, then, these media remained personal networks, sometimes intensely so. A number of researchers have identified a turning towards strangers as well as friends on social media platforms such as Facebook for reassurance in the aftermath of disasters. Howell and Taylor (2011) see social media as sometimes operating as psychological first aid, providing reassurance, compassion and a sense of empowerment, as well as an information resource, passing on information sent out by authorities. Their survey, which included residents in Christchurch as well as those affected by Australian and Japanese disasters, is echoed elsewhere. When reflecting on how they coped with the disaster, these media are frequently mentioned by those in Christchurch as well as by those outside it. One Auckland resident wrote:

> i have between 30–50 friends that live in christchurch and some family members, cousins. i dont know how people survived BEFORE facebook in events like this—with each status update from a friend to say "IM SAFE" i rejoiced and embraced the peace that comes with that knowledge that someone is okay. (Claire 2011).

Further afield, the US social media blog Khou.com noted on the day after the February quake:

> Christchurch is a trending topic worldwide on Twitter. Tweets are pouring in with prayers and well wishes for the victims of the earthquake. Twitter users are also sharing images of the devastation, and tweeting a map showing the aftershocks from the quake. (Khou 2011)

Like the student's father cited above, who sent his daughter a reassuring text message as he was walking out of the city centre for home, many individuals prioritised, in the immediate aftermath, connecting with loved ones. I experienced myself a wave of expressions of support from around the world via Facebook and email, once I was able to get access to the internet. As electricity and telecommunications became widely available again, these channels augmented other forms of support for people.

For some, Facebook continued to provide these spaces of interpersonal, sometimes semi-public support for some time. The manager of the University of Canterbury 'UC Quake recovery' Facebook page wrote of her experience after one night-time aftershock:

> I was worn out, lying in bed with the laptop beside me then there was a big aftershock. I wondered who else was out there and how they were feeling, so I grabbed my laptop and posted something (post: Three doozy aftershocks in a row! Anybody else jump out of bed like I did? Sept 7th 1:28 am). And we got all these responses (47 comments immediately – 51 likes). It helped me as well because I was sitting there scared as well. (Janelle Blythe, quoted in Dabner 2012: 75)

Talking through the trauma of the post-disaster situation and sharing the experience via social media are mentioned so frequently in people's personal accounts of the disaster that they cannot be underestimated as central to the collective experience.

COMMUNITY RESPONSE ONLINE

However, the extent to which that interpersonal support developed into moments of collectivity that would have implications at a wider community or public level is harder to gauge. In terms of gathering and sharing information, social media seem to have served a relatively minor adjunct role to the information provision from governmental and professional media organisations. The website, Geonet, run by the government's geological agency, GNS, became a very heavily used and much discussed website during the thousands of aftershocks in the following months. Bruns and Burgess (2011) also found in a study of the #eqnz hashtag on Twitter that messages from government or infrastructure organisations were a large part of a deluge of tweets they gathered, as people sent on information about boiling water or how to apply to the government's earthquake insurance scheme. By itself, the account of the Canterbury Earthquake Authority (set up after the September 2010 quake) 'is responsible for nearly 2500 tweets during the first fortnight after the 2011 tremor', an average of nearly 180 tweets a day (ibid.).

Some non-governmental websites rose to prominence. Google People Finder and the online auction site TradeMe were widely used as sources for people to find each other and to find out about services in the disrupted city—TradeMe reported 11,000 listings about lost pets, offers of volunteer labour, accommodation and similar topics. For a short time, a social media crowdsourcing initiative provided a well-used site of information. The 'Christchurch Recovery Map', a map based on the Kenyan-developed crowsourcing platform Ushahidi, was created within hours by volunteers involved in Crisis Commons and as many as 10 other volunteer technical communities across the world. The map allowed individuals to upload information via text messages or tweets about such matters as which roads were passable, which petrol stations remained open and where council drinking water tanks had been set up. Leson (2011) reports 70,000 visits to the site within two days of the event, among them a diabetic who thanked the site for helping locate a pharmacy to supply insulin. This information was, it must be said, superseded within days by websites from a host of governmental organisations, including local councils, Civil Defence, social welfare agencies and even the usually reticent defence force, which all appeared to have learnt quickly from the earlier quake about the value of the quick and open provision of information online (Sweeney 2011). Indeed, there was some tension between those running the Recovery Map site and authorities over the risk that people would not know that their tweets or text messages were going directly onto a public website or that people would confuse the site with official information or treat the site as a way of contacting the police or fire service. While the crowd-sourced site remained active for only a short time, it has lead to discussions among authorities about the value of drawing on this kind of volunteer technology in future emergencies.

COMMUNITY RESPONSE FACE TO FACE

These technologically enabled initiatives, however, remain a small part of the story of the community response to the disaster, and there is a risk of overstating them. A large number of other self-help and volunteer initiatives emerged, from the Farmy Army—800 farmers who ferried cooked food from nearby towns, brought in equipment to clear silt and provided drinking water—to spontaneous trips by transport companies whose trucks collected donations from cities and towns across New Zealand to community hubs run by locals that sprang up in churches and schools in the city's suburbs. A central story in both official and individuals' reflections on what took place in the initial weeks after the earthquake is one of large numbers of people taking the initiative to help themselves and each other. News media and researchers alike identified an 'organic community response' where communities organised themselves (Mamula-Seadon et al. 2012). A key, if understandable failing of the official and news media responses to the events, it is widely agreed, was a focus on the central city, where lives had been lost and where the heaviest building damage was. That led to a neglect of the damaged suburbs in the city's east, often places with large numbers of less well-off and less mobile people. Self-organised groups of residents in these areas proved highly effective. Community group CanCern, for example, divided whole suburbs into cells of 12–18 houses, each with a coordinator.

The official response was similarly thin in the port town of Lyttelton, closest to the quake's epicentre. The civil defence team had been called into Christchurch city and public buildings such as council offices and the police station were closed because of damage. Instead, the volunteer information centre, already a community focus through the 'Timebank' scheme it facilitated, where residents traded their skills in a cashless economy, filled many gaps in the official response. Its volunteers coordinated requests to the local fire brigade for chimneys to be made safe or roofs made watertight, found accommodation for those left homeless, and distributed official information through the Timebank email network and an information whiteboard on the footpath (Everingham 2012: 29).

What appeared to characterise the most successful initiatives was people's relationships as neighbours and community members. That is, established and already valued community networks, through which knowledge had accumulated and could then be shared, came to the fore. Knowledge about who lived where, which neighbours had particular skills or where accommodation might be available, as well as the legitimacy and trustworthiness of local organisers, were identified by many community members in their reflections afterwards. Daniel Aldrich is one of a number of scholars who has identified the same trend in the aftermath of disasters across the world. He argues (e.g., Aldrich 2010) that communities with the highest social capital, that is, the richest social networks and social resources, not only respond best to disaster but recover most quickly because their residents provide each other with the 'informal insurance' of information, assistance and financial help and because they have more confidence in the ability to rebuild.

Social media–led initiatives such as the Quake Recovery Map and the Student Volunteer Army drew upon the particular cultural knowledge of technical volunteering communities and of a student network, which seeded wider, if temporary, social action. Elsewhere in the city, it seems that the strong networks of established communities and relationships fostered around the portable toilets and water tanks placed at street corners were more significant sites of information sharing and collective action.

FINAL THOUGHTS

My own experiences of living through the events discussed above has led me to want to place the way networked media were used within a wider context. Mobile phones, portable computer devices and social media platforms have become part of our communication expectations and are reached for, in many different ways, to solve problems after a disruptive event. Dramatic cases can also be found in which there has been a transformational effect of these media on community and public life. The Student Volunteer Army looms large in the story of the Christchurch quake and deservedly so: it powerfully indicates a change in the expectations among young people in particular about how to get things done. However, the facts that at the same time swathes of the city were in the dark and that much of the community response to the quakes happened through the quite different medium of trusted community leaders and neighbours talking face-to-face, act as a reminder to keep any claims within proportion. Facebook and Twitter allowed people to cope and stay in touch, but become less significant at the public level. It is possible also that, as official disaster planners prepare themselves better to meet people's information needs after such events, the value of socially shared media will become still more constrained.

The longer-term picture also suggests that the public power of networked and citizen action may fade after the initial moments of dramatic citizen images and Ushahidi maps. As the city has moved into recovery mode, the control of large institutional actors, such as the government's Canterbury Earthquake Recovery Authority (a body with extraordinarily wide powers), the insurance industry, property developers and large construction companies, has grown, in ways that have rarely been tempered by the voices of citizens. A year after the February 2011 quake, the city council launched a large consultation exercise about how the city should be rebuilt. Along with a community expo, a roadshow and workshops, the council ran a website called 'Share an idea', where people could post their vision of the city's future. Across all these initiatives, 106,000 different ideas were gathered together, netting the council an award from the Netherlands-based Co-creation Association (CCC 2011). Yet these ideas have languished since. Civic discussion has largely been replaced by commercially sensitive negotiations along with a command-and-control model of recovery. Some political scientists (e.g. Hayward 2012) talk of a political quake succeeding the physical one, in which democratic institutions were damaged. What perhaps characterises the story of participatory or citizen media most here is that it was rarely politicised, in contrast to the grass-roots politics of some of the neighbourhood action. A tool in people's pockets, they quickly went back into the pocket.

REFERENCES

Aldrich, Daniel P. (2010). 'Fixing recovery: Social capital in post-crisis resilience. *Journal of Homeland Security* 6: 1–10.

———. (2012). *Building resilience: Social capital in post-disaster recovery*. Chicago: University of Chicago Press.

Allan, Stuart (2013). *Citizen witnessing: Revisioning journalism in times of crisis*. Cambridge: Polity Press.

Booker, Jarrod (2012). 'Christchurch quake third most expensive disaster ever—insurer'. *New Zealand Herald*, 29 March. Available at http://www.nzherald.co.nz/nz/news/article.cfm?c_id=1&objectid=10795342

Bruns, Axel, and Burgess, Jean E. (2012). 'Local and global responses to disaster: #eqnz and the Christchurch earthquake'. In Sugg , Peter (Ed.) *Disaster and Emergency Management Conference, Conference Proceedings*, pp. 86–103. Brisbane: AST Management.

Christchurch City Council (CCC) (2011). '"Share an Idea" wins international award', http://www.ccc.govt.nz/thecouncil/newsmedia/mediareleases/2011/201111241.aspx, 24 November.

Claire (2011). '@stickyjesus: Demystifying Facebook 11 & Twitter 12'. One passion, one devotion, 30 May. Available at http://onepassiononedevotion.wordpress.com/2011/05/30/stickyjesus-demystifying-facebook-11-twitter-12

Couldry, N. (2006). *Listening beyond the echoes: Media, ethics and agency in an uncertain world*. London: Paradigm.

Dabner, Nicki (2012). '"Breaking Ground" in the Use of Social Media: A Case Study of a University Earthquake Response to Inform Educational Design with Facebook.' *Internet and Higher Education*, 15(1), 69–78.

Everingham, Wendy (2012). 'Lyttelton's Grassroots Response,' in *Tephra—Community resilience: Case studies from the Canterbury earthquakes*. Vol. 23, November, 26–31.

Flanagan, P. (2011). 'Reflections on the Christchurch earthquake. *ISBT Science Series*, 6:2, 350–353.

Hayward, Bronwyn (2012). Canterbury's political quake'. *The Press*, 30 March, 7.

Howell, Gwyneth V.J., and Taylor, Mel (2011). 'When a crisis happens who turns to Facebook and why?' *Asia Pacific Public Relations Journal* 12:2, 1–8.

Khou (2011). Social media plays key role after New Zealand earthquake. Khou.com, 22 February. Available at http://www.khou.com/news/Social-media-plays-key-role-after-New-Zealand-earthquake—116676449.html

Ingram, Mathew (2012). 'Hurricane Sandy and Twitter as a self-cleaning oven for news', *Gigaom*, http://gigaom.com/2012/10/30/hurricane-sandy-and-twitter-as-a-self-cleaning-oven-for-news/, 30 October.

Johnson, Sam (2012). 'Students vs. the machine'. *TEPHRA 23: Community resilience: Cases studies from the Christchurch earthquakes*, 18–22. Wellington: MCDEM.

Kivikuru, Ullamaja (2006). 'Tsunami communication in Finland: Revealing tensions in the sender–receiver relationship'. *European Journal of Communication* 21:4, 499–520.

Leson, Heather (2011). 'How the Eq.org.nz site came about to help with the Christchurch earthquake'. *Crisis Commons* blog, 24 February. Available at http://crisiscommons.org/2011/02/24/how-the-eq-org-nz-site-came-about-to-help-with-the-christchurch-earthquake

McCrone, John (2011). 'Relish the power of a new generation'. *The Press*, 25 June, C4–5.

McLean, Ian, Oughton, David, Ellis, Stuart, Wakelin, Basil, & Rubin, Claire B. (2012). 'Review of the Civil Defence Emergency Management Response to the 22 February Christchurch Earthquake', *Ian McLean Consultancy Services Ltd*, Director of Civil Defence and Emergency Management.

Mabbett, Charles (2011). 'The latest Christchurch earthquakes told on social media platforms'. Socialmedianz.com, 22 February. Available at http://socialmedianz.com/opinion2/2011/02/22/the-christchurch-earthquake-told-on-social-media-platforms

Mamula-Seadon, Ljubica, et al. (2012). 'Exploring resilience'. *TEPHRA 23: Community resilience: Cases studies from the Christchurch earthquakes*, 18–22. Wellington: MCDEM.

Muralidharan, S., et al. (2011). 'Hope for Haiti: An analysis of Facebook and Twitter usage during the earthquake relief efforts'. *Public Relations Review* 37:2, 175–77.

Palen, L., et al. (2010). 'A vision for technology-mediated support for public participation and assistance in mass emergencies and disasters'. *Proceedings of ACM-BSC Visions of Computer Science*. Available at http://www.bcs.org/content/conWebDoc/35016

Radio NZ (2012). 'No one in charge of CTV operation, inquest told.' Radio NZ, 1 Nov. Available at http://www.radionz.co.nz/news/canterbury-earthquake/119579/no-one-in-charge-of-ctv-operation,-inquest-told

Rosen, Jay (2008). 'A most useful definition of citizen journalism.' PressThink, Available at http://journalism.nyu.edu/pubzone/weblogs/pressthink/2008/07/14/a_most_useful_d.html

Sarah (2011). 'What has changed for me in Christchurch'. Quake Stories, 4 September. Available at http://www.quakestories.govt.nz/111/story

Sophia (2011). 'I was in my final year…'. Quake Stories, 4 September. Available at http://www.quakestories.govt.nz/111/story

Sutton, Jeanette et al (2008). Backchannels on the front lines: Emergent uses of social media in the 2007 Southern California wildfires'. *Proceedings of the 5th International ISCRAM Conference*. Available at http://www.iscramlive.org/portal/taxonomy/term/4

Sutton et al (2012). 'When online is off: Public communications following the February 2011 Christchurch, NZ, earthquake'. *Proceedings of the 9th International ISCRAM Conference*. Available at http://www.iscramlive.org/portal/taxonomy/term/11

Sweeney, Kevin (2011). 'Supporting earthquake efforts in Christchurch'. New Zealand Geospatial Strategy blog, 2 March. Available at http://www.geospatial.govt.nz/supporting-earthquake-efforts-in-christchurch

CHAPTER EIGHT

Hurricane Sandy AND THE Adoption OF Citizen Journalism Platforms

TREVOR KNOBLICH

Citizen journalists have demonstrated a variety of creative solutions to communications challenges when documenting large-scale crisis events. These solutions have included adapting emerging social media platforms, organizing information in new ways, and even shifting the use of any given platform as access becomes limited. Hurricane Katrina and the 7 July 2005 London transit bombings were captured vividly on Flickr and YouTube. The series of protests in Tunisia, Egypt, Libya and elsewhere, beginning in December 2010, prompted global conversations with those affected using Twitter and Facebook. In October 2012, as Hurricane Sandy bore down on the East Coast of the United States, citizens began to adapt yet another tool for journalistic purposes: the mobile photo sharing application Instagram (Taylor, 2012).

As new tools for capturing text, photographs and videos emerge, citizens around the world have increasing opportunities to participate in documenting, sharing, and providing nuance to breaking news events. In turn, each crisis event illustrates that public familiarity with a platform, as well as persistent access, largely shapes how citizens choose to share information. In other words, when sharing important information, citizens tend to gravitate toward popular, easy-to-use platforms with which they feel most comfortable. Furthermore, they will likely turn to other platforms only when access to their platform of choice is diminished or unavailable.

This chapter presents a qualitative view of citizen journalism coverage of Hurricane Sandy in the United States. In particular, I focus on how the continuing emergence of social media platforms increases citizen participation in news. I will also explore why newsrooms should understand evolving public opinion about social media sites, as well as the manner in which citizens choose to engage with a given platform. Finally, the chapter will conclude with a discussion of the emerging challenges around data ownership, which can be critical information for newsrooms seeking contributions from citizen journalists during a crisis.

DOCUMENTING THE STORM

Hurricane Sandy was one of the worst storms in the 2012 hurricane season, causing damage from islands throughout the Caribbean then north along the eastern United States. The storm affected Cuba, the Dominican Republic, Haiti, Jamaica and other Caribbean islands, killing dozens and causing hundreds of millions of dollars in damage to property and infrastructure (Pierre-Pierre, 2012).

In the United States, Hurricane Sandy became recognized as one of the most destructive storms in history. By the time the storm subsided, it had affected the entire eastern seaboard of the United States, from Florida to Maine. The worst effects occurred in the Mid-Atlantic region, where Hurricane Sandy merged with another storm moving in from the west. This process intensified the effects of both storm systems, causing some media outlets to dub the massive weather pattern "Superstorm Sandy" or "Frankenstorm."

As Hurricane Sandy struck the Mid-Atlantic region, more than 8.5 million homes lost power. The affected area was vast, with the largest storm surges occurring in New York, New Jersey and Connecticut. The National Hurricane Center estimated that Sandy was responsible for 159 deaths. Some of these deaths resulted from direct causes, like the storm surges, while others resulted from indirect causes, such as hypothermia. Total damages from the storm reached an estimated $50 billion, making it the second costliest storm in US history, following Hurricane Katrina (Blake et al., 2013).

As evidenced elsewhere in this volume, major crisis events inspire a significant amount of citizen reporting, largely through social media platforms. Hurricane Sandy was no different. Twitter estimates that the storm sparked 20 million tweets, using the words "Sandy" or "hurricane," or the related hashtags, #sandy or #hurricane (Shih, 2012). Facebook calculated that Sandy was the second-most discussed event on the platform in 2012, after the Super Bowl[1] (Griggs, 2012).

CITIZEN PARTICIPATION IN NEWS

The capabilities for citizens to participate in breaking news events have been increasing dramatically over the last decade. This can be attributed to a variety of technical factors including increasing ownership of smartphones with high-quality cameras, a variety of user-friendly photo and video applications, the popularity and variety of social media platforms, and persistent access to broadband internet.

By early 2012, before Hurricane Sandy struck, approximately 46 percent of American adults owned a smartphone, according to a Pew internet study (Smith, 2012). Photo sharing applications, such as Flickr and Instagram, were also widely used and deeply integrated with social media platforms, including Facebook and Twitter.

As major news events arise, citizens often turn to these types of platforms to coordinate information, share updates, and post videos and photos. For instance, a Pew research study indicates that approximately 34 percent of tweets with the hashtag #Sandy were related to sharing news and information. Another 25 percent of tweets shared photos or videos, "speaking to the degree to which visuals have become a more common element" of Twitter (Pew, 2012). Both figures signify a willingness among citizens to participate in breaking news events, using social media platforms to coordinate and share information.

Many journalists also noted the emergence of Instagram as a popular platform for citizens sharing photographs of the storm. Hurricane Sandy was among the first major news events to gain coverage in the United States since Instagram's rise to popularity. By October 2012 when the storm arrived, Instagram claimed nearly 100 million users, making it a large-scale, mainstream platform in the United States.

This variety of platforms for sharing information makes it increasingly easier for users to participate in news events as they occur. Users are willing to generate text, videos, or photos with tools readily available in their pockets, and hashtag and search features simplify the ability for people to see what others are posting and share that within their own networks.

TWITTER CONTENT: "GLOBAL" TO LOCAL

Twitter users looking to share news and information during Hurricane Sandy demonstrated a specific and unique behavior on the platform. As the storm approached, Twitter users largely relied on broad hashtags for organizing information related to the hurricane. Early hashtags included #superstorm, #sandy, and #hurricanesandy. These are helpful in communicating about the storm in general, but less helpful when discussing its effects on a particular region, city or neighborhood—the type of nuanced perspective that serves as a hallmark of citizen journalism.

In fact, the use of hashtags shifted over time, from the initial hashtags simply describing the storm itself to more locally focused ways of organizing information as the storm reached specific communities. In some cases, the storm name was combined with an additional hash tag for location-based relevance, such as #sandy #eastcoast or #sandy #nyc. Over time, the hashtags began to unify and became even more specific. For example, #sandyNYC, #sandynj, and #sandyDC were used to report on the storm's effects on New York City, New Jersey, and Washington, DC, respectively.

Citizens providing these types of location details may prove valuable for media outlets or others trying to understand massive crisis events in real time. For instance, the social media marketing company Social Flow created a map based on tweets mentioning power outages during the course of the storm. Gilad Lotan, Vice President of Research and Development at Social Flow, argues that geolocated hashtags could present crisis responders and media outlets with a real-time picture of such outages as they occur. "Think of generating these types of maps for different scenarios—power loss, flooding, strong winds, trees falling," Lotan writes. "While basing these observations on people's Tweets might not always bring back valid results (Someone may jokingly tweet about losing power), the aggregate, especially when compared to the norm, can be a pretty powerful signal" (2012).

Many people have suggested both the "global" and hyperlocal nature of Twitter were useful. For example, following Hurricane Sandy, Choire Sicha, founder of New York City news site *The Awl*, was quoted in *The New York Times' Media Decoder* blog as saying, "Twitter was phenomenally useful microscopically—I was literally finding out information about how much flooding the Zone A block next to me was having, hour by hour—and macroscopically, too—I didn't even have to turn on the TV once the whole storm," he wrote (cited in Carr, 2012). Citizens, then, can create an aggregate but still highly nuanced sense of ground-level realities in breaking news events.

In addition to location, the tone of information being shared via Twitter shifted during the course of the storm as well. Since citizens had ample warning of the slow-moving storm, there were many hours before the storm's arrival where citizens were reporting on preparations by individuals and government officials. Some early tweets even expressed a sense of humor. User @andrewcochranx referenced the viral video "Gangnam Style," a popular dance song by Korean pop star Psy: "What if gangnam style was actually just a giant rain dance and we brought this hurricane on ourselves? #sandy" (Pew, 2012). That tweet was sent at 13:23 on 29 October, before the storm reached its peak effects in the United States.

As the storm worsened, tweets grew more serious in tone. And as the storm itself subsided, citizen journalists began providing the type of information and local context that is often so valuable in a crisis. For example, people used #njpower to report specific locations where power outages were occurring, or #njgas, to share the specific location of local gas stations that had power and were open to the public (Saffer, 2012). Others began sharing information about how to donate to the Red Cross, for those who wanted to provide additional assistance to people affected by the storm.

PLATFORM ADOPTION IS THE CHIEF DRIVER OF CITIZEN PARTICIPATION

As new crises emerge, citizen reporting is often driven by the popularity of a given social media platform. Adoption of a social application, then, can mark a significant consideration for those looking to study or use data generated by citizen journalists. Flickr, for example, became a critical mechanism for organizing and sharing photos following the December 2004 Indian Ocean tsunami. This was remarkable in part because the platform had only debuted earlier that year.

Twitter, created in 2006, did not exist during the tsunami, nor in 2005 when both Hurricane Katrina and the London public transit bombings marked high profile moments for citizen journalism. But by early 2011, when a number of Arab nations experienced citizens protesting their governments, Twitter had more than 500 million registered users. Along with Facebook, the platforms represented a significant opportunity for citizen reporters to share content globally. Twitter in particular became an important tool for professional journalists attempting to make sense of various events, allowing them to interact with citizen reporters and activists.[2]

The launch, and perhaps more importantly, broad adoption of a platform, drives participation by citizens during large-scale crisis events. This is made even more apparent by the uptake of Instagram. The photo-sharing mobile application, launched in 2010, claimed nearly 100 million users by the time Hurricane Sandy reached the United States in October 2012. The storm represented a significant moment for Instagram—and arguably for user-generated news content—because it was the largest ever single use of the platform at the time. Users posted 1.3 million photos using the hashtags #sandy, #hurricanesandy, or #frankenstorm. At peak times, users were uploading 10 storm-related photos per second (Taylor, 2012).

This behavior was not unique to citizen journalists. *Time* magazine sent its own photographers out to document Hurricane Sandy via Instagram for a photo essay (Bercovici, 2012). Kira Pollack, *Time*'s director of photography, later told *Forbes* that, "We just thought this is going to be the fastest way we can cover this, and it's the most direct route. It wasn't like, 'Oh, this is a trend, let's assign this on Instagram.' It was about how quickly we can get pictures to our readers."

Time's actions demonstrate that media outlets, like citizen journalists, may have similar motivations when selecting a platform to record major events. Speed and comfort with a given platform are major drivers of use in a crisis.

Instagram was newsworthy for another reason too: in the United States, at least, Instagram had supplanted Flickr as the most popular photo-sharing site for organizing crisis information. This differed from the previous major storm event, Hurricane Katrina, in which Flickr was a critical tool for sharing images.

In part, this shift is clearly due to the ease by which photos from a mobile phone could be shared to the internet in the Instagram platform. It is also an interesting story of numbers. By the time the storm arrived, Instagram had nearly 100 million users, while Flickr had roughly 87 million users. Instagram recorded approximately 1.3 million photos related to Hurricane Sandy, while Flickr had a mere 134,000.[3]

The lessons are two-fold: audience adoption is a critical consideration when engaging with citizen reporters, and popular platforms for sharing and organizing information can change dramatically within a few years' time.

Citizens may choose to share photos, videos, and updates on popular platforms because it increases the chances of their friends or family seeing their updates, as well as the general public. Adoption also provides a sense of familiarity, and may be where people inherently turn in a time of crisis. No doubt many users will prefer to quickly post information on a platform they understand well, with an interface they are comfortable with, rather than attempt to register with and learn a new platform with more limited reach.

PLATFORM ACCESS

Access to a platform is just as critical as adoption in terms of volume of user-generated content. The more ability a citizen has to access his or her platform of choice, the more likely he or she is to participate in that platform during a crisis event. Evidence of the importance of access can be seen in numbers looking at the adoption of mobile technology. Nielsen and NM Incite, a joint venture between Nielsen and McKinsey, published a study in December 2012 that indicates that mobile is becoming a critical driver of the amount of time people spend accessing social media (Perez, 2012). According to the research, nearly half of all social media users say they access their platform of choice on a smartphone. Furthermore, the most popular platforms experienced a significant growth in the amount of time users interacted with them in 2012, with mobile apps and mobile websites accounting for 63 percent of this year-over-year growth. Mobile access, then, is a large part of what drives adoption and use of products like Instagram.

Conversely, a lack of mobile access limits user behavior. Tech blogger Mat Honan argues that this is the case for Flickr; the perceived difficulty of uploading photos to Flickr via a mobile device in 2012 meant users were turning to other applications when they only had their mobile phones available. He wrote:

> As a result of being resource-starved, Flickr quit planting the anchors it needed to climb ever higher. It missed the boat on local, on real time, on mobile, and even ultimately on social—the field it pioneered. And so, it never became the Flickr of video; YouTube snagged that ring. It never became the Flickr of people, which was of course Facebook. It remained the Flickr of photos. At least, until Instagram came along. (Honan, 2012)

Other factors may also affect access to a platform. In a crisis, the popularity of a given platform may ultimately serve as its downfall. A major crisis can overwhelm server requests to major social networking sites. Twitter was especially prone to this type of failure early on, given how quickly it rose in prominence as a social network. Server failures can be caused by a variety of other reasons, including power outages or disruption of internet networks during a natural disaster. In some extreme cases, such as during the 2011 Egyptian revolution, governments can manually attempt to shut down access to the internet and to mobile towers. Such shutdowns forced users to abandon coordination efforts on social media platforms and coordinate via other means, such as SMS (text messaging).

Under other circumstances, governments or other entities may even request that citizens self-impose limits on communications channels. During both Hurricane Sandy and the aftermath of the Boston Marathon bombings, government officials were concerned that voice calls would overwhelm network capacity. They specifically directed citizens to coordinate with friends and family via SMS, as the service is less taxing on mobile networks and in some cases works even when towers are busy or partially damaged.

In instances where citizens lose access to their first platform of choice, they will likely turn to the next platform that best serves their needs. It stands to reason that people do turn to SMS when access to mobile networks is limited, though this data is typically difficult to come by due to the relatively private, one-to-one nature of SMS. Still, there is some evidence that coordination was happening via SMS during Hurricane Sandy. One team established an SMS system that allowed people seeking relief, or offering assistance, to text keywords to a single number. This allowed coordinators to begin connecting volunteers with people overlooked by other support mechanisms. The fledgling service received more than 100 requests for help within a week, often from underprivileged citizens in remote areas (Strochlic, 2012).

There was evidence of this behavior among media outlets themselves during Hurricane Sandy. Severe flooding, caused by the storm, damaged servers at Datagram, Inc. in New York City. This in turn brought down websites for several media outlets, including *The Huffington Post*, *BuzzFeed*, and Gawker Media—many of which were providing critical information during the course of the storm. Adrienne LaFrance, writing for *Digital First Media*, noted that many of the websites that went down turned to Tumblr, a site famous for quickly allowing its users to start sharing web content.

> BuzzFeed began publishing section-by-section content to the blogging platform Tumblr. It also worked with Akamai, a content distribution network, so that some article pages would still load even if the core BuzzFeed site could not. Gizmodo and Jezebel each set up Tumblr blogs, "Sandy 2012 Emergency Site" and "Post-Hurricane Emergency Blog," respectively. Gawker, too, relied on the easy-to-use Tumblr as a back-up. While Tumblr enabled practically seamless connectedness amid the chaos, the resulting aesthetic was a bit of a throwback to the spartan design of the internet's earlier days. (LaFrance, 2012)

Other services were focused on using SMS as a vehicle for updating popular sites. As the storm approached, Twitter promoted its own service through which SMS can be used to generate tweets, without internet access. Likewise, *Mashable* posted an article, specifically in advance of the hurricane, titled "How to Update Twitter and Facebook Without the internet" using SMS (Franceschi-Bicchierai, 2012). Again, it is difficult to determine to what extent these services were used, but it is safe to say that many people recognized the potential for accessing SMS, even as the storm limited access to other platforms.

PERCEIVED CULTURAL CONTEXT FOR SOCIAL MEDIA

Instagram presented a particular challenge for media outlets because of the ease with which users can manipulate photos. To briefly summarize: Instagram users typically take a photo with a smartphone, then crop the photo and apply a "filter" that changes the image in terms of light balance, saturation, contrast and color tint. Instagram itself states on its website that the filters "transform the look and feel" of any given photo. At the time of publication, the platform does not easily lend itself to adding layers or combining images, the way Adobe Photoshop might. But for the tools available, photos can be altered with extreme ease. This stands in contrast to Flickr, previously among the most common sites for journalists to seek citizen-generated content, which has fewer features for altering photos. Of course, any site runs the risk of a photo, video or text being significantly manipulated before being posted to a platform, but that is beyond the scope of this particular discussion.

The ease of image manipulation within Instagram raised a question with a number of observers over ethics and taste related to user-generated content. Some critics argued that Instagram - at least prior to Hurricane Sandy—had largely been affiliated with taking playful photos, then giving them a fun, nostalgic, de-saturated tone. These critics argued that altering photos with the platform's basic tools diminished the tragedy of the storm by applying a whimsical filter to the final image (Bosker, 2012). Others argued that because the tools for altering the photos are largely limited to cropping, light, and color balance, or the types of tools used in photography in newsrooms anyway, that the use of Instagram photos presented no ethical problem, and may have made the tragic images more compelling for the audience.

This sort of argument reflects subjective issues of taste and ethics. The ease with which a photo or video can be altered relates specifically to a newsroom's ethical policy regarding the use of images. How a platform is perceived, or its social context, relates to audience taste. News outlets will need to consider both issues as smartphone adoption increases worldwide, photo sharing and visualization tools become increasingly mobile, and the ability to alter images becomes increasingly user-friendly.

DATA CORRECTION AMONG CITIZEN JOURNALISTS

In nearly any ongoing news event, misinformation is rampant. Exaggerated or outright false information can spread quickly over social media. In some cases, such misinformation is then broadcast by major media outlets. Any spread of misinformation during a crisis potentially creates public confusion during critical rescue times. That said, many analysts have also pointed out how rapidly online communities, taken in aggregate, have the ability to self-correct bad information.

Hurricane Sandy represented many instances of self-correction by online communities. Early in the storm's progression, many fake messages and photographs were shared via social media, most notably Twitter and Facebook. However, users on those platforms quickly pointed out the fake photographs, creating a public channel for correcting misinformation.

In some cases, the misinformation appears to have been an earnest mistake. In the early hours of the storm, many Facebook and Twitter users shared a photograph of soldiers guarding the Tomb of the Unknown Soldier in Arlington National Cemetery in the midst of significant wind and rain. Many people assumed that the photo related to Hurricane Sandy. However, the photo had actually been taken more than a month prior to Hurricane Sandy, in a different storm. The Old Guard Association, which supports the infantry group responsible for watching the tomb, worked hard to circumvent the confusion, explaining on Twitter when the photograph was taken. They even included a new, albeit less visually compelling, photograph of the soldiers standing watch as Hurricane Sandy approached. The Old Guard did note that its members regularly endure difficult weather conditions, since they are in position 365 days a year (Laird, 2012).

In a more infamous example, one Twitter user appeared to intentionally spread false information during the storm, including that the New York Stock Exchange was under three feet of water. This information was potentially dangerous, as it could have affected global financial markets. CNN even mentioned the tweet during a live television feed. Over time, Twitter users began to question the information and eventually started sharing news that the report was bogus. CNN also later acknowledged that the report was false. The Twitter user that apparently started the rumor was eventually named by *BuzzFeed* and subsequently fired from his job as a hedge fund analyst.

The speed with which the false information was shared was remarkable, but so too was the speed with which it was corrected. *Gigaom*'s Matthew Ingram wrote: "What's interesting isn't that there was fake news, it's how quickly those fakes were exposed and debunked, not just by Twitter users themselves, but by an emerging ecosystem of blogs and social networks working together" (2012). Citizen journalists, then, serve a critical role in helping fact-check information, even as it is broadcast quickly around the internet.

CHALLENGES OF PUBLIC DATA

Public data can be extremely helpful in collecting and sharing news stories, correcting misinformation, and adding context to major crisis events. Of course, news outlets also face a number of challenges in using the "wisdom of the crowd," which is another way of saying "a large and publicly populated data set."

One issue is that public data sets are incomplete, and the participants are self-selecting. As such, they can bias news stories in a number of ways. Kate Crawford (2013), a principal researcher at Microsoft Research, wrote about this challenge in the aftermath of Hurricane Sandy:

> Data and data sets are not objective; they are creations of human design. We give numbers their voice, draw inferences from them, and define their meaning through our interpretations. Hidden biases in both the collection and analysis stages present considerable risks, and are as important to the big-data equation as the numbers themselves.
>
> For example, consider the Twitter data generated by Hurricane Sandy, more than 20 million tweets between October 27 and November 1. A fascinating study combining Sandy-related Twitter and Foursquare data produced some expected findings (grocery shopping peaks the night before the storm) and some surprising ones (nightlife picked up the day after—presumably when cabin fever strikes). But these data don't represent the whole picture. The greatest number of tweets about Sandy came from Manhattan. This makes sense given the city's high level of smartphone ownership and Twitter use, but it creates the illusion that Manhattan was the hub of the disaster. Very few messages originated from more severely affected locations, such as Breezy Point, Coney Island and Rockaway. As extended power blackouts drained batteries and limited cellular access, even fewer tweets came from the worst hit areas. In fact, there was much more going on outside the privileged, urban experience of Sandy that Twitter data failed to convey, especially in aggregate. We can think of this as a "signal problem": Data are assumed to accurately reflect the social world, but

> there are significant gaps, with little or no signal coming from particular communities. (Crawford, 2013)

Again, Crawford's points address the issue of access. If there are communities that are not accessing social media platforms, reports from citizen journalists may not be included in professional reporting or analysis of an event.

Another challenge beginning to emerge is one of data ownership, or the rights to use information that perhaps was never intended to be broadcast to a large audience. Even as news outlets begin looking at this type of data in aggregate, the role of data ownership may affect what can be used and how. The terms of service or privacy policies within a social media platform largely govern this issue. Some platforms require users' permission before text or images can be used. Media outlets posed permission questions directly to Instagram users during the course of Hurricane Sandy requesting the use of certain images (Lacy, 2012). Media outlets, then, will need to be aware of the ever evolving terms of service agreements users make with various platforms that might be used by citizen reporters, and in particular question whether they will need user permission to share information.

CONCLUSION

Citizen journalists will take to a variety of platforms and services, often gravitating toward those with which they feel most comfortable. Media outlets or others seeking to engage with citizen journalists need to be aware of emerging platforms, as well as the uptake and access to those platforms. The increase in access to mobile phones worldwide, and the continued uptake of smartphones, should further propel global access to many platforms, as well create increasingly easier adoption of emerging social applications. Instagram, coupled with the expanding US smartphone market, served as a prime example of this trend during Hurricane Sandy.

That said, journalists should continue to be aware of Kate Crawford's warnings about gaps in media coverage, or "signal problems," and consider who might not be participating in a given news story. It begs some important questions: Who are the people for which access to popular platforms was diminished in some way? Are there citizens who never had access to begin with? Who are they and what mediums are they using to share information? Are there platforms, such as SMS, that may be useful in collecting information relevant to covering a particular story for otherwise excluded citizen journalists? Furthering access to underserved groups may provide invaluable data as communities, media outlets and governments seek to understand events in times of large-scale crisis, as well as their aftermath.

NOTES

1. It is worth noting that the 2012 US Presidential Election was broken down by specific events, such as "First Presidential Election Debate," for Facebook's purposes. In aggregate, the election coverage was by far the most discussed event, followed next by the Super Bowl and then Hurricane Sandy.
2. Facebook was often a critical platform for people sharing news with one another. However, since it is largely a more private platform than Twitter, fewer professional journalists were using Facebook to organize information.
3. I used the search terms "hurricane sandy" and "Frankenstorm" for Flickr, but dropped the search term "Sandy," as it generally turned up individuals and pets rather than storm-related photos.

REFERENCES

Bercovici, J. (2012). "Why *Time* Magazine Used Instagram to Cover Hurricane Sandy." *Forbes* blog, 1 November.

Blake, Eric S., Kimberlain, Todd B., Berg, Robert J., Cangialosi, John P., & Beven, John L. II (2013). *Tropical Cyclone Report Hurricane Sandy*. National Hurricane Center, 12 February.

Bosker, B. (20120 "When Instagram Makes Disaster Beautiful: Hurricane Sandy Destroyed NYC, But Instagram Sure Made It Pretty." *Huffington Post*, 1 Nov.

Carr, D. (2012). "How Hurricane Sandy Slapped the Sarcasm out of Twitter." *New York Times Media Decoder Blog*, 31 October.

Crawford, K. (2013). "The Hidden Biases in Big Data." *Harvard Business Review Blog Network*, 1 April.

Franceschi-Bicchierai, L. (2012). "How to Use Facebook and Twitter without the internet." *Mashable*, 29 October.

Griggs, B. (2012). "Sandy Is Year's No. 2 Topic on Facebook." *CNN*, 30 October.

Honan, M. (2012). "How Yahoo Killed Flickr and Lost the internet." Gizmodo, 15 May.

Lacy, S. (2012). "Could Sandy Be Instagram's Big Citizen Journalism Moment?" *PandoDaily*, 29 October.

Laird, S. (2012). "Incredible Viral Soldier Pic Debunked by *Military*." Mashable, 29 October.

LaFrance, A. (2012). "Hurricane Sandy Washed Out Servers, Knocked Down Major Websites." *The Denver Post*, 1 November.

Lotan, G. (2012). "#Sandy: Social Media Mapping." *Social Flow* blog, 5 November.

Perez, S. (2012). "Mobile Drives Adoption of Social Media in 2012: Apps & Mobile Web Account for Majority of Growth; Nearly Half of Social Media Users Access Sites on Smartphones." *Tech Crunch*, 3 December.

Pew Research Center's Project for Excellence in Journalism (2012). "PEJ New Media Index: Hurricane Sandy and Twitter."

Pierre-Pierre, G. (2012). "Hurricane Sandy: It Hit the Caribbean Too, You Know." TheGuardian.com, 2 November.

Saffer, M. (2012). "#HurricaneSandy: How Twitter and Hashtags Helped New Jersey." PCG Digital Marketing blog, 20 November.

Shih, G. (2012). "Over 20 Million Tweets Sent as Sandy Struck." Reuters, 2 November.

Smith A. (2012). "Nearly Half of American Adults are Smartphone Owners." Pew internet & American Life Project, 1 Mar.

Strochlic, N. (2012). "Text In to Help Hurricane Sandy Victims." *The Daily Beast*, 16 November.

Taylor, C. (2012). "Sandy Really Was Instagram's Moment: 1.3 Million Pics Posted." *Mashable*, 5 November.

CHAPTER NINE

Live Reporting Terror: Remediating Citizen Crisis Communication

EINAR THORSEN

> "I have to admit I had no idea that news got so real—so fast in social media. [...] It's not news. It's real life. It's rawer, uncensored. If you where [sic] Norwegian on Twitter today you got it uncensored whether you wanted it or not."
>
> —HANNAH AASE, PERSONAL BLOG, 22 JULY 2011

On 22 July 2011 Norway was marred by two devastatingly violent attacks, carried out by a right-wing extremist, Anders Behring Breivik. The first attack involved a car bomb detonated outside government buildings in Oslo at 15:25, killing 8 people and injuring a further 98. The second attack, reported to police at 17:24, was a shooting spree at the AUF youth camp on the Utøya island—69 of the 564 people on the island were eventually confirmed shot, and more than 60 injured. Most of those shot were 18 or younger, two of whom only 14 years old. Breivik was eventually arrested on Utøya at 18:34, the two attacks having lasted just over three hours and killing 77 people.

People who were caught up in the Oslo attack used mobile phones and social media to document events as they unfolded, providing for others a raw, uncensored and immediate account of what they were witnessing. Victims on Utøya used their mobiles as emergency communication tools—publishing cries for help, confirming they were alive and seeking information about what was going on. In both instances ordinary citizens were contributing to public understanding of the crisis as it unfolded. Such was the prominence of social media during and after the attacks that established news organisations not only used them as sources, but even published online articles based entirely on victims' Twitter feeds and blog posts.

Drawing on a larger study into social media and news reporting of the 22 July 2011 attacks in Norway, this chapter will provide a critique of breaking news in times of crisis. My primary focus will be the role of citizen crisis communication from victims that were publishing eyewitness accounts of events as they unfolded. Attention will also be on a comparative analysis of international news organisation's live blogging of the attacks—highlighting firstly the use of citizen eyewitness accounts, and secondly the changing news frames from the initial assumption that the attacks were grounded in "international terrorism" to the confirmation of Breivik as a domestic right-wing extremist.

RESEARCHING THE 22 JULY ATTACKS

The Oslo and Utøya attacks have attracted considerable public scrutiny, both through the trial of Anders Behring Breivik and the official Norwegian 22 July commission (Gjørv et al., 2012). The comprehensive report delivered by Gjørv et al. (2012) also considers the role of national media and indeed comments in passing on the prominence of social media. There is also a growing body of scholarly work analysing the two attacks from a range of perspectives—ones concerning social media cover, for example: Twitter in relation to emergency responses (Perng et al., 2013), use of personal blogs by Utøya survivors (Thorsen, 2013a), a Master thesis on the role of social media for Labour Youth Party members on Utøya (Johnsen, 2012), and a Master thesis analysing public sentiment in relation to sharing traumatic experiences using the #oslo hash tag (Eriksson, 2013). Much less attention has been given to the contribution social media and citizen journalism during the two attacks made to public information and breaking news, which is the focus of this chapter.

This study is based on the author's personal archives of online news and social media from a 48-hour period covering the attacks, as they were unfolding on 22 July 2011 and their immediate aftermath. The study involved a systematic textual analysis of: 1) Twitter messages and associated audio-visual material (e.g., YouTube or TwitPic),[1] and 2) live blogs from international news organisations. Finally, the study encompassed a close reading of both Norwegian and international news websites, and public reports into the events.

Twitter messages captured by the author were verified and cross-referenced against NRK's public archive of 250,000 tweets from 22–25 July 2011 (http://nrk.no/terrortwitter/). In addition to this, the social analytics search engine, Topsy (http://topsy.com), was used to examine tweets in the same time period with relevant hash tags or from select Twitter users to reveal their timeline and interactions with others. This combined resource enabled a detailed, chronological reconstruction of how ordinary citizens caught up in the two attacks contributed to public information about the crisis.

Social networking data was then compared against the live blogs of four international broadcasters—Al-Jazeera, BBC News, CNN and Sky News—plus two British broadsheet newspapers, *The Guardian* and *The Telegraph*. These were analysed to understand how quickly the international media reacted to the news as it unfolded and what, if any, citizen journalism news organisations relied upon in their online live reporting. For the purpose of this chapter only live blogs published on 22 July 2011 were analysed, with a cut-off point for those continuing over midnight to 05:00 on 23 July.

Table 1. Live blogs covering the 22 July 2001 crisis

Site	First post	First Utøya mention	Updates	Words
Al-Jazeera	17:28	18:09 (ref Tweet by @Made-Chris)	44	1,762
BBC News	16:34	18:19 (ref Reuters)	146	2,857
CNN	16:00	18:48 (ref Reuters)	41	2,176
Guardian	16:30	18:16 (ref Neil Perry (reader) who ref Stoltenberg on TV2)	48	1,813
Sky News	16:36	18:06 (ref unconfirmed reports, ref police reports at 18:17)	98	1,536
Telegraph	16:00	18:19 (ref Police reports)	110	3,670

Live blogging has become increasingly popular among established news organisations as a way of tracking significant news stories as they unfold (Thorsen, 2013b; Thurman and Walters, 2013). Journalists updating a live blog curate reports from a wide range of sources and publish these as time-stamped updates in a single post. The short updates are presented in reverse chronological order, usually integrating audio-visual material, links and third-party content. Live blogs document the life of a story, often finding room for material that would otherwise not be published. Indeed live blogs sometimes interweave commentary about how the reporting is evolving and interactions with audiences. To this end, live blogs provided a unique way of tracking how international news organisations reported on the 22 July attacks in Norway. It also enables an analysis of what sources are reported and the extent to which non-domestic news was making use of social media.

"HOLY CRAP. DID OSLO JUST EXPLODE?"

On 22 July 2011 when the car bomb went off in Oslo it destroyed the Prime Minister's offices as well as other buildings in the surrounding area. This included government buildings and the headquarters of one of Norway's leading tabloid newspapers, *Verdens Gang* (VG), and the daily newspaper, *Dagsavisen*. Coupled with the unexpected nature of the attacks, this impact on both state power and news organisations inevitably contributed to a chaotic information picture in the hours following the attacks. Immediately following the Oslo explosion were uncertainty and contradictory messages about what had happened; even pinpointing the exact time of when the bomb went off proved difficult. Media variously reported the impact of the blast as around 15:20–15:30, and the police for a long time operated with an official timing of 15:26.

Commercial radio station, P4, appeared to be the first to report the blast on air at 15:27, with NRK mentioning it on their website at 15:30 and VG publishing it online at 15:34 (Gjørv et al., 2012:22). Yet, public reports were already circulating on social networking sites. One of the first was this Twitter post by Gunnar Tjomlid (@CiViX): "Holy crap. Did Oslo just explode?"—posted at 15:25, only seconds after the bomb went off and before the police's official estimate of the impact time (see Figure 1).

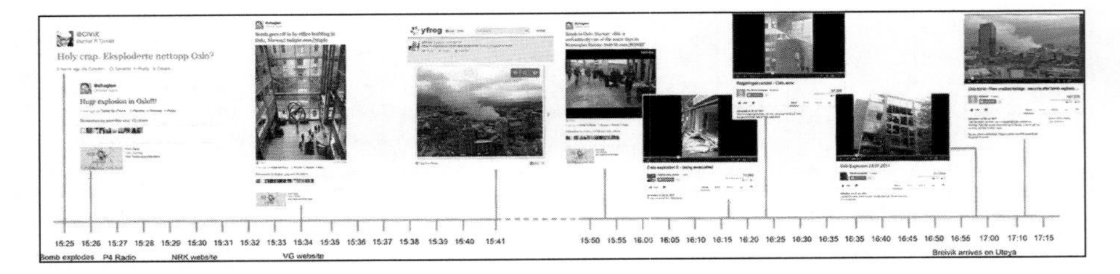

Figure 1. Oslo timeline on 22 July 2011

Caught in the heart of the affected area was Christian Aglen (@chaglen). At 15:26 he tweeted: "Huge explosion in Oslo!!!" followed at 15:34 by "Bomb goes off in by [sic] office building in Oslo, Norway!" and a photograph from inside his building showing people escaping and shattered windows. Upon exiting the building Aglen started filming and continued to take photographs as he walked through the chaotic streets. In a tweet at 15:53 he described it as "undoubtedly one of the worst days in Norwegian history." The 21-second video that accompanied this message was shaky hand-held footage, panning side-to-side as he walked up the street, giving viewers a first-person view of what it was like. Sounds of people walking on broken glass, burglary alarms and sirens, made the film almost as iconic for its diegetic sound as its vivid imagery. He posted an image of people helping an injured person at 15:57, with the caption: "Injuries, possible deaths as a result of this tragic event. #Oslo #Bomb #Explosion."

Whilst Aglen himself was not a professional journalist, he worked in a news environment as the development manager at E24.no (Norway's leading financial news website). This may have contributed to his instinctive documentation and publishing of eyewitness reports. It also meant being connected to influential sources that would remediate his reports further. His first picture, for example, was retweeted more than 350 times and the video 215 times. Indeed the video was played on loop by international news channels during the early parts of the crisis. Such was the pressure on Aglen to confirm requests for permission or to appear as an eyewitness interviewee, that he twice felt compelled to post general public responses:

> @chaglen: I am receiving many media requests now. Unable to respond to all of them right now unfortunately… (16:15)

> @chaglen: To the media—I see your requests and will try to respond to them. A bit overload now. I will answer your questions as soon as possible… (20:15)

Other eyewitnesses received similar attention for material they published on the internet. Geir-Olav Goksøyr (@geirolav), for example, tweeted one of the first public images of the explosion at 15:41. Taken from a tall building afar, the picture showed a plume of smoke coming from central Oslo and helped establish the general location of the blast. According to Goksøyr he took the photograph some 20 seconds after the bomb had detonated, and shortly after he also posted a one-minute video of the smoke. International news organisations soon picked up on the images and he received eight requests for permission to use his material posted on his Twitter and yfrog accounts, including from ABC News, the BBC, *The New York Times* and Storyful. YouTube user kuleland posted a similar video of the bellowing smoke from afar at 17:13, entitled "Oslo bomb—Raw unedited footage—seconds after bomb explosion—Terror attack 22.07.2011." In the description, kuleland explained the gut reaction to record the act of witnessing:

> I felt the blast—so loud—as if it happened right outside my building. Then the smoke cloud shot up in the sky. I ran to get my camera, and this is what I saw.

Prominent YouTube eyewitness videos were also posted by Joakim Vars Nilsen (joakimvnilsen) at 16:17 showing people being evacuated, Per Øivind Eriksen (pereriks1) at 16:24 containing close-ups of the blown-out government buildings, and Stardestroyer65 at 16:57 capturing the destroyed VG headquarters opposite the government buildings. Whilst posted at different times, the videos were mostly recorded 10–15 minutes after the blast and subsequently uploaded to social networking sites.

Further citizen eyewitness videos continued to emerge in the hours and days after the attack, and were often distributed via news organisations that had taken care to conceal graphic images or the identity of those injured. Andreas Helgesen worked in a nearby photography shop, FotoVideo, and was one of the first eyewitnesses to reach the government building. His video shows a completely blown-out building, rubble and broken glass everywhere. The absence of other people provides an eerie complement to the sound of alarms and Helgesen, out-of-breath, shouting "hello, anyone need help?" There is no response. An edited version of the video was published by Norwegian broadcaster NRK at 16:39 on 22 July and was quickly picked up internationally (the original, longer version contained images of a dead body that were edited out).

Garnering most attention some days after the attack was a 16-minute continuous video shot by estate agent, Johan Christian Tandberg. He was driving through a nearby tunnel at the time of the blast and started filming as soon as he could get out of the car. The film shows Tandberg running through the streets towards the government buildings, stopping occasionally to speak to others and passing himself off as a rescue worker—purportedly to avoid having to answer questions about what he was doing. Uniquely he also entered one of the government buildings and continued to film as he ran through the corridors, shouting for survivors, warning people to get out and asking for status reports from others. In one of many interviews after the video was published, Tandberg reflected:

> It is the worst I have ever seen in my whole life. It looked like a war zone. But I figured it was not war, but terror, so I stopped the car and started filming. I thought if anyone wanted to destroy my country, it would be important to document for posterity (cited in VG Nett, 27/07/2011).

Asked why he entered the building, Tandberg replied "to document, and help get injured out" (ibid.). Describing how he acted on "auto-pilot", Tandberg claimed he provided commentary to the viewers "so they can get a realistic impression of how awful this was [...] I spoke to the camera to document" (cited in NRK, 27/07/2011). While he studied journalism at the University of Utah in the 1990s, his initial reaction had been to film material that might be of use to the police investigations. Yet, he acknowledged afterwards the filming had developed in a way that would have a much broader interest and that the event represented something he had "not experienced before."

Alongside the eyewitness accounts of the Oslo bomb, social networking sites and news media were also rife with speculation and amplification of rumours. This included, for example, the possibility of further bombs, the associated uncertainty being considered a continued threat. Similarly, there were people seeking to explain the bomb attack by connecting it with past events, thereby beginning to frame peoples' understanding of possible perpetrators and their motivations. References here included previous suicide bombings in Sweden and Finland. More problematic was the reaction by many to connote a "terrorist attack" with foreign assailant(s). Parallels were frequently drawn to the US September 11, 2001, attack, with tenuous reasons offered for why Norway might have been a target—including Norway's involvement in Afghanistan, a Norwegian newspaper's reprinting of the Danish Muhammed cartoons and so forth.

Citizen reporting of the Oslo explosion was rich in photographic and video material, providing an on-the-ground illustration of the wild disarray that was unfolding. Whilst eyewitness material was remediated by regional, national and international news organisations, many had first learned details of what was happening on social networking sites. This was also the case for the second attack, the shooting spree on Utøya.

"DEAR GOD THERE ARE PEOPLE SHOOTING ON UTØYA"

In the trial of Breivik it was ascertained that he boarded the ferry to Utøya island at 16:57, and after arriving on the island he walked up towards the main building before opening fire at 17:21. The Norwegian emergency services were notified about the shooting at 17:24, and Breivik was captured at 18:34. However, the first public messages about the shooting spree were not from official sources, but posted on social networking sites by friends and relatives of those on the island.

Twitter user Ikkekatarina (@unnimi) had been called up by her best friend who was on Utøya, and at 17:36 posted: "DEAR GOD THERE ARE PEOPLE SHOOTING ON UTØYA AT THE AUF-CAMP" (see Figure 2). She then engaged in a conversation with Jørgen (@jvik), who was curating information from tweets and news reports relating to the Oslo explosion. In her replies to him she clarified how she knew about the shooting and provided insight into her jumbled emotions:

> @unnimi: @jvik received a phonecall from someone there, they have jumped in the sea, there are people shooting people there!!!!! (17:38)

> @unnimi: @jvik I can't understand it (17:39)

> @unnimi: @jvik it must be someone messing around. it can't be real. (17:40)

In a separate post at 17:41, the user Tiril (@FrkTiril) posted: "There is shooting on Utøya, my little sister is there and just called home!" Seconds later @uvette tweeted "There are reports of gunfire on Utøya. Ambulances and police are on their way."

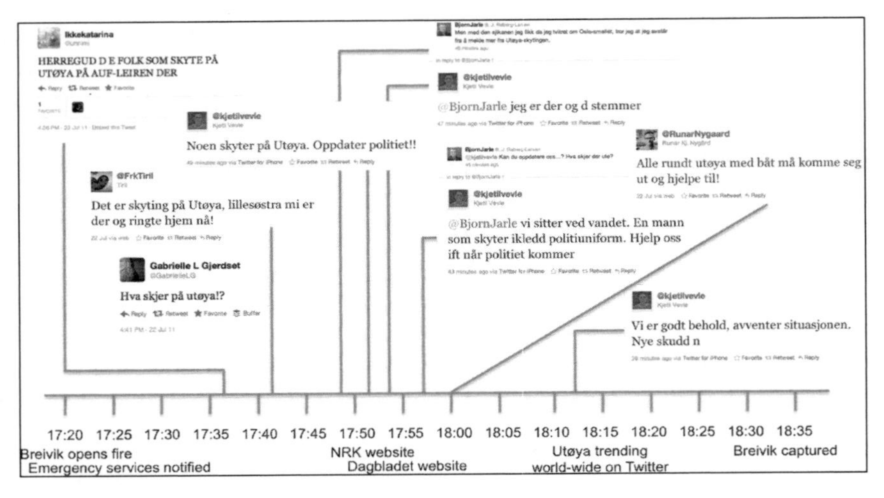

Figure 2. Utøya timeline on 22 July 2011

At the same time Gabrielle L. Gjerdset (@gabriellelg) tweeted "What is happening on Utøya!?" and replied to queries about what she meant by encouraging people to search social networks for information. Minutes later at 17:49, Bjørn Tore Hansen (@BTHansen) tweeted: "Reports about shooting on Utøya. Just told now. Police on its way." followed by Magne Mellem Enoksen (@magneme) at 17:51: "Anyone heard about shooting on Utøya? My partner's little sister just phoned to say, and a Twitter search shows others reporting it too."

Shortly after people who were themselves present on the Utøya island also started posting public messages about what was happening. One of the first came from Kjetil Vevle (@kjetilvevle) at 17:52: "Somebody is shooting on Utøya. Update the police!!" After initially reporting gunfire, Bjørn Jarle Rødberg Larsen (@BjornJarle) posted scepticism about the shooting and the following exchanges ensued:

> @BjornJarle: Oi! Gunfire on Utøya. AUF's summer camp evacuating. (17:48)
>
> @BjornJarle: But given the abuse I received following my tweets about the Oslo explosion, I think I'll decline to report anything more about the Utøya shooting. (17:52)
>
> @kjetilvevle: @BjornJarle I'm there and it is happening (17:54)
>
> @BjornJarle: Receiving several confirmations that there IS gunfire on Utøya, and that police are on their way. (17:55)
>
> @chreriks: @BjornJarle Are you on Utøya? Do you know anything more about the situation? (17:55)

@HaugBente: @BjornJarle Please share if you have information. We have several youths on Utøya we are concerned about. (17:56)

@BjornJarle: @kjetilvevle Can you update us? What is happening out there? (17:56)

@kjetilvevle: @BjornJarle we are sat by the water. A man is shooting dressed in police uniform. Help us with regards to when the police arrives (17:58)

@kjetilvevle: We are ok, awaiting the situation. New shots n (18:13)

The last two tweets were retweeted 235 and 140 times respectively, indicating news on social networking sites was not only read by posters' immediate friends and followers, but also redistributed to reach even larger audiences. Similar messages were also posted in private Facebook exchanges, and some people even took to reposting Facebook status updates on Twitter from victims on the island.

Reports about the Utøya shooting first began emerging on online news websites at 17:53 (*Dagbladet* first to go public, later followed by NRK), some 17 minutes after the first Twitter message about it. Indeed several people were tweeting bemused messages about the silence from national newspapers and broadcasters about what was evidently taking place. Oda Rygh (@OdaRygh) summed up the frustration shared by many others: "Utøya-shooting not confirmed, but from several sources. None of them media." (17:57). Prime Minister, Jens Stoltenberg, confirmed knowledge of a critical situation on Utøya at around 18:00, but referred to police for further information.

Ordinary citizens also got involved in spreading emergency support messages, including encouraging anyone with boats to help pick up victims attempting to escape by swimming to the mainland, and warning people against contacting anyone on the island as the ring-tone or vibration might reveal their hiding place:

@RunarNygaard: Everyone around Utøya with boats must get out to help! (18:00)

@RunarNygaard: @KyrreJansen have spoken to friends on Utøya via phone who whispered "someone is shooting, have to hang up"

@fmellem: shooting on #Utøya. Don't call anyone there. They are hiding in bushes. (18:17)

Others tweeted pleas for people nearby to avoid using mobile phones altogether, to keep the networks free for victims attempting to contact relatives, or encouraged people to give blood. Several people also took to reposting crisis support telephone numbers, updates from emergency services and the police, and confirmed details from official sources and news reports. Such was the level of activity that by about 18:15 the term Utøya was trending on Twitter world-wide. With details about Breivik emerging whilst the shooting was still on-going, some people began to counter the prevailing assumption by analysts and commentators about a foreign terrorist attack:

> @Tore_Sveen: If Utøya-reports are correct it would seem there are Norwegian speaking people behind the attacks at least (18:22)

Indeed further information about the perpetrator, including his identity and links to his online profiles and manifesto, were circulating publicly long before it was reported by established news organisations (initially on VG's website, quickly followed by others). In the evening of 22 July, anyone following claims circulating on social networking sites would have garnered a very different picture to that reported by national or international news organisations—or indeed official spokespeople who were slower at confirming updates. Even a simple search on Twitter or Facebook would reveal direct or remediated eyewitness accounts proving that the number of dead was much greater than what was being confirmed by official sources and reported in news bulletins. Such challenges were borne out in international reporting of the attacks too, which will be explored in the final section below.

LIVE BLOGGING TERROR ATTACKS

The reasons why news about both the Oslo and Utøya attacks was first reported by ordinary citizens online proved significant. Whilst national media were quick to begin coverage of the unprecedented events in Oslo, they were comparatively slow to react to the news emerging from Utøya. Journalists were clearly preoccupied with what was already considered the worst attack on Norwegian soil since World War Two. With all the resources focussed on the capital and the rumour-mill running rampant, journalists seemed reluctant to spread further unconfirmed reports. Indeed, verification was more challenging for the second attack since reports emerging from Utøya were all text based—victims using social networking as a desperate plea for help in their fight for survival rather than attempting to record or document the killings. Such challenges were even more pronounced for international news organisations, which were similarly ill-prepared for an attack on Norway of this magnitude. The live blogs analysed as part of this study provide an insight into their sourcing practices, especially where citizen journalism and eyewitness accounts are concerned, as well as the lag in reporting compared to social networking sites and Norwegian national media.

The first news organisations to establish live blogs for the Oslo attack were CNN and *The Telegraph* at 16:00, some 35 minutes after the explosion (see Table 2 below). BBC News, *The Guardian* and Sky News all set up live blogs around 30 minutes later, with Al-Jazeera a comparatively late arrival at 17:28. Whilst the BBC provided the largest volume of updates, *The Telegraph* was by far the longest live blog when it concluded. Sky News was the first to report the Utøya shooting, referring to "unconfirmed reports" at 18:06 and following this up with "police reports" at 18:17. Having been slow to establish a live blog presence, Al-Jazeera was then quick to report the second attack at 18:09—uniquely it was also the only news organisation that publicly referred to a tweet as the source of the information (by @MadeChris). *The Guardian* referred to an audience email from Neil Perry, described as a former colleague living in Oslo, who claimed to have watched Prime Minister Stoltenberg confirm an incident on Utøya when interviewed by TV2. BBC News, *The Telegraph* and CNN all followed later, and only once the second attack had been confirmed by Reuters or the police.

All the live blogs embedded or referred to citizen eyewitness accounts posted on social networking sites. Overall BBC News made the most references to eyewitness accounts, audience emails or social media reports from citizens (34 mentions), though CNN had the greatest proportion of such material in its live blog overall (37% of its updates). Christian Aglen's TwitPic images, for example, were used by CNN and Sky News, while BBC News cited one of his tweets. Similarly, Geir-Olav Goksøyr's picture capturing an aerial view of the plume of smoke was embedded in the live blogs of Al-Jazeera, BBC News (without attribution), and Sky News. Both CNN and Al-Jazeera also made use of @TPB_Stun's images, and several YouTube videos recorded by eyewitnesses were also embedded in the live blogs.

Eyewitness material sourced from social networking sites was limited to the early part of most of the live blogs, however. Once the crisis scenario was established, the live blogs increasingly relied on official sources or "eyewitness reports" (presumably obtained by their own journalists or through wire services). These eyewitness accounts were providing emotional descriptions about their feelings when the bomb went off and tales about the chaotic aftermath. *The Telegraph* also embedded a large amount of tweets from officials and other journalists—e.g., Rune Håkonsen (@runehak) from NRK, and Swedish foreign minister Carl Bildt (@carlbildt). Sky News was also prolific in its embedding of tweets, although chiefly using this as a way of incorporating reports from its own correspondents, such as live tweeting the Prime Minister's late evening speech.

BBC News and *The Guardian* were the only two live blogs that used audience emails as sources. In the case of the BBC live blog, emails consisted almost exclusively of eyewitness accounts from people in Oslo. These contained information about the location of the blast, the scale of the devastation, explanations for why so few people were in the building ("the public sector leave work early on Friday's during the summer"), police notices and evacuation procedures, and the emotions victims were experiencing. There was only one email concerning Utøya, from Bjørn Magne Slinde, warning others not to call people on the island for fear of revealing their location. Half of *The Guardian* emails were from Neil Perry, with the rest seemingly from ordinary citizens. Unlike the eyewitness accounts in the BBC's audience emails, *The Guardian*'s audience emails were all referring to other established news organisations—typically Norwegian media such as NRK, *Aftenposten* and *Dagbladet*. In other words, *The Guardian* was acknowledging not only the source of the information, but the citizens who provided them with the tip-off.

Table 2. Live blogs covering 22 July 2011 attacks

Live blog	First post	First Utøya mention	Updates	Words	% EW / AE / CJ	Number of updates containing Eye-witness	Audience emails	SM - citizens	SM - official
Al-Jazeera	17:28	**18:09 (ref Tweet by @MadeChris)**	44	1762	20%	5	0	4	0
BBC News	16:30	18:19 (ref Reuters)	146	2857	23%	15	16	3	0
CNN	16:00	18:48 (ref Reuters)	41	2176	37%	12	0	3	0
Guardian	16:34	18:16 (ref Neil Perry (reader) who ref Stoltenberg on TV2)	48	1813	29%	3	9	2	5
Sky News	16:36	18:06 (ref unconfirmed reports, ref police reports at 18:17)	98	1536	21%	16	0	5	18
Telegraph	16:00	18:19 (ref Police reports)	110	3670	15%	10	0	6	21

Live blogs also provide an opportunity to track how the respective news organisations incorporated rumours about "international terrorism" as responsible for the attacks (see Figure 3). *The Telegraph* first cited claims that the Oslo bomb could be retaliation for Norwegian involvement in Afghanistan at 17:45. It was the first live blog to cite claims from Will McCants, who studies militant Islam and was perceived by many news organisations as a reliable source. At 19:03 *The Telegraph* linked to McCants' claim that Ansar al-Jihad al-Alam appeared to have claimed responsibility for the attacks via "elite jihadi forum" Shmukh, allegedly "in response to the occupation of Afghanistan and insults to the Prophet Mohammed." At 19:45 they cited Twitter user @finansakrobat as claiming "the person arrested in Utoya was reportedly Blonde and 'Norwegian-looking.'" Further speculation about the perpetrator's nationality was posted at 20:00, citing a UK risk consultancy suggesting "it is likely he was ethnically Norwegian," yet still maintaining that "the Labour Party would be a favourable target for Islamist groups." It was not until 20:08 that *The Telegraph* posted a tweet from McCants indicating he was "now backtracking on Twitter on his earlier claims." After citing eyewitness accounts at 20:17 describing in detail how the perpetrator "had a Norwegian look and spoke in a common eastern dialect," the paper dropped any further mention of a connection with international terrorism.

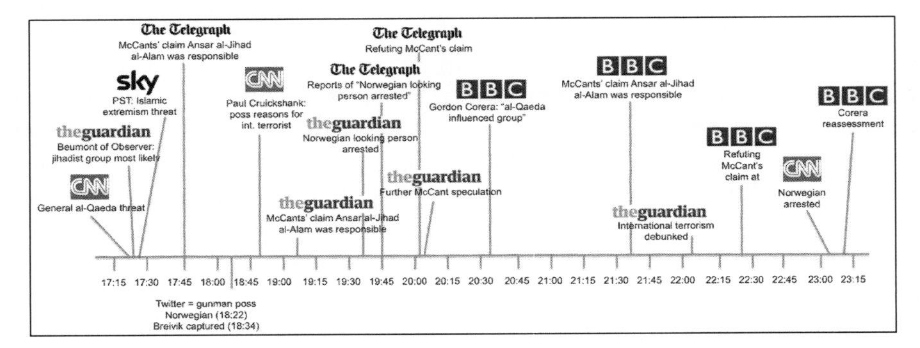

Fig 3. Timeline of selected live blogs and key updates on 22 July 2011

At 20:32 the BBC's security correspondent, Gordon Corera, was cited in the Corporation's live blog as believing "an al-Qaeda influenced group is most likely behind the attack." McCants' claim was also indirectly referred to by the BBC live blog, when it cited a report in *The New York Times* at 21:39. The BBC live blog provided a rebuttal from a terrorism expert at 22:24, and confirmation that the man arrested was Caucasian at 21:56. Corera's own reassessment of the erroneous al-Qaeda link was only posted at 23:11. Meanwhile, *The Guardian* live blog had at 17:25 cited a lengthy analysis by Peter Beaumont, foreign affairs editor of its sister publication *the Observer*, explaining why "a jihadist group is most likely behind the blast." At 19:09 it cited McCant's Ansar al-Jihad al-Alam allegations, using both NRK and Jihadica as a source. Whilst *The Guardian* reported at 19:40 that "the attacker in Utøya, now arrested, looked Norwegian," it still continued to explore the Ansar al-Jihad al-Alam claim at 20:11 by citing *The New York Times* article. The "international terrorism" link was first debunked at 22:07, citing tweets from @runehak of NRK, and final confirmation at 22:48 when the justice minister confirmed the suspect was Norwegian.

CNN made a reference towards a general al-Qaeda threat at 17:23, and a cautious analysis at 18:48 by CNN terrorism analyst, Paul Cruickshank, outlining possible reasons why the group might have targeted Norway. No references were made to McCant's Ansar al-Jihad al-Alam allegations, the broadcaster instead preferring to cite claims that it was too early to attribute responsibility. It confirmed "a Norwegian man was arrested" at 23:07. Sky News adopted an even more cautious approach to the al-Qaeda link, making only a passing reference at 17:27 that "Norway's intelligence police agency (PST) said in February that Islamic extremism was a major threat to the country." No other such references were made until 22:59 when @alexrossiSKY was live tweeting from the Prime Minister's press conference, Stoltenberg refusing to be "drawn on speculation [the perpetrator] was inspired by radical Islam or [the] far right." Finally, Al-Jazeera did not make a single reference towards an al-Qaeda link, McCant's allegations, or any other rumoured claims to or accusations of responsibility.

There are two important aspects to highlight from these timelines: firstly, that the complexity of verifying source claims on the internet are not limited to eyewitnesses or citizen journalists, but evidently also those of seemingly reputable analysts. Secondly, and perhaps more worryingly, the slow pace at which certain framing conventions are relinquished in light of contradictory evidence. *The Telegraph*, BBC News and *The Guardian* all continued to report on al-Qaeda or jihadist links long after Breivik had been arrested—indeed even after their own reports that the perpetrator was most likely or confirmed as Norwegian. Here the live blogs reflect a broader trend in the immediate reporting of the 22 July 2011 attacks, whereby many international news organisations were quick to adopt "international terrorism" as a way of making sense of the unexpected—and then unable to let go of this conventional framing when it was proven to be false. The lack of attention to citizen reports, that were indicating early on that the perpetrator must have been Norwegian, was staggering. Citizens were a useful resource, in other words, when able to provide audio-visual material or eyewitness portrayals of easily verifiable claims (i.e., a depiction of devastation or police instructions would be expected after a confirmed bomb explosion). Most live blog journalists seemed reluctant, however, to offer the same epistemic value to text reports or citizen's judgements about Breivik's identity.

CONCLUSION

Today's news cycle is condensed and the challenges to journalistic practice—be they about verification or ethical concerns about who to contact or what to publish—are being played out in real time. News of both attacks on 22 July 2011 spread rapidly on Twitter and Facebook, with citizens engaging in different forms of crisis communication and rapid remediation of fellow citizens' eyewitness reports. In the case of the Oslo bomb these were: firstly, eyewitness accounts or practical information; secondly, rife speculation about the potential for further bombs and the amplification of those rumours; and thirdly, people seeking to explain what happened by connecting them with past events and thus beginning to frame our understanding of possible perpetrators and their motivations. Victims on the Utøya island, however, were posting citizen distress calls—effectively struggling for their lives in public. This was less a case of documenting what was taking place (in a journalistic sense), and more an attempt by victims to attract attention or help to ensure their survival. Yet victims, their friends and relatives were making information public ahead of established news media, which in turn were ahead of official accounts from emergency services.

Indeed both national and international news organisations were found to draw on citizen eyewitness accounts in their coverage—in particular during the initial phases of the Oslo attack. Several national news websites also referenced tweets from Utøya in their coverage, some even publishing stories based entirely on the role of social media during the attacks as they unfolded—with detailed profiles of people like Kjetil Vevle. This escalated further when, in the days following the Utøya attack, many survivors turned to blogging as a way of publishing their personal experiences. Whilst these blogs might have served as cathartic experiences for the writers, Utøya survivor blogs also appeared to appease both news organisations' and audiences' hunger for an unmediated realism with intensely emotional, frequently gory details about the attack (see Thorsen, 2013a).

The Oslo/Utøya attack serve as a useful reminder about two important aspects concerning information flow in contemporary crisis and disasters: firstly, how citizens caught up in crisis events are increasingly self-publishing their experiences and eyewitness accounts; secondly, how people tangentially or connected with the event contribute to the creation of a shared understanding of events and seek to aid the relief effort through communicative acts. The ability for eyewitnesses to self-publish their observations and experiences, especially during fast-paced breaking news events, is changing the dialogic relationship between sources, journalists and indeed emergency services. Citizen journalism, in all its different forms and guises, has become such a normalised component of today's media landscape that it is almost a naturalised expectation of any crisis or disaster event. The often vivid, intensely emotional and captivating nature of raw eyewitness accounts offers an irresistible resource to journalists seeking to capture the humanitarian side of crisis as they unfold. More importantly, ordinary citizens are through their own journalistic efforts helping to document crisis events and participating in the co-creation of our collective memory of their significance.

NOTES

1. All non-English material translated from Norwegian by the author. All timestamps converted to Norwegian local time for ease of comparison.

REFERENCES

Eriksson, M. (2013). *To the youth: Social sharing of sentiments on Twitter after the 2011 Norway attacks.* (Master Thesis). Uppsala universitet.

Gjørv, A., B., Auglend, R. L., Bokhari, L., Enger, E. S., Gerkman, S., Hagen, T., Hansen, H. B., et al. (2012). *Rapport fra 22. juli-kommisjonen* (NOU No. 2012:14). Norges offentlige utredninger. Oslo: Departementenes Servicesenter.

Johnsen, C. (2012). *Bruken av sosiale medier i en krisesituasjon: En kvalitativ studie av hvilke roller sosiale medier spilte for ungdommer i AUF, i forbindelse med terrorangrepet på Utøya 22. juli 2011.* (Master Thesis). Norges teknisk-naturvitenskapelige universitet.

Perng, S. Y., Büscher, M., Wood, L., Halvorsrud, R., Stiso, M., Ramirez, L., & Al-Akkad, A. (2013). Peripheral Response: Microblogging During the 22/7/2011 Norway Attacks. *International Journal of Information Systems for Crisis Response and Management (IJISCRAM)*, *5*(1), 41–57.

Thorsen, E. (2013a) Survivor blogging: Reporting emotion and trauma after Utøya. In Cooper, G. (ed) *The Future of Humanitarian Reporting*, City Centre for Law, Justice & Journalism, City University London.

Thorsen, E. (2013b) 'Live blogging and social media curation'. In Fowler-Watt, K. and Allan, S. (eds) *Journalism: New Challenges*, Centre for Journalism and Communication Research, Bournemouth University, Available at: http://microsites.bournemouth.ac.uk/cjcr/publications/journalism-new-challenges/

CHAPTER TEN

Eyewitness Images as a Genre of Crisis Reporting

METTE MORTENSEN

Some of the past years' most widely disseminated, most debated, and most spectacular images have been produced and distributed by non-professionals. Since the last turn of century, a new era of crisis reporting has emerged as a result of ubiquitous camera phones and increasingly blurred boundaries between media producers and media audiences. Citizens or participants create otherwise unobtainable stills and videos of unfolding events, which are featured regularly in the mainstream news media. The amateur pictures gaining broad public attention have often undermined the conventionally tight regulation of images from areas of disaster and conflict. This is most striking in relation to eyewitness images from regimes with limited or no freedom of press such as Burma, Iran, Syria, and Libya. However, they also provide counternarratives in democratic countries, as was the case with the revelations of the pictures from the Abu Ghraib prison in 2004.

In the competitive, commercialized global news circuit giving priority to being first with the latest and live reporting, amateur footage remains the only user-generated content to be routinely presented in a manner similar to professional content (Pantti and Bakker 2009, 485). Especially in connection with breaking news events, non-professional imagemakers frequently become the initial link in the chain of news as they are on the spot and able to document events as they play out. Eyewitness images circulated in the established news are often associated with an exclusive insider perspective and proximity in time and space to events. Yet, even if they are habitually considered authentic on account of their urgency, immediacy, and handheld aesthetics, non-professional images still put the norms, editorial routines, and professional self-perception of journalism to the test, because they rarely live up to conventional journalistic ideals such as objectivity and autonomy (see e.g., Andén-Papadopuolos and Pantti 2011a, Wall 2012, Allan 2013, Kristensen and Mortensen 2013). They are mostly produced and mobilized by people involved on some level or other in the situations depicted. Moreover, the standard journalistic procedure of source criticism is complicated by the frequent anonymity of the producers, as well as the challenges involved with verifying facts and extracting a coherent narrative from the clips.

This chapter contributes to the understanding of the role played by amateur visuals in today's convergent and digitalized media landscape by defining eyewitness images as an emergent genre of crisis reporting in the mainstream news. Proceeding in four sections, the chapter first provides a brief overview of amateur images in the news. The second section engages critically with the vocabulary deployed to designate amateur images and argues for the term 'eyewitness images'. In the third section, five defining traits of eyewitness images are outlined, 1) auto-recorded; 2) subjectivity; 3) media institutional ambiguity; 4) participation and documentation; and 5) decontextualization. Fourth and finally, the conclusion outlines questions for further research.

EYEWITNESS IMAGES IN THE NEWS

Even though taking pictures has gradually developed into a ritualized response to crisis during the latter part of the 20th century, the advent of digital technologies has obviously rendered amateur images much more widespread in the public realm. The various ways in which eyewitness images enter the media coverage of terror, war, conflict, natural disaster, accidents, and other situations of tension or unrest can of course not be exhausted in this chapter. Instead, this section offers a brief overview of eyewitness images in the context of terror and war/conflict.

The multitude and significance of amateur images first obtained public notice in connection with the terrorist attacks on September 11, 2001. As Kirschenblatt-Gimblett laconically remarks, this event constituted 'the ultimate Kodak moment' (2003, 14). Numerous amateur stills and moving images circulating in the news media and memorial culture contributed greatly to turning 9/11 into 'the most photographed disaster in history' (Stubblefield 2011). Subsequently, eyewitness images have played a prominent role in the coverage of terror attacks such as the ones in London in 2005 (Reading 2009, Allan 2012), Mumbai (2008), and Boston (2013). The bombings during the Boston Marathon highlight the multiplicity of purposes and functions assigned to eyewitness images in the wake of disastrous events in a complex, communicative web of actors and platforms. Seeing as the bombs exploded in the final stretch of the marathon, onlookers had produced much footage of the scene prior to, during, and after the attack. The vast amount of videos and stills was not only used intensely in the media coverage, the FBI and the police also examined the material for possible evidence about the identity of the terrorists and the course of the event, just as media users engaged in collaborative efforts to crowd-source the eyewitness images with the aim of partaking in the criminal investigation, for instance on the social media Reddit (see also chapters by Allan and Meikle, respectively, in this volume).

With regard to the context of war and conflict, the Abu Ghraib photographs probably constitute the most famous example of amateur images (see e.g., Danner 2004, Hersh 2005, Grusin 2010). The snapshots were attributed news value on account of the authentic glimpse they appeared to offer into the 'dirty reality' (Birchall 2008) of war. Due to censorship and embedded reporting, the Iraq War did not seem to yield any 'memorable professional images' as Struk (2011, 147) contends, and the audience was left with 'a pervading sense that traditional photojournalism was inadequate and outdated.' Abu Ghraib forms an exceptional case in this overview of eyewitness images in so far as the snapshots were not taken with the aim of public dissemination. On the lines of many other amateur images intended for private purposes, they draw on various visual traditions, e.g., pornography, trophy shots, tourist snapshots, and colonial photography, and hence fall in-between performative and documentary genres. The perpetrators furthermore radicalized the dual position of documentation and participation by standing behind the camera themselves, and the act of photography at one and the same time documented the abuse of prisoners *and* contributed to the abuse of prisoners. In this sense, the Abu Ghraib images are reminiscent of the controversial photographs by German conscripts on the East Front of World War II disclosing their crimes of war (see e.g., Heer and Naumann 1995, Struk 2004, Heer, et al. 2008, Allan 2011, Struk 2011), but they do not point forward to the following years' typical amateur pictures in the news. The eyewitness images entering the media circuit tend to be more straightforward and direct in their communication of events, just as they have been taken with the explicit goal of public distribution, for instance in connection with the anti-government uprisings in Myanmar (Burma) (2007), Iran (2009) (Andén-Papadopoulos and Pantti 2011b, Mortensen 2011a), and in a number of Northern African and Middle Eastern countries from 2011 (Kristensen and Mortensen 2013, Pantti 2013).

The different strategies deployed by the news media in their use of eyewitness visuals may be illustrated by two paradigmatic and oppositely directed examples. During the anti-government revolts in Iran following the assumed electoral fraud in 2009, on-site activists turned into essential sources of information for the international news media on account of the government's prohibition against the foreign media reporting from within Iran. However, despite the huge number of amateur sources at hand, the news media took a familiar path and presented the footage of the killing of the woman Neda Agha Soltan as a centralized, symbolic icon to illustrate the complex political situation (Mortensen 2011a). The reverse strategy was taken in connection with the death of former president Muammar Ghaddafi in Libya in 2011. Speed appeared to outweigh verification, and the first phase of the coverage to a large extent consisted of passing on incoming, unconfirmed amateur videos as they became available in a seemingly unfiltered manner (Kristensen and Mortensen 2013).

NON-PROFESSIONAL IMAGES: CONCEPTS AND DEFINITIONS

Various terms are used to describe non-professional images in the news circuit: 'citizen journalism', 'citizen photojournalism', 'citizen images', 'citizen eyewitness images', and 'eyewitness photography'. This chapter proposes the term 'eyewitness images' as an alternative to the umbrella concept citizen journalism, which is problematic for three main reasons (see also Mortensen 2011b).

First, 'citizen journalism' and 'citizen photojournalism' are used indiscriminately. 'Citizen journalism' became a widespread denomination in the aftermath of the tsunami in South Asia in December 2004, when pictures produced by tourists on location were recognized as an extraordinary contribution to mainstream news coverage (Allan 2009, 18). Covering a broad field, 'citizen journalism' holds no critical sensibility to capture the specificity of images, for example their circulability across regional and linguistic borders.

Second, simply employing the concept 'citizen photojournalism' does not solve the problem as both 'citizen' and 'photojournalism' suffer from imprecision. 'Citizen' stems from a notion of a subject in a nation state carrying rights and obligations, which is hardly universally applicable—especially not in a number of totalitarian states with flourishing cultures for eyewitness images. Even if we bracket this reservation, many producers hardly seem to be ordinary 'citizens'; a point to which we shall return.

Third, 'photojournalism' does not convey a proper impression of the variety of formats in questions such as still photos, videos, and music accompanied slide shows. Referring to the different visual practices as 'citizen photojournalism' might also give the wrongful impression that they subscribe to the more narrow tradition of photojournalism, i.e., photographs to illustrate news stories, typically living up to the humanistic ideals established in the golden age of this genre from the 1930s to the 1960s.

To establish a more generic vocabulary, this chapter uses the expressions 'eyewitness images'/ 'eyewitness image producer' (Mortensen 2011b, see also e.g. Allan 2013, Andén-Papadopoulos and Pantti 2013). Apart from shunning 'citizen' and 'photojournalism' as problematic designations, 'images' and alternatively 'pictures' or 'visuals' are broad terms encompassing diverse visual and audiovisual forms. This choice of vocabulary also holds the advantage of relating current practices to the wider historical perspective of media, politics, and culture brought in play by the eyewitness and witness testimonies. As it goes beyond the scope of this chapter to relay in detail for the centrality of witnessing in law, religion, culture, and journalism (see e.g., Felman and Laub 1992, Ellis 2000, Peters 2001, Zelizer 2007, Frosh and Pinchevski 2009), suffice it here to briefly point out how the contemporary non-professional image-producer might be seen as a predecessor to the figure of the eyewitness. Today's amateur photographer operating in the world's hotspots has, most importantly, inherited the eyewitness' physical presence and privileged firsthand access to events. Eyewitness images also share the traditional witness testimony's subjective viewpoint and the partial, fragmented narration due to the 'annihilation of perspective' (Ashuri and Pinchevski 2009, 140) involved in the personal investment and proximity to events.

The major novelty in relation to the proliferation of digital technologies remains, of course, that eyewitnesses do not just make appearances in the courtroom or in the media as sources of information, but are capable of creating and distributing media content themselves. In 2001, John Durham Peters writes in his seminal essay 'Witnessing' (722) that most eyewitnesses do not think of themselves as such until after the closing of an event. This no longer holds true per se. Witnessing has evolved into an individual choice, a recurring option, and a mass phenomenon. In other words, eyewitnesses have become self-mediated by incorporating digital media technologies into their practice and adapting to the logics of the current media system.

DEFINING EYEWITNESS IMAGES

Different approaches and methods could be applied to define eyewitness images. In line with this chapter's focus on eyewitness images as a genre within the mainstream news media's crisis reporting, emphasis is put on the most important structural conditions and formal characteristics arising from actors outside the institutionalized media contributing to the news circuit. On this basis, five distinguishing traits are singled out in the following: 1) *Auto-recorded*; 2) *subjectivity;* 3) *media institutional ambiguity*; 4) *participation and documentation*; and 5) *decontextualization*.

1. Auto-Recorded

Eyewitness images are usually recorded and distributed by the very same person. Compared to the traditional witness, testimonies have changed profoundly by this figure moving behind the camera and holding the option of uploading pictures almost instantaneously. This becomes clear when comparing eyewitness images to spoken acts of witnessing, which Peters divides into the two distinctive phases of 'seeing' and 'saying':

To witness thus has two faces: the passive one of *seeing* and the active one of *saying*. In passive witnessing an accidental audience observes the events of the world; in active witnessing one is a privileged possessor and producer of knowledge in an extraordinary, often forensic, setting in which speech and truth are policed in multiple ways. What one has seen authorizes what one says: an active witness first must have been a passive one. Herein lies the fragility of witnessing: the difficult juncture between experience and discourse. The witness is authorized to speak by having been present at an occurrence. A private experience enables a public statement. But the journey from experience (the seen) into words (the said) is precarious (2001, 709–710, italics in original).

Witnessing traditionally involves an act of translation between a passive, private experience and an active, public performance. This passage between 'seeing' and 'saying' is difficult to cross, following Peters. While events are messy and hard to manage in the here and now of their unfolding, stories are expected to be coherent and smooth (Peters 2009, 44–45). The media have traditionally counterbalanced the eyewitness' subjective disposition by offering a role governed by rules and rituals (Thomas 2009, 101), which turns 'seeing' into 'saying' by framing and structuring the act of witnessing. Eyewitness images, by contrast, collapse the distance in time and space between 'seeing' and 'saying'. The situation is already experienced in a mediated form as it proceeds; mentally by the individual's decision to transform observation to representation in situ and physically by the split attention between the event in real life and the mobile phone screen's reproduction of the event. Whereas the junction from 'seeing' to 'saying' used to be 'precarious' and 'difficult' as Peters writes in the quote above, digital recording devices now easily cover this distance. The testimonies themselves have in turn become 'precarious' and 'difficult' with their unedited, fragmented, and subjective form. For the same reason, eyewitness images may give the impression of bringing the viewer close to the scene of the event—at least that would form a parallel to how spoken witness testimonies tend to be perceived as more authentic and intimate when they are still incoherent and unstructured, i.e., when 'seeing' has not yet quite been converted into 'saying'.

2. Subjective

Second, a subjective point of view characterizes images created by an eyewitness—or I-witness', to use a catchphrase sometimes deployed.[1] Eyewitness images in this way feed into the overarching individualization and the omnipresence of personal confessions in today's media and culture. For instance, reality television and the performative staging of the self on social media also take their point of departure in the subjective viewpoint of 'ordinary' people. A public statement by a person regarding a privately experienced event always contains an element of self-representation (Thomas 2009, 100–102), because the account is communicated from a subjective, individualized perspective. In this way, first-person testimonies invariably address the fundamental question of how private sensations might be transformed into intelligible and meaningful public statements (Peters 2001, 711).

Eyewitness images may, or they may not, hold the potential for giving voice to collective concerns by putting larger political, social or cultural issues into perspective. Along the same lines, the literature on the Holocaust witness has accentuated the paradox inherent in testifying (see e.g., Felman and Laub 1992): On the one hand, eyewitness testimonies stand out as singular and unique accounts for the specific experiences of an individual. On the other hand, witnesses address collective experiences as they 'speak *for* and *to* others' (Felman 1992, 3). This tension between the singular and the collective often manifests itself in the reception of eyewitness images. In connection with the Abu Ghraib photographs, for instance, debates ensued as to whether they represented the point of view of the individual military employees, or epitomized the spirit and behaviour of the US military in a more generalized way. Similarly, many eyewitness images linger ambiguously between private and public recollection, when they have been created for personal purposes, at the same time as they play a part in the public mediation or documentation of crisis.

3. Participation and Documentation

As a third characteristic, amateur images in the news blur the boundaries between participation and documentation. In some of the most high-profile media cases generated and/or illustrated by amateur images, photographers were on-site by virtue of their occupation, e.g., as cargo worker (the pictures of caskets with fallen American soldiers from 2004) or security guard (the bootleg video of the hanging of Saddam Hussein in 2006). With the same dual perspective of observer and participant, soldiers and activists comprise two other significant suppliers of non-professional images. Even though amateur photographers on sites of natural disasters are not involved in the same sense, they are nonetheless affected as victims or witnesses to the suffering and distress of others.

Regarding weaker degrees of participation, theoreticians in the field have brought forward the important argument that bearing witness per definition entails a crossing of the line between observation and action (e.g., Peters 2009, Wagner-Pacifici 2005). Every act of witnessing is tied to a 'transformation' (Thomas 2009, 96). Put differently, taking on the part of the eyewitness constitutes a form of participation in itself. This applies especially to the auto-mediated eyewitness, whose attendance is more self-reflexive and active than the classical eyewitness. Witnessing becomes part of the event, which changes both the witness' perception of the event, the retrospective reconstruction of the event, and at times even the course of the event itself. The mere physical presence of the camera is often enough to exert an influence, most dramatically in regimes interpreting filming in public spaces as a sign of dissent, but also on a smaller scale when people become conscious of their actions being recorded. Events take place in front of the lens, which would otherwise not have happened in that way, or would perhaps not have happened at all.

4. Media Institutional Ambiguity

Formerly 'institutionally powerless' (Cottle 2006, 2) individuals are able today to produce and distribute content outside the realms of the traditional media institutions. Still, to distinguish simply between institutionalized and non-institutionalized media practices would be misleading. Eyewitness images are often created and circulated in a manner which cuts across the two.

As amateur images enter the established news media, the producers seem to become more and more professionalized. Activists raising awareness for their cause constitute a conspicuous example. To a still greater extent, they operate on a semi-professional level to produce and distribute images targeted at established media institutions, which they also collaborate with in implicit and explicit ways. Some visual, digital activism in Burma, for example, emanates from a tightly structured network of so-called 'video journalists' on the ground, supplying footage for the Oslo-based organization Democratic Voices of Burma, which e.g., succeeded in thrusting images of anti-government protests into the international news media in 2007.

From the other side of this traditional divide, amateur practices are increasingly welcomed and facilitated by the mainstream news media, even if they are also ambivalent with regard to their validity and legitimacy as sources. For instance, the 'Assignment Desk' at CNN's iReport encourages the audience to sign up for reporting on topical subjects or events with the promise that 'Your iReport could be featured in a CNN story' (http://ireport.cnn.com 2013). Seen from a media institutional perspective, amateur pictures have the advantage of low production costs, just as they solve some of the traditional obstacles of crisis reporting concerning available sources of information and access to areas of combat or disaster. The appeal of non-professional visuals is possibly enhanced by the way citizens and others producing and distributing images are inclined to adjust to or internalize the logics, requirements, and norms of today's news industry, for instance concerning what is deemed newsworthy information, offering identification, proximity, immediacy, drama, authenticity, a sense of urgency etc. (Kristensen and Mortensen 2013, see also Andén-Papadopoulos and Pantti 2011a). Further, amateur images prompt a reciprocal effect; as they become a standardized feature in crisis reporting, they in turn change the criteria for the content and form of newsworthiness (see also Pantti 2013, 203).

5. Decontextualization

Eyewitness images tend to be less didactic, informative, and self-explanatory than professional press photographs owing to the decontextualized content and transmission of the material. Producers often remain unnamed or assume pseudonyms, e.g., for safety or legal concerns, and/or their precise identity is not considered important by the media circuit. Moreover, information about the actors, circumstances, locations etc. represented is frequently neither volunteered, nor easily available through journalistic research. This lack of factual information and of a steady, identifiable figure or institution responsible for the images naturally makes it difficult to test the reliability of the source.

On account of their decontextualized form, eyewitness images lend themselves to be contextualized anew. Related to this, Struk writes that amateur images rarely '*tell stories*', but rather '*prompt stories*' (2011, 150, italics in original). They often enter into event-driven journalism as opposed to source-driven journalism (Chouliaraki 2010, see also Livingston and Bennett 2003) in connection with breaking news events, when news networks have a propensity to present the information and material at hand rather than offering a prioritized, synthesized, and balanced account.

CONCLUSION

According to the five defining traits of eyewitness images, they are recorded and distributed by individuals with exclusive first-hand experience, who mostly reside outside or are only informally affiliated with a media institution. However, the question of institutional ties is not an easy one, due to the conflation of the borderlines between professional and non-professional media production. Eyewitness images accordingly contain a particular viewpoint, which is at once anonymous and subjective. They are often conveyed in a decontextualized form, and are to be considered partial, if not biased, due to the merging of participation and documentation.

This chapter has predominantly paid attention to eyewitness images in the context of manmade disaster in accordance with the greater part of the literature on the subject. By contrast, research is less developed into the impact of non-professional visuals in media representations of natural disasters (but see e.g., Liu Palen et al. 2009, Chouliaraki 2010, Pantti et al. 2012). This is somewhat surprising considering the frequency with which amateur images have been featured in coverage of natural catastrophes such as the tsunami in the Indian

Ocean in 2004, hurricane Katrina in 2005, the earthquake in Haiti 2010, the tsunami in Japan in 2011, and many others. The role performed by amateur images in relation to media coverage of these events would be a highly interesting and relevant area for future research.

In addition, this chapter has focused on defining eyewitness images in the news, i.e., typically footage entering the institutionalized news media from social media or the established news media's outlets for citizen journalism. The definition in this way takes as its implicit condition the convergence between different media systems. Another subject of interest for future research would be to expand this circle and look into the way in which eyewitness images become an integral part of larger communicative networks, including also official crisis communication, written citizen journalism, communication between users spurred by the images etc. Moreover, field studies of the different functions performed by the images would also form a valid contribution to the research area, e.g., it would be highly relevant to look into when and how they are used by activists to document events, create awareness and sympathy for their cause, to recruit new members, communicate to the media etc. A final topic worth more scrutiny would be how different media institutions are developing editorial norms and routines for selecting, verifying, and framing eyewitness images.

NOTES

1. See, for instance, http://i-witness-news.com/.

REFERENCES

Allan, S. (2009). Histories of Citizen Journalism *Citizen Journalism: Global Perspectives*. S. Allan and E. Thorsen. New York, Peter Lang: 17–31.

Allan, S. (2011). Amateur Photography in Wartime: Early Histories. *Amateur Images and Global News* K. A.-P and M. Pantti. Bristol/UK and Chicago, Intellect/University of Chicago Press: 41–59.

Allan, S. (2012). Citizen Journalism and the Rise of "Mass Self-Communication": Reporting the London Bombings. *Citizen Journalism: Valuable, Useless, or Dangerous?* M. Wall. New York, International Debate Education Association: 11–26.

Allan, S. (2013). *Citizen Witnessing: Revisioning Journalism in Times of Crisis*. Cambridge, Polity Press.

Andén-Papadopoulos, K., and M. Pantti (2011a). *Amateur Images and Global News*. Chicago/Bristol, UK, University of Chicago Press/Intellect Press.

Andén-Papadopoulos, K. and M. Pantti (2011b). Transparency and Trustworthiness: Strategies for Incorporating Amateur Photography into News Discourse. *Amateur Images and Global News*. M. P. and K. Andén-Papadopoulos. Chicago/ Bristol, UK, University of Chicago Press/Intellect Press: 97–112.

Andén-Papadopoulos, K., and M. Pantti (2013). Re-imagining Crisis Reporting: Professional Ideology of Journalists and Citizen Eyewitness Images" *Journalism: Theory, Practice and Criticism*: 1–18 (online first).

Ashuri, T., and A. Pinchevski (2009). Witnessing as a Field. *Media Witnessing. Testimony in the Age of Mass Communciation*. P. F. a. A. Pinchevski. Basingstoke, Palgrave Macmillan: 133–157.

Birchall, D. (2008). Online Documentaries. *Rethinking Documentary. New Perspectives, New Practices*. T. A. a. W. d. Jong. Maidenhead, Open University Press.

Bruns, A. (2008). *Blogs, Wikipedia, Second life, and Beyond: From production to produsage*. New York, Peter Lang.

Chouliaraki, L. (2010). Ordinary Witnessing in Post-Television News: Towards a New Moral Imagination "*Critical Discourse Studies* 7(4): 305–319.

Cottle, S. (2006). *Mediatized Conflict: Developments in Media and Conflict Studies*. Maidenhead, Berkshire, England; New York, NY, USA, Open University Press.

Danner, M. (2004). *Torture and Truth: America, Abu Ghraib, and the War on Terror*. New York, New York Review Books.

Ellis, J. (2000). *Seeing Things: Television in the Age of Uncertainty*. London ; New York, I.B. Tauris.

Ellis, J. (2009). Mundane Witness. *Media Witnessing. Testimony in the Age of Mass Communication*. P. a. P. Frosh, Amit. Basingstoke Palgrave Macmillan: 73–88.

Felman, S., and D. Laub (1992). *Testimony: Crises of Witnessing in Literature, Psychoanalysis, and History*. New York; London, Routledge.

Felman, S. (1992). Education and Crisis, or the Vicissitudes of Teaching. *Testimony: Crises of Witnessing in Literature, Psychoanalysis, and History*. s. a. L. Felman, Dori. New York; London, Routledge: 1–56.

Frosh, P., and A. Pinchevski, Eds. (2009). *Media Witnessing. Testimony in the Age of Mass Communciation*. Basingstoke, Palgrave Macmillan.

Gillette, S., et al. (2007). Citizen Journalism in a Time of Crisis: Lessons from California Wilderfires. *The Electronic Journal of Communication* 17(3 & 4).

Grusin, R. A. (2010). *Premediation: Affect and Mediality after 9/11*. Basingstoke; New York, Palgrave Macmillan.

Heer, H., et al., Eds. (2008). *The Discursive Construction of history: remembering the Wehrmacht's War of Annihilation*. Basingstoke; New York, Palgrave Macmillan.

Heer, H., and K. Naumann (1995). *Vernichtungskrieg: Verbrechen der Wehrmacht 1941–1944*. Hamburg, Hamburger Edition.

Hersh, S. M. (2005). *Chain of Command: The Road from 9/11 to Abu Ghraib*. New York, Harper Perennial.

iReport, CNN (2013). http://ireport.cnn.com

Kirschenblatt-Gimblett, B. (2003). Kodak Moments, Flashbulb Memories. Reflections on 9/11. *The Drama Review* 47(1): 11–48.

Kristensen, N. N., and M. Mortensen (2013). Amateur Sources Breaking the News, Metasources Authorizing the News of Gaddafi's Death. New Patterns of Journalistic information Gathering and Dissemination in the Digital Age. *Digital Journalism*, online first: http://www.tandfonline.com/doi/full/10.1080/21670811.2013.790610.

Liu, S. B., et al. (2009). Citizen Photojournalism during Crisis Events. *Citizen Journalism: Global Perspectives* S. Allan and E. Thorsen (eds) New York, Peter Lang: 43–64.

Livingston, Steven and Bennett, W. Lance (2003). 'Gatekeeping, Indexing, and Live-Event News: Is Technology Altering the Construction of News?', *Political Communication*, 20(4):363–380.

Mortensen, M. (2011a). When Citizen Photojournalism Sets the News Agenda: Neda Agha Soltan as a Web 2.0 Icon of Post-Election Unrest in Iran *Global Media and Communication* 7(1): 4–16.

Mortensen, M. (2011b). The Eyewitness in the Age of Digital Transformation. *Amateur Images and Global News* K. a. P. Andén-Papadopoulos, Mervi. Bristol/Chicago, Intellect Press/University of Chicago Press: 61–76.

Pantti, M., and P. Bakker (2009). Misfortunes, Memories and Sunsets: Non-professional Images in Dutch News Media *International Jounal of Cultural Studies* 12(5): 471–489.

Pantti, M., et al. (2012). *Disasters and the Media*. New York, Peter Lang.

Pantti, M. (2013). Getting Closer? Encounters of the National Media with Global Images. *Journalism Studies* 14(2): 201–218.

Peters, J. D. (2001). Witnessing. *Media, Culture & Society* 23(6): 707–723.

Peters, J. D. (2009). An Afterword: Torchlight Red on Sweaty Faces. *Media Witnessing. Testimony in the Age of Mass Communciation*. P. a. P. Frosh, Amit. Basingstoke, Palgrave Macmillan: 42–48.

Reading, A. (2009). Mobile Witnessing: Ethics and the Camera Phone in the 'War on Terror'. *Globalizations* 6(1): 61–76.

Struk, J. (2004). *Photographing the Holocaust: Interpretations of the Evidence*. London, I.B. Tauris.

Struk, J. (2011). *Private Pictures: Soldiers' Inside View of War*. London; New York, I. B. Tauris.

Stubblefield, T. (2011). Does the Disaster Want to be Photographed? Reconsidering the Camera's Presence at Ground Zero. *Afterimage* 39(1/2): 9–12.

Thomas, G. (2009). Witness as a Cultural Form of Communication. *Media Witnessing. Testimony in the Age of Mass Communication*. P. a. P. Frosh, Amit. Basingstoke, Palgrave Macmillan: 89–111.

Wagner-Pacifici, R. (2005). Witness to Surrender *Visual Worlds*. B. S. a. L. T. B. John R. Hall. London and New York, Routledge.

Wall, M. (2012). *Citizen Journalism : Valuable, Useless, or Dangerous?* New York, International Debate Education Association.

Zelizer, B. (2007). On 'Having Been There': 'Eyewitnessing' as a Journalistic Key Word." *Critical Studies in Media Communication* 24(5): 408–428.

CHAPTER ELEVEN

Reformulating Photojournalism: Interweaving Professional AND Citizen Photo-reportage OF THE Boston Bombings

STUART ALLAN

If everyone with a smartphone can be a citizen photojournalist, who needs photojournalism? This rather flippant question cuts to the heart of a set of pressing issues, where an array of impassioned voices can be heard in vigorous debate. Closer inspection reveals an evolving continuum of perspectives. At one end are voices confidently prophesying photojournalism's impending demise as the latest casualty of internet-driven imperatives of convergence, while at the other end are those heralding its dramatic rebirth under the rippling banner of citizen journalism. Voices situated in-between these stark polarities include those calling for a new economic model to sustain its integrity as a professional craft, even though the means by which such a model is to be secured remain frustratingly elusive. Regardless of where one is situated along this continuum, however, it is readily apparent that photojournalism is being decisively transformed across fluidly uneven digital contexts in ways that raise important questions for civic engagement in public life.

Taking as its point of departure a telling example of the perceived attenuation of photojournalism in the United States, this chapter proceeds to identify and critique a range of factors currently recasting its professional ethos. A startling number of newsrooms, under intense financial pressure to trim expenditure wherever possible, evidently regard the resources vital for a dedicated photo desk to be a luxury increasingly difficult to justify. Some fear that photojournalism's viability risks being compromised as a result, its guiding principles on the verge of collapsing in a climate of managerial indifference, if not outright neglect. Against this backdrop, the figure of the 'citizen photojournalist' will be examined, with specific reference to news coverage of the bombing of the Boston marathon on 15 April 2013. Drawing on a critical analysis of public commentaries concerning the respective roles of professional photojournalism and the improvised contributions made by ordinary bystanders during the crisis, I shall examine competing claims regarding photojournalism's capacity to thrive or perish with ever-greater citizen involvement in newsmaking. Illuminated by the challenges confronting photojournalism that day, I shall argue, is a basis for envisioning new opportunities to realign it with its publics in a manner at once more transparent and accountable while, at the same time, encouraging a more openly inclusive news culture committed to fostering dialogue across diverse communities.

CHALLENGING COMMITMENTS

Former chief photographer for the *Chicago Sun-Times*, Bob Kotalik, began his career in 1942. A teenager at the time, his part-time job was to clean out the loft built for homing pigeons on the newspaper building's roof—the pigeons having been trained to return clutching a roll of film shot elsewhere in the city, not least at sporting events. Over the course of the years to follow, Kotalik rose through the ranks of *Sun-Times* photographers, earning a reputation for fearless determination. 'He had a knack for news and an extra sense of anticipating what was going to happen,' former colleague John H. White later recalled. 'He knew that you could never publish an excuse, so he always produced' (cited in Golab, 2013). White was one of several photographers quoted in the *Sun-Times*'s obituary for Kotalik published on 30 May, 2013, following his passing at the age of 87. 'He was a good person to work for because he had been on the street for quite a number of years,' photographer Gene Pesek added. 'He knew what to expect.'

A professional life well lived. In what would be recognised to be a bitter twist of irony, however, Kotalik's obituary appeared on the *Sun-Times* website the same day startling news was breaking that would call into question the daily's longstanding commitment to photojournalism. Earlier that morning, each of the 28 members of the photography staff had received an email instructing them to attend a 9:30 am meeting. Managing Editor Jim Kirk took less than a minute to tell them that the 'tough decision' had been taken to terminate their employment, the latest in a series of cost-cutting measures. The newspaper executive's public statement, released to the Associated Press later that day, read in full:

> The Sun-Times business is changing rapidly and our audiences are consistently seeking more video content with their news. We have made great progress in meeting this demand and are focused on bolstering our reporting capabilities with video and other multimedia

> elements. The Chicago Sun-Times continues to evolve with our digitally savvy customers, and as a result, we have had to restructure the way we manage multimedia, including photography, across the network (cited in Marek, 2013).

Meanwhile the assembled group of stunned photographers and photo editors—amongst them Pulitzer Prize winner John H. White, quoted in Kotalik's obituary—were told to complete the paperwork for their 'layoff packages,' return their company-owned camera equipment, and surrender their access badges. White commented afterward that it was 'only the second meeting with the new managers' the photographers had attended. It was 'intimidating,' with 'a toxic and unkind spirit in the office' (cited in Irby, 2013).

Initial reactions from across the mediascape ranged from the incredulous to the appalled, soon giving way to the scathing. The Newspaper Guild expressed its outrage that the entire *Sun-Times* photography staff was being summarily dismissed without notice, but expressed little hope of reversing management's action. 'I have learned time and again how the eye of a professional photographer can see and express things that I can't,' Bernie Lunzer, president of the Guild-CWA stated. 'Apparently, some accountant/manager can see and express things that I can't understand. Because this makes no sense' (cited in AFP, 2013). Elsewhere rumours were circulating on social media sites that *Sun-Times* reporters had received a memo from Managing Editor Craig Newman informing them that they would be undergoing 'mandatory' training in 'iPhone photography basics' in order to supplement the work of freelance photographers wherever possible. 'In the coming days and weeks, we'll be working with all editorial employees to train and outfit you as much as possible to produce the content we need,' it explained (Newman, 2013). Summing up the situation, one commentator from the photography community observed:

> Out the door went 28 people, and decades of experience and skill. All at once the paper emptied a deep reservoir of photojournalistic talent. Before Thursday, it had a staff of professionals with broad knowledge of a great city, with the hard-earned ability to tell stories with pictures—the not-so-easy thing that newspaper photographers do every day. Now it has some freelancers and reporters toting cheap cameras with their notebooks and pens (Downes, 2013).

Several commentators pointed out that there was much more to photojournalism than equipping journalists with cameras, a point underscored by Dan Mitchell (2013), writing in *Fortune Tech*:

> Reporters, it should be noted, are in general terrible at taking pictures. Photographs snapped on iPhones by photographically inept reporters who are also trying to gather information at an accident scene, for example, are not going to impress anyone, digitally savvy or not. [This is...] pushing the idea of 'multimedia journalism'—that is, having reporters take photos and shoot video [...]. Most often, this results in nothing more than one person

> doing three jobs poorly rather than doing one job well. It also tends to sabotage the notion that all of these are professional endeavors and to strengthen the false notion that anybody could perform any of them equally well. This reveals a shocking level of disrespect for both journalists and readers (Mitchell, 2013).

Photojournalism's 'death spiral' was gaining momentum, several commentators warned, with its status as a professional craft in danger of unravelling. 'Nobody is actually saying it,' Chicago blogger Samuel Smith (2013) pointed out, 'but I'm also willing to bet that they'll be "crowdsourcing" more "content" from "citizen journalists" with camera phones.' While images captured by amateur bystanders were often impressive, maintained Richard Cahan, former *Sun-Times* picture editor, such shots did not amount to photojournalism, in his view. Wryly conceding he would accept the concept of 'citizen journalists' when people were ready for 'citizen surgeons,' he underscored a crucial difference when remarking that amateurs without journalistic training 'are not being asked to go to the widow of a policeman who has been shot' (cited in Lydersen, 2013). Alex Garcia, photojournalist at the crosstown rival *Chicago Tribune*, concurred. To 'reduce photojournalism down to pressing a couple of buttons on the point and shoot' is to 'take the journalism out of photojournalism,' he argued. 'It's about experience, perspective and contacts and your specialty and it's about seeing in perception' (interview transcript, Deacon, 2013).[1]

Opposing views over the *Sun-Times* decision, and the wider repercussions its sudden implementation sparked, attracted considerable debate in the days to follow. The ensuing crisis of confidence in photojournalism's future was illustrative of a deeper malaise in the eyes of some critics, where audiences were effectively blamed for embracing 'free' news on digital devices, rather than being prepared to pay for ink-and-paper alternatives driven to capitalise (in every sense of the word) on their attention for advertisers. For photojournalists acutely aware of the competing demands made on the revenue necessary to sustain their livelihoods, attempts made to 'enhance' or 'complement' their work with citizen imagery risked sowing the seeds of their role's destruction. Such a blurring of boundaries invites awkward questions regarding what counts as 'real photojournalism' for dedicated practitioners, and yet evidence is readily available to suggest that audiences expect to see impromptu contributions from amateurs alongside the work of professionals. It is not unusual to read comments in feedback sections on news webpages, for example, where citizen imagery is praised for the gritty rawness of its authenticity, the very disruption of unspoken rules of composition, framing or technique being upheld as virtuous in its own right. Still, to the extent photojournalism is deprofessionalised, critics reasoned, these same audiences become increasingly vulnerable to visual misrepresentation (deliberate or otherwise), with accustomed protocols of impartiality being rewritten by those intent on advancing personal, frequently non-journalistic priorities.

The profusion of citizen imagery documenting vital aspects of breaking crisis events throws into sharp relief the extent to which news organisations strive to narrativise conflicting truths, a process of mediation valorised, in part, on the basis of news photographers' privileged claim to expertise. Citizens feeling compelled to generate firsthand, embodied forms of visual reportage—such as cell or mobile telephone imagery, digital photographs or camcorder video footage shared across social networks—raise searching questions for the epistemic certainties of professional norms, values and protocols (Allan, 2013a, b; see also Liu et al., 2009). Few would dispute, however, that should quantity be confused with quality in the heat of the moment, photojournalism's standards will be jeopardised. In the next section, our attention turns to a tragic example where these tensions came to the fore in the weeks leading up to the *Sun-Times* controversy. Specifically, on April 15 at approximately 2:50 pm, two pressure-cooker bombs were detonated near the crowded finishing line of the Boston Marathon, killing three people—amongst them an eight year-old boy—and injuring 264 others, many suffering broken lower leg bones and shrapnel wounds engendered by nails and ball bearings packed into the devices. The two suspects believed by the FBI to be responsible, brothers Tamerlan and Dzhokhar Tsarnaev, were soon identified with public assistance, the latter currently in a federal prison awaiting trial.

'AND THEN WE HEARD THIS EXPLOSION'

'I was covering the finish line at ground level at the marathon,' *Boston Globe* photojournalist John Tlumacki (2013a) later recalled. 'Everything was going on as usual. It was jovial—people were happy, clapping—and getting to a point where it gets a little boring as a photographer. And then we heard this explosion'. What had been an ordinary day in Tlumacki's more than 30-year career photographing city events was shockingly transformed into an extraordinary one. The percussion of the bomb blast threw his camera gear into the air, yet he barely hesitated in his response. 'My instinct was…no matter what it is, you're a photographer first, that's what you're doing. I ran towards the explosion, towards the police; they had their guns drawn. It was pandemonium. Nobody knew what was going on.' Amidst the turmoil swirling around him—'the first thing I saw were people's limbs blown off'—Tlumacki did his best to document the scene while keeping his emotions in check. Rendered temporarily speechless, eyes 'swelling up behind my camera,' his shoes 'covered in blood from walking on the sidewalk taking pictures,' he persevered best he was able. At one point in the confusion, he remembered, 'a cop came to me, grabbed me, and said: "Do me a favor. Do not exploit the situation." And that resonated with me. I can't think about it—I gotta keep doing what I'm doing.'

Tlumacki was one of several professional photojournalists who found themselves abruptly pressed into service to capture the grisly horrors of what breaking news coverage was calling a 'war zone' erupting in the city. Amongst them were fellow citizens, seizing the moment to bear witness from their vantage point along the street, their responses and images—relayed via social networking sites such as Twitter, Facebook, YouTube, Flickr, Instagram and Vine—representing personal, impromptu contributions to real-time reportage. PR consultant Bruce Mendelsohn, for example, had been enjoying a celebratory party in an office above the finishing line when the first bomb detonated. 'The building shuddered. I saw smoke; I smelled cordite,' he told NPR in an interview. 'I'm a veteran, so I know what that stuff smells like, I know what that stuff sounds like' (cited in NPR, 2013). Having been knocked from his seat on a couch by the concussion, he hurried downstairs to reach the scene, where he helped to unite a mother with her lost child before providing medical care to several of those injured ('I'm not a medic or anything, but I pressure-treated wounds') before being moved along by the police fifteen minutes later. 'What I saw was more equivalent to newspaper reports of Baghdad than to Boston,' he later recalled (cited in Dinges, 2013). Returning to the third-floor office, he took an image of the street's carnage and posted it on Twitter, where it was discovered by the Associated Press and widely distributed.

Hundreds of those amongst the race's assembled spectators had similarly maintained the presence of mind to engage in spur-of-the-moment, improvised forms of what may be termed 'citizen witnessing' (Allan, 2013a). College student Daniel Lampariello, situated some 200 feet from the finish line—there to cheer on his aunt and uncle running in the marathon—found himself making a precipitous decision to proffer firsthand images and observations via Twitter. 'We thought maybe it was fireworks at first, but when we saw the second explosion we definitely knew that something was wrong,' he told ABC News afterward (cited in Bhattacharjee, 2013; Effron, 2013). Lampariello's photograph of marathon runners continuing to run as the second bomb detonates was uploaded to his Twitter account within a minute of being shot, its geo-location details recording the time as 2:50 pm. Spotted by Reuters' social media editor Anthony De Roas minutes later, it was promptly retweeted to his extensive followers. Eventually heralded as one of the most 'iconic images' of the crisis, its extensive usage by news organisations was facilitated by Reuters having moved swiftly to secure the exclusive license for its distribution. Close analysis of the image revealed a lone figure on the roof of an adjacent building, inviting intense speculation across the social mediascape. A tweet asking 'Who is that guy on the roof?' went viral, while 'Boston Marathon roof' was soon trending worldwide—such was the public interest in delving through crowdsourced imagery in search of clues about the crime's perpetrators (some of those involved were promptly dubbed 'amateur photo sleuths' by the press, although fears were also expressed about 'online vigilantes,' 'digital witch-hunts,' and 'conspiracy nuts').

The sheer scope and diversity of eye-witness citizen reportage of the Boston attack was recurrently described as a critical turning point in media commentaries, some of which contended that this was the 'first atrocity to be covered in real-time for a mass audience on social media' (BBC News, 2013). Not surprisingly in the ensuing deluge of material, however, evidence of the perils engendered by misleading rumours, lapses in judgement and outright disinformation attracted severe criticism. 'This is one of the most alarming social media events of our time,' media academic Siva Vaidhyanathan warned. 'We're really good at uploading images and unleashing amateurs, but we're not good with the social norms that would protect the innocent' (cited in Bensinger and Chang, 2013). Other critics complained of 'me-first journalism,' 'unsubstantiated amateur footage,' 'mass photo dumps,' or 'unreliable crowd-sourced material,' amongst other, more colourful objections. 'The chaos of breaking news is no longer something out of which coverage arises,' Poynter analyst Jason Fry (2013) observed, 'it's the coverage itself.' Pointed ethical questions arose regarding how some news organisations were striving to compete with their citizen media rivals to be first with a 'scoop' or fresh angle, including with respect to their 'ripping' of imagery without independent verification (the *New York Post* frontpage story 'Bag Men: Feds seek these two pictured at Boston Marathon' being one of the more notorious examples of this rush to judgement, the two young men depicted in the full-page photo being innocent bystanders).

In the eyes of some, however, the very legitimacy of citizen photo-reportage was morally problematic. 'You're not a journalist just because you have your smartphone in your pocket and can take pictures of someone who has just had their leg blown off and their life shattered' was one telling Facebook comment prompting debate in the blogosphere (cited in Geleff, 2013). Further criticisms revolved around assertions made about the callousness of individuals too busy taking images of victims to lend assistance, the prospect of media celebrity allegedly proving impossible to resist. Some worried about the emotional affectivity such disturbing imagery might engender amongst vulnerable publics, while others expressed concerns that such depictions of carnage and panic were fulfilling the perpetrators' narcissistic desire for notoriety, possibly even inviting 'copy-cat' responses as a result. Evidence in support of these and related claims was seldom cited beyond hearsay, nor did the individual participants' relative investment in self-identifying as journalists tend to be made apparent. In any case, such graphic imagery was firmly defended by others, perhaps most resolutely by those counterpoising it against visually sanitised treatments proffered by corporate media. 'Reporters have been normalizing the abnormal for so long that they've created well-worn catastrophe templates to convey their stories,' Jack Shafer (2013) of

Reuters argued, hence the importance of these amateurs—'instant Zapruders'—working alongside professionals to create and share 'unfiltered' news as a vivid alternative to the repetitive sameness of template-centred coverage.[2]

The issue of filtering proved contentious in further ways, perhaps most markedly with regard to certain perceived transgressions of photojournalism's normative limits—typically expressed in a subjunctive language of 'good taste,' 'public decency' or 'personal privacy'. A case in point was imagery widely regarded to be iconic—*The New York Times*'s Tim Rohan (2013) called it 'a searing symbol of the attacks' – showing spectator Jeff Bauman in a wheelchair being rushed from the scene, the best part of both legs ripped away. Shot from different angles by Charles Krupa of AP and Kelvin Ma for the Bloomberg Photo Service, respectively, this grisly depiction of Bauman's injuries posed awkward questions for news organisations. Several went with a cropped version to exclude the appearance of protruding bone matter from what was left of his legs. 'You did not need to see the rest of the picture,' *Times* photography editor Michele McNally insisted. 'The legs actually distracted you from seeing the intense look on his face, the ashen quality that suggested how much blood had been lost' (cited in Sullivan, 2013). Other news organisations imposed a black bar to conceal the more gruesome aspects, whilst still others employed digital pixilation for cautionary purposes.

No decision was immune from criticism. *The Atlantic*'s InFocus column initially posted one of the Bauman images without pixilation on its webpage, but fifteen minutes later changed tack: 'An earlier version of this gallery featured this photo with the graphic warning but without the image blurred,' a note to readers stated. 'We have since decided to blur the subject's face out of his respect for privacy.' Bob Cohn, digital editor for *The Atlantic*, explained: 'We thought it was such an honest and powerful representation of the tragic impact of the bombings. [However, he] obviously was in a very vulnerable situation. He was fully identifiable' (cited in Haughney, 2013). For those news organisations opting to pixilate Bauman's legs instead, it was because the graphic nature of the damage was deemed too upsetting to disclose. Freelance photographer Melissa Golden was amongst those expressing their concern about censorial implications. While conceding the gallery image in question was 'horrifying,' she nevertheless wondered: 'Since when do legitimate print journalism outfits modify photos like this? Run it or don't, but don't enact a double standard for Americans when we're totally cool running unadulterated photos of bombing victims from foreign lands' (cited in Murabayashi, 2013). For Ma (2013), writing on Bloomberg's blog afterward, these types of negative comments were understandable. 'But as a professional witness,' he countered, 'I don't know how else to show not only the evil of the world, but also the compassion and humanity that ultimately overcomes it.' Here he had in mind 'the actions of the first responders and volunteers who dove headfirst into the smoke to save so many lives on Monday.'

A related concern with the depiction of victims arose where the use of digital manipulation was concealed, thereby calling into question image integrity at the level of perceived truthfulness. One such instance occurred when a reader, sports designer Andy Neumann of Louisville, spotted a discrepancy between one of John Tlumacki's photographs of wounded people on the street (posted on the *Boston Globe*'s 'Big Picture' blog) and the version of it published by the *New York Daily News* on its front page the next day. Close scrutiny revealed that the image's record of a woman's broken leg had been carefully adjusted to obscure the broken, bloodied bone emerging above her ankle. 'Looks to me like somebody did a little doctoring of that photo to remove a bit of gore,' visual editor Charles Apple (2013) remarked on his blog, credited with calling attention to Neumann's acumen. 'If you can't stomach the gore, don't run the photo. Period.' Numerous commentators weighed into the emergent debate about the acceptable limits of Photoshop. The ethics code of the National Press Photographers Association was widely quoted, namely that: 'Editing should maintain the integrity of the photographic images' content and context.' Images are not to be manipulated 'in any way that can mislead viewers or misrepresent subjects.' Initially refusing to comment on its editorial decision-making, the *Daily News* eventually released a statement. 'The Daily News edited that photo out of sensitivity to the victims, the families and the survivors,' a spokesperson explained. 'There were far more gory photos that the paper chose not to run, and frankly I think the rest of the media should have been as sensitive as the Daily News' (cited in Pompeo, 2013).

Comments posted online by readers were typically forthright in their appraisal of the issues at stake. Amongst the tensions brought to the fore, it is worth noting the extent to which contending views regarding the verisimilitude of visual journalism revolved around public trust. So-called 'real news' was frequently perceived to be at risk, including where the misappropriation of the 'ordinary submissions by cellphone-wielding citizens' were concerned, by 'MSM' (mainstream media) advancing their own editorial interests. Implicit here is the apparent authenticity of the ordinary citizen's near-instant uploading of imagery via social media, effectively contrasted with the routines of institutional processing—intrinsically biased or ideologically motivated, in the eyes of some—enacted by their professional counterparts.

For those inclined to consider avowed commitments to objective reportage contrived or self-serving, spontaneous citizen engagement in unapologetically subjective newsmaking—and its impassioned distribution via retweets and shares—may well be deemed to provide a truer image of what is really happening on the ground. Moreover, in praising members of the public for gathering, interpreting and sharing visual evidence, particularly when so much of it proved distressing, is to recognise in such activities a reportorial role that invites further reflection about the professional's corresponding responsibilities during this type of crisis.

'It's haunting to be a journalist and have to cover it,' Tlumacki (2013b) of the *Boston Globe* observed. 'What newspapers and professional journalists need to realize, and the world has to realize, is that we are news photographers, not somebody out there with an iPhone and a camera, jumping over people to put images on YouTube.' Despite the ubiquity of citizen imagery, he insisted, the news photographer's specialist role remains vital. 'I'm so sick of citizen journalism, which kind of dilutes the real professionals' work. I am promoting real journalism, because I think that what we do is kind of unappreciated and slips into the background.' Fellow photojournalist Alex Garcia (2013a) of the *Chicago Tribune* evidently holds similar views regarding 'the need for professionals in this age of de-professionalization of the news industry.' When 'spectators with cameras' were fleeing the explosions, he maintained, it was the photojournalists who 'headed towards the madness.' This deep conviction of journalistic purpose, informed by 'professional instincts, training and mission,' produced a record that conveyed 'the horror, confusion, and fear of the moment' in a manner as calmly detached as it was publicly relevant.

Disputes over what counts as photojournalism—and who qualifies to be a photojournalist—are hardly new, of course, but there is little doubt the Boston crisis highlighted the extent to which photo editors found themselves relying upon imagery shot by non-professionals. The very amateurness of citizen imagery tempers normalised conventions of journalistic authority, its up-close affirmation of presence, 'I am here' and this is 'what it means to be there,' intimately intertwining time, space and place to claim an emotional, often poignant purchase. Managing this proliferation of imagery invited fresh thinking about how to best perform a curatorial role, one consistent with professional standards and procedures while, at the same time, benefiting from the news value associated with the raw, visceral immediacy of citizen witnessing. Photo editors scrambling to figure out the guiding imperatives of this role would be wary of the reputational risk for their news organisations posed by decisions hurriedly made under seemingly incessant pressure to push ahead of the competition. Safely ensuring a professional's ostensibly credible image was trustworthy, its captioning accurate, or its placement properly contextualised demanded close and methodical scrutiny, yet sifting through citizen documentation in search of deeper understanding was a curatorial challenge of an altogether different order. Moreover, where the imagery in question captured the explosions and their brutal aftermath, compassion for the victims further complicated editorial judgements about explanative significance. Selecting the most appropriate one to tell this violent story was a balancing act, Michael Days of the *Philadelphia Daily News* maintained. 'You want people to feel, you want people to feel a bit of the horror, you want them to feel a bit of the terror, without crossing the line that would make people turn away' (cited in Scott, 2013). Here it almost goes without saying, of course, that transgressions of this dividing line make it easier to discern, subtly disclosing tacit norms shaping what is permissible to see—and, it follows, to remember—as a shared experience in the fullness of time.

RECASTING ROLES AND RESPONSIBILITIES

'If I don't go to the action and shoot it, then who will?,' asked Boston photojournalist Michael Cummo, all too aware that when almost everyone else was running away from the scene of the bombings, people like him were racing toward it. 'You are human before a photographer but there is nothing you could have done to stop what happened,' Cummo's colleague Scott Eisen added. 'Your job as a journalist is to keep documenting it' (cited in Hamedy, 2013). In contrast with much of the 'accidental journalism' of spectators situated near the finishing line, professionals were knowingly putting themselves in harm's way in pursuit of images to help convey a story in all of its dreadful complexity. At the same time, however, the Boston crisis provides evidence to suggest that many of the ordinary individuals finding themselves on the scene felt a personal obligation to engage in their own form of citizen witnessing. While their relative investment in journalistic intent may have been hesitant or tentative, perhaps the compulsion to record and share a traumatic experience by connecting with distant others being a stronger motivation in many cases, time and again their self-reflective comments revealed a sincerity of purpose when quoted or broadcast in the news media coverage. The reportorial value of their contributions tended to attract grudging professional recognition at best, however, with examples of outright disparagement easily found in the blogosphere.

News organisations willing to revision their role anew, namely by making the most of this potential to forge collaborative relationships between professionals and their citizen counterparts, will secure new opportunities to reinvent photojournalism at a time of considerable scepticism about its future prospects. Which returns us to the plight of the *Chicago Sun-Times* photography staff discussed above, the closure of their department having taken place six weeks after the Boston crisis sharply underscoring why this reinvention of the craft is so important. The *Sun-Times*'s 'knee-jerk reaction' to financial pressures, as it was characterised by some critics, appears to be consistent with a growing pattern to 'outsource' photographic responsibilities in order to better ensure the viability of news organisations under threat of closure by anxious investors. 'It's not common, but it's not unprecedented either,' Kenny Irby of the Poynter Institute observed. 'This is part of an ongoing trend that has been happening for the last 10 years or so in American newsrooms, with the downsizing and devaluing of professional photojournalism' (cited in Marek, 2013). The price such organisations are paying is proving to be considerable, not least with regard to sustaining a reputation—or 'brand' in managerial discourse—based upon public trust to inspire loyalty amongst readers. 'While our reporters are doing the best they can to take photos with their iPhones and still trying to deliver quality stories, visually, the story has taken a big hit,' Beth Kramer of Chicago's Newspaper Guild told ABC News two months after the *Sun-Times* decision. Reporter Maureen O'Donnell echoed the point with regard to her and colleagues' newfound responsibility on the daily. 'Our photos just can't compare,' she confirmed. 'There's a soul and a life and a vibrancy that the photographers bring,' which is sorely missed (cited in Jordan, 2013).

Photojournalists help to connect news organisations to their audiences, reaffirming the terms of a social contract of visual reciprocity from one day to the next. Trained to skilfully convey a sense of emotion and intimacy with their images, Alex Garcia (2013c) of the *Chicago Tribune* maintains, they make this relationship vibrant and relevant. In John H. White's case, it was a 35-year career on the *Sun-Times* that came to an abrupt end, yet he made every effort to remain positive. 'There's no place in my heart for anger. I'm hurt, sure, I'm human, I'm disappointed, but I don't curse darkness, I light candles,' he said in an interview afterwards (cited in Riley and Hampton, 2013). Photojournalism was not about him but rather about helping others, he believed, particularly those struggling to cope with everyday hardships, such as in impoverished inner-city neighbourhoods. 'We're part of, and we cover, the heartbeat of humanity,' he continued. 'That's going to suffer' (cited in Ryssdal, 2013). Photographers' personal presence in these local communities, their informal contacts and networks giving shape and direction to *Sun-Times* coverage, is vitally important—and yet recurrently overlooked when managers focus their attention on cost structures quantified on financial balance sheets. 'It was as if they pushed a button and deleted a whole culture of photojournalism,' White surmised. 'Humanity is being robbed,' he added, 'by people with money on their minds' (cited in Irby, 2013).

To close, then, this chapter has aimed to highlight vantage points from which to assess several challenges confronting photojournalism as it evolves in contested circumstances. Contrary to certain pessimistic assessments about its relative viability, some news organisations are finding that the demand for visual storytelling through photography is intensifying—indeed, many newspapers are enlarging spaces for pictorial reportage, while online slideshows and galleries are frequently credited with being key drivers of web traffic. This is particularly apparent where breaking news is concerned, when ad hoc citizen contributions can effectively supplement—but not supplant—images secured by professionals, as the Boston crisis demonstrated so profoundly. Very few of citizen photojournalism's most passionate advocates, in my judgement at least, would contend that it is a replacement for the role of the professional. And yet, its merits must embolden photojournalists to speak with clarity and conviction about what this role means today, and to revisit questions regarding how it may be enhanced in the name of public service. We need to be alert to the prospects for refashioning photojournalism so as to create spaces for citizen witnessing while, at the same time, recognising how these opportunities will be conditioned by negotiated compromises wrought by institutional pressures and priorities. In seeking to move debates about how best to enliven photojournalism's future beyond the soaring rhetoric of advocates and critics alike, the importance of developing this collaborative, co-operative ethos of connectivity becomes evermore pressing.

NOTES

1. Equipping reporters with iPhones to step into the breech is hardly desirable in the eyes of editors, one can safely presume, yet likely to be rationalised as a regrettable necessity in such circumstances. In a reader's comment on *The New York Times* website, Frank Caramelli (2013) of Los Angeles weighed into the debate. 'As a video editor in a large market news department; I struggle daily with the idea that an I-phone is just as good as an expensive camera in the hands of a seasoned professional. It isn't. In addition, it also is NOT easier, faster, or more convenient. It takes me more time to download (limited) garbage off the internet than it does to have a good photographer shoot exactly what I need.' Alex Garcia (2013b), writing on his blog Assignment Chicago, observed: 'Most Sun-Times photojournalists I knew, because of their decades of experience, were unsung journalists more than photographers. They knew how things worked and what made communities tick. They found stories and passed them on. They helped to shape stories, correct misperceptions and convey understandings that have deep resonance with readers. I am sure that many of their reporter colleagues would attest to this. I would also bet that some reporters will continue to call them, hoping to get a little help here and there.'
2. The phrase 'instant Zapruders' refers to Abraham Zapruder, whose 8mm 'home movie' footage of the assassination of US President John F. Kennedy in 1963 is frequently cited as an early exemplar of citizen journalism (for a discussion, see Allan 2013a). Elsewhere I have contrasted the photo-reportage of the marathon bombings with a further 'terror attack' taking place the following month, namely the killing of a British soldier in Woolwich, southeast London (see Allan, 2014).

REFERENCES

AFP (2013). 'iPhones replace US paper's photo staff,' Inquirer Technology.net, 6 June.

Allan, S. (2013a) *Citizen witnessing: Revisioning journalism in times of crisis*. Cambridge: Polity Press.

Allan, S. (2013b) 'Blurring boundaries: Professional and citizen photojournalism in a digital age,' in M. Lister (ed) *The photographic image in digital culture*, second edition. London and New York: Routledge, 183–200.

Allan, S. (2014). 'Witnessing in crisis: Photo-reportage of terror attacks in Boston and London,' *Media, War & Conflict*, 7(2); in press.

Apple, C. (2013). 'I hate to make an accusation here, but…,' Apple.Copydesk.org, 16 April.

BBC News (2013). 'Jamie Bartlett: Social media is "big rumour mill",' BBC News.co.uk, 22 April.

Bensinger, K., and Chang, A. (2013). 'Boston bombings: Social media spirals out of control,' *The Los Angeles Times*, 20 April.

Bhattacharjee, R. (2013). 'Meet Dan Lampariello, the college student whose Boston Marathon photos went viral,' News.MSN.com, 16 April.
Caramelli, F. (2013). Reader comment. Taking note: The editorial page editor's blog. *The New York Times*, 5 June.
Crabbe, L. (2013). 'Boston Marathon snapshots take on new meaning,' Connect.DPReview.com, 18 April.
Deacon, G. (2013). 'Photojournalism still matters, thank you.' Transcript of interview with A. Garcia. Assignment Chicago, 4 July.
Dinges, T. (2013). 'Boston explosion eyewitness describes "carpet of glass," a dozen bodies,' *The Star-Ledger*, 15 April.
Downes, L. (2013). 'Do newspapers need photographers?' Taking note: The editorial page editor's blog. *The New York Times*, 31 May.
Effron, L. (2013). 'Mystery "Man on the roof" sparks Boston Marathon chatter,' *Technology Review*, 16 April.
Fry, J. (2013). 'Boston explosions a reminder of how breaking news reporting is changing,' Poynter.org, 16 April.
Garcia, A. (2013a). 'Tragedy and the role of professional photojournalists,' Assignment Chicago, 16 April.
Garcia, A. (2013b). 'The Idiocy of Eliminating a Photo Staff,' Assignment Chicago, 30 May.
Garcia, A. (2013c). '10 Responses to the Sun-Times debacle,' Assignment Chicago, 4 June.
Geleff, A. (2013). 'Citizen journalism and social media in 2013: Is there a "too much" or is it just what we need?,' ByteNow.net, 17 April.
Golab, A. (2013). 'Bob Kotalik, former chief photographer for Chicago Sun-Times, dies at 87,' *Chicago Sun-Times*, 30 May.
Greenwald, G. (2013). 'The Boston bombing produces familiar and revealing reactions,' *The Guardian*, 16 April.
Hamedy, S. (2013). 'Boston University photojournalists reflect on marathon mayhem,' NPPA.org, 26 April.
Haughney, C. (2013). 'News media weigh use of photos of carnage,' *The New York Times*, 17 April.
Irby, K. (2013). 'John White on Sun-Times layoffs,' Poynter, 31 May.
Jarvis, J. (2013). 'To the dauntless lensmen,' BuzzMachine, 31 May.
Jordan, K. (2013). 'Iconic images on display as Sun-Times photographers protest layoffs,' ABC 7 News, WLS-TV/DT, 30 July.
Liu, S.B., Palen, L., Sutton, J., Hughes, A.L., and Vieweg, S. (2009). 'Citizen photojournalism during crisis events,' in S. Allan and E. Thorsen (eds) *Citizen journalism: Global perspectives*, New York: Peter Lang, 43–63.
Lydersen, K. (2013). 'Why a picture is worth a thousand words: Laid-off photojournalists defend their field,' *In These Times*, 8 July.
Ma, K. (2013). Bloomberg photographer captures image seen around the globe. Blog.Bloomberg.com, 19 April.
Marek, L. (2013). 'Chicago Sun-Times cuts entire photography staff,' Crain's Chicago Business.com, 30 May.
Mitchell, D. (2013). 'Chicago Sun-Times fires its entire photo staff,' *Fortune*, 31 May.
Murabayashi, A. (2013). 'A blurry double standard? A photo from the Boston Marathon bombing,' PhotoShelter.com, 16 April.

Newman, C. (2013). 'Memo to Sun-Times editorial staff from Managing Editor Craig Newman,' posted on the Facebook page of Robert Feder, 30 May.
NPR (2013). 'Eyewitness, Special series: Explosions at Boston Marathon,' NPR.org, 15 April.
Pompeo, J. (2013). '*Daily News* doctored front-page photo from Boston bombing,' CapitalNewYork.com, 17 April.
Riley, M., and Hampton, I. (2013). 'Fired photographers picket Sun-Times building,' *NBC Chicago*, 6 June.
Rohan, T. (2013). 'Beyond the finish line,' *The New York Times*, 7 July.
Ryssdal, K. (2013). 'Photojournalist John H. White on layoffs, 35 years at Chicago Sun-Times,' Marketplace.org, 4 June.
Scott, M. (2013). 'Flood of graphic images after Boston blasts raises concerns,' Newsworks.org, 16 April.
Shafer, J. (2013). 'Terror and the template of disaster journalism,' Reuters.com, 15 April.
Smith, S. (2013). 'Cost over quality: Chicago Sun-Times fires its photo staff, and journalism's death spiral continues,' Scholars and Rogues.com, 31 May.
Sullivan, M. M. (2013). 'A model of restraint in the race for news,' *The New York Times*, 20 April.
Tlumacki, J. (2013a) 'Tragedy in Boston: One photographer's eyewitness account,' LightBox, *The New York Times*, 15 April.
Tlumacki, J. (2013b) Interview with K. Irby, Poynter.org, 22 April.

CHAPTER TWELVE

Citizen Journalism, Sharing, AND THE Ethics OF Visibility

GRAHAM MEIKLE

At 2.49pm local time on Monday 15 April 2013, two home-made bombs improvised from pressure cookers were detonated near the finish line of the Boston Marathon—a major annual US public sporting event, which attracts tens of thousands of spectators. Three people were killed in the explosions, including an eight-year-old boy, and hundreds more were injured. For the next four days, under very intense public and media scrutiny, the police and FBI hunted the suspected perpetrators, who were identified on Friday 19 April as brothers Tamerlan (26) and Dzhokhar Tsarnaev (19). The investigation reached what seemed to be a climax in the shooting of a police officer and the subsequent shooting dead of Tamerlan Tsarnaev in an exchange of gunfire with police in the early hours of Friday 19 April. But with one of the brothers still at large, events then moved to a bizarre plateau when Boston was shut down for the full Friday—the public transport system, businesses, schools, colleges and shops were closed, as people obeyed instructions to stay at home until the eventual apprehension on Friday evening of Dzhokhar Tsarnaev. Many people shared in the manhunt in real time by following the Boston police department scanner and sharing responses to its updates through social media.

The Boston Marathon bombing and subsequent manhunt took place in an always-on news environment characterised by rolling news channels, push notifications, and the non-stop update service that is Twitter. There is a spatial dimension to this environment, characterised above all by the imperative for reporters to be present at a scene. But its dominant dimension is temporal—speed is the perceived cardinal virtue, as news accelerates from telling us what has happened, through telling us what is happening now, to speculating about what might be about to happen next. The Boston events exposed some appalling shortcomings in this established breaking-news paradigm, recalling Elihu Katz's criticism of rolling news as news that 'almost wants to be wrong' (1992: 9). Some news organisations reported the story cautiously, with NBC's coverage attracting much acclaim for its accuracy and restraint. But others presented the story in ways which revealed the limitations of a live news approach in which being first counts more than being right.

Very shortly after the bombing, the *New York Post* announced that a Saudi suspect had been 'caught' and was 'under guard' in a Boston hospital, none of which was correct. On Wednesday 17 April, CNN's John King and Wolf Blitzer told their live audience that a suspect had been arrested, King using the phrases 'a dark-skinned male' and 'we got him', and the network repeatedly boosting their own 'exclusive reporting' until investigators stepped in to deny the report as baseless ('it was exclusive', Jon Stewart would later declare on *The Daily Show,* 'because it was completely fucking wrong'). The Associated Press also tweeted at 7.02pm on that day that a suspect was in custody and expected in court—an error retweeted by thousands of others to an unknowable cumulative audience. The next morning, the *New York Post* came back for another go, running a front-page picture of two entirely innocent individuals with the headline 'BAG MEN: Feds seek these two pictured at Boston Marathon'. The *Post* defended this potentially defamatory labelling by saying the image had been 'distributed' by the FBI—but so had other images of innocent bystanders, to many other media organisations, which did not run them on page one.

If the established news media struggled to cope with this story, it would be heartening to claim that citizen journalists and networked non-professionals had done better. But that wouldn't be true. The Boston bombing case also revealed the limitations of a crowd-sourced citizen journalism in which networked individuals come together on social media platforms to share and make visible ideas. Pierre Lévy's concept of 'collective intelligence' is often invoked in discussions of networked collaboration: Levy observes that 'No one knows everything, everyone knows something' (1997: 13–14). But the limitations of this for real-time citizen journalism are exposed by the Boston bombing case—*sometimes no one knows anything.*

This chapter discusses the use of the social media platform Reddit by networks of individuals who attempted to crowd-source the identities of possible bombing suspects by sharing images and speculation. The chapter focuses on three key aspects of this. First, it situates these events within the frame of citizen journalism. Second, the chapter considers the centrality of sharing to social media and its uses for non-professional journalism and related forms of collaborative information provision. And third, it argues that such uses of social media for citizen journalism reveal the need for an ethics of visibility.

REDDIT AND *FINDBOSTONBOMBERS*

In response to the bombing, thousands of users of social media platforms began sharing photos of crowds of spectators near the marathon finish line before and after the explosions, in an effort to isolate images of the killers. Images from Flickr were redistributed and edited with graphics programmes to paint red circles around the heads of bystanders who variously had backpacks large enough to have contained a pressure cooker, or who were running away from the explosion, or, in the case of the individual who became known online as Blue Robe Guy, were wearing an oversized fleece; some of these identifications appeared to rest on nothing but crude racial profiling by uninformed amateurs. These edited images were then in turn shared across Facebook, across Twitter, and on 4Chan, where they were mixed with flakes of information and misinformation shared by users listening to the Boston police radio scanner. Rumours spread that the police had named two suspects over the scanner, one of whom was a university student who had previously been reported as missing (Kang 2013, Madrigal 2013), and who would later be found dead in circumstances entirely unrelated to the marathon bombings. On Thursday 18 April, investigators released pictures of the Tsarnaevs, in part to reduce the impact of misidentifications through social media (Montgomery, Horwitz & Fisher 2013). This chapter focuses on such activity within one forum on Reddit, *findbostonbombers,* http://www.reddit.com/r/findbostonbombers.

Reddit is a social news platform, which brands itself as 'the front page of the internet'. One 2013 study by a reputable organisation claimed that 6% of all online adults in the US use Reddit (Duggan & Smith 2013). Users can share links, images or text posts to which others can respond by commenting and by offering a single up or down vote; the cumulative score of these votes determines the prominence of each post within the site. As with precursor sites such as Slashdot or Digg, Reddit combines the affordances of user-generated or curated material with community voting on the interest or importance of that material. Anyone can create a free account under any username, and the site discourages the posting of personal information or links to identifiable non-famous individuals such as their Facebook profiles. The platform is divided into thousands of smaller forums or communities called subreddits, each moderated by its creator or by other volunteers independent of Reddit the company.

The subreddit *findbostonbombers* was created on 17 April, two days after the bombings, by a Redditor using the name 'oops777', who was to later delete that account after giving an interview to the *Atlantic* (Abad-Santos 2013) and participating in a Reddit AMA (Ask Me Anything) Q and A session. By 23 April, the entire forum had been removed from the site, placed behind a page that denies access. However, a substantial sample is still viewable through the internet Archive's Wayback Machine, which captures 'snapshot' copies of significant web pages for archiving (http://archive.org/web/web.php). For the crucial days of Thursday 18 and Friday 19 April there are three and ten snapshots respectively, each of which offers an archive of the top 25 posts to the subreddit at a given time, along with comments and stats on user numbers. This captures essential dimensions of the subreddit's activity that week, although we should note that some of the content had been deleted by those who had posted it before the archival snapshots could capture it, and that the subreddit's moderators also intervened to delete posts which misidentified a missing student as a bombing suspect. There were good reasons for the eventual decision to remove the whole subreddit from public view: some posts appear likely to have been defamatory, others likely to have been distressing to identifiable individuals or to those who know them. So while this chapter quotes some posts from the archived sample of the subreddit, no further usernames are given below.

While contributors to the subreddit at times appeared to imagine they were having a small private discussion, there were actually thousands of people on the forum at any one time (still more than 7500 shortly before midnight on the Friday, for example, by which time Tamerlan Tsarnaev was already dead and Dzhokhar in custody), and many of those were further distributing material from the forum through other platforms such as Twitter or 4Chan; still others, of course, were writing articles about the discussion, which was reported in a large number of established news media outlets, both across the US and around the world, as part of their own coverage of the hunt for the suspected bombers.

FINDBOSTONBOMBERS AND THE LIMITS OF CITIZEN JOURNALISM

The phrase *citizen journalism* is at once obvious (journalism is always bound up with citizenship) and shocking (*anyone* can do this now?). It's a phrase which tries to capture in just two words a complex of shifting and developing capacities and expectations. The emerging capacities of networked digital media bring new technological affordances, social opportunities and cultural possibilities; these are met by the upturned expectations brought by eroding revenues and investment, vanishing readerships, new kinds of competition for attention, and disintegrating trust in journalism. Changing technological capacities for audiences, journalists and news organisations alike bring altered expectations about voice and participation, feedback and response, access and ubiquity, instantaneity and visibility. Changing industrial capacities, by no means all changes for the better, involve established news media trying to do more with less, and trying to invent, embrace, enhance or rip off newer innovations at the same time as expecting more from less investment and fewer staff. The convergent media environment brings new social and cultural capacities for news, built around networked connectivity, and emerging expectations of news as a networked set of relations rather than a hierarchical one. The news doesn't just talk to us, but we can now talk back; and perhaps more importantly, we are also now all the more likely to talk to each other about what we're hearing. The news—always too important to be left to the news media alone—is now something that its users can *do* as well as watch, listen to or read.

In what sense was the *findbostonbombers* Reddit activity citizen journalism? After all, the contributors were not writing a news story. They were not collaborating with a particular news organisation or outlet. They were not preparing a report for publication. And a substantial number of the comments on many posts were trolling or snark. Indeed, much of the discussion in the forum turned on the question of whether or not it was itself part of the media: recurring themes in the subreddit included not only this question, but also those of the ethical dimensions to public discussion of identifiable people, not least in an environment in which established news media may regard social media interactions as source material.

The subreddit's description states that:

> This is nothing more than one single place for people to compile, analyse, and discuss images, links, and thoughts about the Boston Bombing

and the page explicitly tries to distance the subreddit from journalism:

> IMPORTANT r/FindBostonBombers is a *discussion forum*, not a journalistic media outlet. We do not strive, nor pretend, to release journalist-quality content for the sake of informing the public.

But this disclaimer was not an accurate account of how the discussion was being used, even if it were an accurate description of its creator's expectations for the forum. The subreddit's users were responding to and extending a major news story (and in turn themselves became a part of the story). The participants were trying to identify a potential killer and in this were contributing to the story by generating original research—it just turned out to be hopeless, worthless, and to some extent harmful research.

One of the most highly rated posts is one from the day of the subreddit's creation and posted by its creator with the title:

> Media Outlets, please stop making the images of potential suspects go viral, then blaming this small subreddit for it. And read the rules we've imposed before calling us "vigilantes".

The post claims:

> Until the media got involved, none of the images were going anywhere but to the FBI.

The rules referred to above are listed prominently in a sidebar:

> 1) We do not condone vigilante justice.
> 2) DO NOT POST ANY PERSONAL INFORMATION.
> 3) Any racism will not be tolerated.
> 4) Theories are welcome, but make sure you fact check your sources.
> 5) Remember, we are only a subreddit. We must remember where helping ends and the job of professionals begins.
> 6) Do not make any images viral. Limit reposting images outside of this sub.
> 7) Finally keep in mind that most or all of the "suspects" being discussed are, in all likelihood, innocent people and that they should be treated as innocent until they are proven guilty.

But each of these rules is flouted at every turn, and as the first comment points out in relation to the claim that the images were not going anywhere until 'the media' became involved:

> You have to admit, they're actually going everywhere.

And as another adds:

> If you want people to stop reporting on this you should shut it down. And you should shut it down because it's a terrible idea.

In contrast, some other commenters defend the subreddit by attacking the more established news media:

> The media is scared because they are behind us. They want to attack our credibility, because we are undermining theirs. Also, we are doing what they are doing better, faster, and for free. How can CNN expect to make a profit if we will do their job for them better, faster, and for free?

But not everyone involved is prepared to accept this distinction between 'the media' and their own large communication platform, 'the front page of the internet':

> Reddit *is* the media.

And:

> Uh—like it or not, this IS a media outlet

And:

> Reddit: Now no better than CNN.

In response to such criticism, the subreddit's creator weighs in to argue that:

> If anything we're trying to clear the names of the people who the mainstream media just found images of and made go viral.

Some of the longest threads in this subreddit return again and again to tensions about whether or not what the participants are doing should be considered journalism, and whether or not the space in which they are doing it is part of the media. Braun and Gillespie capture something of this tension in their analysis of the difficulties of integrating the affordances of social media into the websites of established news organizations:

> … news organizations now finding that part of their mission includes hosting an unruly user community that does not always honor the norms of journalism; and media platforms and social networks now finding that the user-generated content being shared is often much like news, some of which violates their established content policies. (Braun & Gillespie 2011: 385).

This characterisation of an unruly community not honouring the rules of journalism can serve as a very good description of what went on in the Reddit forum. In a networked digital media environment characterised by convergence—of technologies, industries, texts, users, modes of communication—such invisible moments of shared response are made visible to new kinds of networks. A platform such as Reddit allows new kinds of ad hoc, shifting coalitions to come together to share their responses to a news event with others whom the platform makes visible to them and to whom they are made visible in turn. It allows for particular modes of interaction built around *sharing*.

SHARING THE NEWS

News is social. It is a collaborative process of making meaning from events. 'The first typical reaction of an individual to the news', observed Robert Park as long ago as 1940, 'is likely to be a desire to repeat it to someone' (1967: 42). While Park was writing in a very different news environment, his insight gains renewed force with the daily affordances of networked digital media. Social media platforms enable their users to enter into flows of public communication, and to add a phatic dimension to the public quality of news discourse (Miller, 2008; Crawford, 2011). News is not just something that is *distributed to* its audiences by news organisations—it is also something that is *shared by* those audiences with others.

The word *share* is at the heart of social media. It appears as an imperative verb under every Facebook post, every YouTube video, every story on *The New York Times* website. Social media tools from Twitter to the BBC iPlayer highlight possibilities of connection and sharing. Instead of simply watching, listening or reading, we share ideas and images, information and entertainment, stories and songs with self-selected networks of friends, contacts and our own personal audiences. From Spotify playlists to Tumblr blogs, from BitTorrent to the *Guardian* opinion page, we are encouraged to communicate, cooperate, collaborate and *share*. Sharing is what is social about social media.

These affordances of sharing extend and augment long-established principles of public communication (see, for example, Raymond Williams's short essay on *communication* in his *Keywords*, 1983). James Carey's discussion of what he called the *ritual view* of communication includes this paragraph, which emphasizes the common roots of *communication*, *community* and *communion*:

> The ritual view of communication, though a minor thread in our national thought, is by far the older of those views—old enough in fact for dictionaries to list it under "Archaic." In a ritual definition, communication is linked to terms such as "sharing," "participation," "association," "fellowship," and "the possession of a common faith." This definition exploits the ancient identity and common roots of the terms "commonness," "communion," "community," and "communication." A ritual view of communication is directed not towards the extension of messages in space but toward the maintenance of society in time; not the act of imparting information but the representation of shared beliefs." (1989: 18).

The word *share* operates discursively in many different ways. To *share* can be to separate and divide, or to copy and multiply. On social media, it can variously be about sharing images, links and ideas, or meanings, opinions and emotions. Sharing may be phatic communication, a moment of ritual or communion, or the performance of versions of self. It may be to affirm someone, or to take something without paying. All of these may be visible, as others share in our actions in the network. And, of course, what we share may be commodified by the sharing industry of social media firms. All of this demands a new ethics of visibility.

TOWARDS AN ETHICS OF VISIBILITY

The identification and exposure of named individuals, and the circulation of their photos in defamatory contexts across dispersed media networks, point to the need for an ethics of visibility in the social media environment. Questions of visibility are central to thinking about contemporary developments in networked digital media, not least the many manifestations of citizen journalism. While such questions are most often framed in terms of privacy, this is too narrow a frame through which to view what are a rather larger set of concerns and practices. The case of *findbostonbombers* underscores the fact that we need to think not only about privacy, but also about exposure and display, connection and networking, community and communion. The frame of *visibility* captures more of these than does privacy, and brings with it the need for an *ethics* of visibility in relation to social media, witnessing, sharing, and all the other elements that orbit the concept of citizen journalism. Those interactions on such social networks which can be thought of as citizen journalism are not intended to be private in the first place. They're intended to be *shared.* The question is rather with whom we imagine we are sharing, and how the balance between rights and responsibilities is calculated.

The circulation and discussion of misidentified photos, across the networked digital media environment from Reddit to the *New York Post,* highlight the need for a broader and deeper debate about visibility in relation to these converging media platforms. Networked digital media bring with them new kinds of visibility, new opportunities and requirements to monitor and be monitored, to perform and display, and to connect with others who are newly visible to us and to whom we are ourselves in turn made visible. Such affordances are fundamental to the development of the various forms of mediated sharing and collaboration that we class as citizen journalism. But as John Thompson points out, 'mediated visibility is a double-edged sword' (1995: 41).

Some contributors to *findbostonbombers* were clearly mindful of the risks of making others visible. As one post notes:

> You can't mark people as terrorism suspects and then get upset that other media pick it up because of your silly rule about it. This is the internet, surprisingly both good and bad information can go viral.

And as another commenter observes:

> everything you are doing here, including all the images and the discussion/interpretation of the details in those images… is being held on a public forum on a website with billions of pageviews. Everything here is by definition "going everywhere" and moreover, it's going everywhere instantly via google search and various image/post auto aggregators. years from now all this stuff will still be out there, whether debunked or not. It's completely meaningless to have some sidebar disclaimer if you are going to hold discussions and throw accusations in a place as public as this.

A third points to a particular irony of visibility:

> we bemoan the rise of the surveillance state, but then when something like this happens, everyone's more than happy to post pictures all over the internet, drawing big red circles around anyone carrying a backpack

One user posted an image of 'Blue Robe Guy', whose key transgression appeared to be wearing a jacket some people didn't like much, under the headline 'Popped up on my facebook newsfeed… this can't happen'; it showed several images of the individual in the blue jacket together with images of a post-explosion backpack and the caption 'Can you identify this man?', which was being circulated across Facebook.

In the context of this kind of activity on this kind of platform, what might an ethics of visibility mean? In individual daily networked digital life, an ethics of visibility would include taking conscious account of the ways in which we make others visible—when we like, when we retweet, when we tag, when we screengrab, when we share. It would include taking conscious account of other people's rights to manage their own visibility as well as our rights to share and comment. It would include asking which invisible audiences are imagined and unimagined in an interaction online, and considering the connections between unsuspecting others that we make each time we link and tag, like and share, and whom we are making visible to whom through these networked interactions.

In collective social life, including news, an ethics of visibility would include a more careful consideration of who and what is valid for exposure. Neither Reddit's Blue Robe Guy nor the *New York Post*'s Bag Men should have been subjected to such intense levels of enforced visibility. While Rupert Murdoch tweeted his support for the *Post,* Reddit the company issued a public apology for what had happened. It addressed their own policies about visibility:

> A few years ago, reddit enacted a policy to not allow personal information on the site. This was because "let's find out who this is" events frequently result in witch hunts, often incorrectly identifying innocent suspects and disrupting or ruining their lives. We hoped that the crowdsourced search for new information would not spark exactly this type of witch hunt. We were wrong (erik [hueypriest], 2013)

The apology may have been just an attempt at damage limitation, but it is nonetheless encouraging to see a major social media platform publicly address an ethics of visibility, in an environment in which those very social media platforms themselves so often act in a way that pushes their own users towards ever greater visibility and disclosure for commercial exploitation.

CONCLUSION

News is no longer something distributed only by news organisations but is now also redistributed and circulated by its readers, viewers and users, shared and discussed among self-selected networks of friends and contacts. These practices of sharing should be understood as part of what we think of as citizen journalism. The convergence of the professional and the non-professional is as much about networking and connection, about sharing and its resulting visibility, as it is about public writing.

The *findbostonbombers* forum, and parallel activity elsewhere on Reddit, on Twitter, on Facebook and 4Chan, shone a stark light on the contours of mediated sharing in relation to news. At its best, citizen journalism can indeed extend and augment, complement and counterpoint, the practices of the established news media, adding depth, breadth and longevity to discussions otherwise curtailed by the imperatives of news organisations. At its best, citizen journalism is a much needed sharing around of the licence to create non-fiction drama, and can bring both spatial authority (through the actual presence of witnesses) and temporal authority (through the real-time immediacy made possible by networked digital media).

But as *findbostonbombers* shows, networked collaboration through social media can also be as shoddy and corrosive as the worst of the established news media (for which let the earlier examples from CNN and the *New York Post* serve as metonyms). It can become harder to distinguish signal from noise, and there may not be any necessary extra value in the extra labour involved. So this networked digital news environment does not eliminate the need for journalists. Rather, it gives the Fourth Estate role of the news media a renewed applicability. Professional journalists—acting professionally—can analyse and sift raw material, can test evidence and redact details that may endanger named individuals, can offer context to help the reader interpret the material, can access high-status sources of official information, and can shape the data into stories, reports and commentaries that make sense of the material for audiences who lack, of course, the time and expertise to process specialised documents and intelligence for themselves—something that was made apparent in the attempts by visitors to the *findbostonbombers* subreddit to undertake DIY forensic work armed only with photos from Flickr and 4Chan. The paradox of citizen journalism is that rather than rendering professional journalists obsolete, it makes them ever more necessary.

REFERENCES

Abad-Santos, Alexander (2013). 'Reddit's "Find Boston Bombers" Founder Says "It Was a Disaster" but "Incredible"', *The Atlantic Wire*, 22 April, http://www.theatlanticwire.com/national/2013/04/reddit-find-boston-bombers-founder-interview/64455, accessed 26 May 2013.

Braun, Joshua, and Gillespie, Tarleton (2011). 'Hosting the public discourse, hosting the public', *Journalism Practice*, vol. 5, no. 4, pp. 383–98.

Carey, James (1989). *Communication as Culture*, New York: Routledge.

Crawford, Kate (2011). 'News to Me: Twitter and the Personal Networking of News' in Graham Meikle and Guy Redden (eds) *News Online: Transformations and Continuities*, Basingstoke: Palgrave Macmillan, pp. 115–31.

Duggan, Maeve, and Smith, Aaron (2013). '6% of Online Adults are Reddit Users', *Pew Internet Project*, 3 July, http://pewinternet.org/Reports/2013/reddit.aspx, accessed 23 July 2013.

erik [hueypriest] (2013). 'Reflections on the Recent Boston Crisis', *blog.reddit*, 22 April, http://blog.reddit.com/2013/04/reflections-on-recent-boston-crisis.html, accessed 13 May 2013.

Kang, Jay Caspian (2013). 'Should Reddit Be Blamed for the Spreading of a Smear?', *The New York Times*, 25 July, http://www.nytimes.com/2013/07/28/magazine/should-reddit-be-blamed-for-the-spreading-of-a-smear.html, accessed 31 July 2013.

Katz, Elihu (1992). 'The End of Journalism? Notes on Watching the War,' *Journal of Communication*, vol. 42, no, 3, pp. 5–13.

Lévy, Pierre (1997). *Collective Intelligence*, Cambridge, MA: Perseus Books.

Madrigal, Alexis C. (2013). 'It Wasn't Sunil Tripathi: The Anatomy of a Misinformation Disaster', *The Atlantic,* 19 April, http://www.theatlantic.com/technology/archive/2013/04/it-wasnt-sunil-tripathi-the-anatomy-of-a-misinformation-disaster/275155, accessed 13 May 2013.
Miller, Vincent (2008). 'New Media, Networking and Phatic Culture', Convergence, vol. 14, no. 4, pp. 387–400.
Montgomery, David, Horwitz, Sari, and Fisher Marc (2013). 'Police, Citizens and Technology Factor into Boston Bombing Probe', *Washington Post*, 21 April, http://www.washingtonpost.com/world/national-security/inside-the-investigation-of-the-boston-marathon-bombing/2013/04/20/19d8c322-a8ff-11e2-b029-8fb7e977ef71_print.html, accessed 13 May 2013.
Park, Robert E. (1967). [1940] 'News as a Form of Knowledge' in his *On Social Control and Collective Behavior*, (ed. Ralph H. Turner), Chicago: University of Chicago Press, pp. 33–52.
Thompson, John B. (1995). *The Media and Modernity,* Cambridge: Polity.
Williams, Raymond (1983). *Keywords: A Vocabulary of Culture and Society* (revised edition), London: Fontana.

SECTION THREE

Globalising Cultures of Citizen Journalism

CHAPTER THIRTEEN

Citizen Journalism, Development AND Social Change: Hype AND Hope

SILVIO WAISBORD

CONCEPTS AND QUESTIONS

Let's clarify the meaning of citizen journalism, development, and social change before we analyze their linkages.

Citizen journalism (CJ) is a relatively straightforward concept. CJ refers to "random acts of journalism" practiced by ordinary people. Stuart Allan (2013, p. 9) defines it as "first-person reportage in which ordinary individuals temporarily adopt the role of journalists to participate in newsmaking, often spontaneously during a time of crisis, accident, tragedy or disaster when they happen to be present on the scene." In recent years, the concept gained popularity with the spread of easy-to-access digital platforms that allow anyone to report and disseminate information and opinion. CJ is different from "industrial journalism" (Anderson, Bell & Shirky 2012) produced by news bureaucracies—it doesn't follow standard rules, isn't produced by paid labor, and isn't housed in news companies. It's non-professional journalism, understanding professionalism as the ability of an occupation to control a certain social jurisdiction that provides a specific social service (Waisbord 2013). The rise of digital CJ signals cracks in the control that industrial journalism had in the provision of information for mass publics.

Unlike CJ, neither development nor social change can be defined succinctly. "Development" remains the subject of long discussions. As a concept identified with political democracy, market economy, and social indicators, it has been a central category in the analysis of the evolution of modern societies as well as a normative horizon of "human progress." During the postwar years, governments, academics and private foundations embraced the narrative of "development" to characterize world societies based on a particular construction of the Western experience and offer solutions to a host of global problems such as poverty, economic growth, authoritarianism, health and education. As a modernist narrative, "development" was premised on a teleological, optimist vision that anticipated a bright future ahead for the "underdeveloped" world if it embraced the Western historical path and policy recommendations. "Developed," then, became associated with "Western progress," industrial capitalism, liberal democracy, and rising social conditions.

Critical scholars have questioned this line of argument on various grounds (Escobar 1995). In their mind, "developmentalism" offers an ethnocentric perspective that reflects a profoundly misguided and narrow interpretation of human history. By elevating the "Western experience" as the model of "good development," it ignores the complex historical linkages between the North and the global South. "Development" was the product of a long history of colonialism and exploitation of social and natural resources. It is also premised on an externally driven model of human improvement that posits the West as the engine of global change and the repository of appropriate resources and knowledge. In doing so, it sidelines alternative models of social development that reflect local knowledge, experience, and aspirations.

This line of argument pushed the debate to redefine "development" in terms of emphasizing local knowledge and voice, rights, social reforms, and collective action. Although this critique was responsible for important changes in international aid, such as the adoption of participation as central to human improvement (Cooke and Kothari 2001), the discourse of "development" still undergrids global aid and international cooperation. It remains institutionalized in academia ("development studies"), government (departments of "social" and "rural" development), and international aid ("development" agencies).

"Post-development" arguments prompted calls to find substitute concepts unburdened by the ideological baggage of developmentalism. The notion of "social change" has emerged as a plausible alternative endorsed by activists, non-government organizations, scholars, donors, and governments (Servaes 2008). Social change, however, is a semantically diffused concept that lacks clear definitions. It is loosely used to refer to transformations in norms, attitudes, socio-economic structures, policies, beliefs, information, power, and behaviors. Unlike "development," it lacks specific normative implications. It has a positive resonance yet its meanings are in the eye of the beholder. Someone's necessary "social change" is somebody else's unimportant change. Although social change is commonly associated with addressing a range of social ills, it is not obvious which are those problems, or if they fit a broad vision about the "good society." "The social question," to use the idea coined by critics of the social consequences of the industrial revolution in the 19th century, is fragmented in a world of social problems: poverty, discrimination, gender inequity, access to health and education services, public safety, food insecurity, water scarcity, labor slavery, poor sanitation, climate change and so on.

Rather than adopting a "thematic" understanding of social change, my sympathy lies with the notion that social change is about social justice—redressing power inequalities, strengthening individual and social rights, improving and expanding opportunities, particularly for socially excluded populations. Social change shouldn't be viewed in terms of conventional divisions between problems and solutions of developed and developing countries. Certainly, social problems may have different significance and urgency across the world. For example, access to safe water and sanitation, infectious diseases, and child and maternal mortality are significantly bigger problems in the global South than in the West. Also, societies have different institutional and economic resources as well as political opportunities and obstacles to address various social problems. Despite these differences, it is possible to find valuable insights across the world in the way societies identify social problems and implement solutions through collective actions.

Social change shouldn't be confused with its cognate "political change" which refers to transformations in political regimes, structures, actions, and dynamics. This has been the analytical focus of recent discussions about protest movements that have used CJ for information and organization for political action (Earl & Kimport 2011). Soldiers who distribute classified information and pictures that challenge official narratives, citizens who produce visual documents of violence and natural disasters, and activists' reporting about uprisings and protests in the Middle East, the Occupy movement, and the *Indignados* illustrate the contributions of CJ to political change and activism. In this chapter, the focus is not on political change, but on the contributions of CJ to giving visibility to and addressing social problems.

The range of topics discussed in this book demonstrates that there is no single question about CJ. Instead, scholars have examined various issues, such as its relationship with mainstream/legacy journalism, political uses, innovations in news production and distribution, and continuities with journalistic traditions and forms of public expression.

How should we think about the relationship between CJ and development/social change? At first glance, CJ can be correctly considered a manifestation of positive social change. CJ seemingly meets a condition for social change: the expansion of opportunities for citizens' voices. Long before the arrival of Web 2.0, participatory communication theorists emphasized the importance of community spaces and citizens' media (e.g., low-powered radio stations, grassroots video, public access television) for human emancipation and the improvement of social conditions (Rodriguez, Kidd & Stein 2009).

One can reasonably argue that CJ follows this tradition. It empowers ordinary citizens by leveling opportunities to speak, define issues, and make demands in the public sphere. Despite persistent inequalities in digital access in the global South, digital CJ enables incalculable numbers of people to be heard. It redefines "news as active citizenship" rather than passive consumption of pre-digested information. Everyone is potentially a protagonist rather than spectator or audience. Citizens are able to relay information to massive numbers of people through blogs, social media and other digital platforms, bypassing the mediating position of media and journalistic organizations.

It is mistaken, however, to view CJ simply as participatory communication. The questions about public expression in social change are not only who talks or whether citizens have opportunities for speaking up. Other questions need to be asked, too. What kind of public conversations take place? What are they about? Who participates? How are problems defined? How is dialogue linked to decision making? These questions refer to the institutional architecture of information resources and communication actions for addressing social problems. Social change demands propitious conditions for citizens' dialogue and action. It's not merely a matter of individuals voicing ideas into an empty void, but rather, whether different voices have a chance to be heard in conversations and decisions about common affairs.

My argument is that to understand how CJ contributes to social change is necessary to examine its connections (or lack of) to information and communication organizations. Social change is not merely citizens blogging, participating in social media, and uploading information across the vast digital universe. It is about how CJ is linked to the organizational scaffolding of public dialogue and opinion formation, politics, and collective change. Without considering these issues, the analysis easily falls into a celebration of CJ as the crystallization of individual expression divorced from a sociological analysis of communication rights and opportunities for collective actors as well as the institutional contexts for participation and decision making.

My interest in this chapter is to analyze two questions in the relationship between CJ and development/social change. One question is whether CJ offers better opportunities for covering development/social change issues and actions. This is an important question given that CJ may help to address the shortcomings of traditional journalism's coverage of social problems. A second question deals with CJ as a form of collective action driving social change. CJ is more than journalism. It blurs key concepts in media and communication studies such as journalism, participation, interpersonal and mediated communication, public expression and collective action. "Social media" postings and blogging can be legitimately considered actions by which citizens try to influence public awareness about specific issues and demands.

SOCIAL CHANGE NEWS IN A DIVERSIFIED NEWS ECOLOGY

Undoubtedly, CJ offers opportunities for expanding news coverage and bringing public visibility to social problems. Its open-ended architecture, constant innovation, and bottomless capacity offer endless opportunities for news. It challenges the elitist premises of "professional" news by bringing voices from the margins to the center. It situates citizens and communities as central actors rather than journalists as "self-appointed" representatives of the news interests of the people. It undermines the mediating role that journalism monopolized in the past and its power to tell millions what is (and what isn't) news. CJ has swung the doors open for citizens with internet access to produce news and information and provide information and perspectives missing or distorted in industrial journalism. The diversification of information about social problems enriches the information resources of societies, and may help to raise public awareness and knowledge, shape public agendas, and inform policy debates and legislation.

How should we assess the contributions of CJ to news about social change? One approach is to assess whether CJ helps to produce sustained and contextualized news about underreported social problems in the mainstream press. This issue matters for two reasons. First, scholars and activists have long criticized legacy news for their limited and sporadic coverage of social problems, particularly as they affect socially marginalized communities (Thakurta & Chaturvedi 2012). Social issues are more likely to get covered when journalists find standard newsworthy elements. Attention by political elites and celebrities, event-driven news, social and geographical proximity, accidents/disasters and other conventional news values are likely to drive coverage on issues such as poverty, displaced peoples, gender-based violence, environmental problems and others. Second, the impact of CJ on legacy news matters because, although the internet is more open and leveled than the old news system, news usage remains lopsided in favor of traditional news companies that attract the lion's share of traffic. Successful bloggers may have a few thousand followers, but CJ as a whole can hardly match the reach of selected national and global news organizations. Despite examples of amateur content suddenly becoming viral news, CJ can't get the attention that legacy newsrooms attract daily. For example, *The New York Times* columnist Nick Kristof and the newspaper's "Fixes" blog regularly spotlight local and global social problems (from gender-based violence to teenage depression) as well as the "invisible" work of organizations in front of a massive audience. The vast majority of citizen journalists lack a similar perch for digital storytelling. In fact, organizations working on social change typically yearn for attention from news powerhouses to elevate the presence of their causes in news and political landscapes (Waisbord 2008, 2010).

CJ influences mainstream news through formal collaborations between citizens and legacy newsrooms in which the former produce information and facilitate access to sources, or informally as citizens act as "crowd-sources" and feed RSS and Twitter entries about social problems that are eventually picked up by the mainstream press. In so doing, citizens enrich the work of "professional" journalists and help to improve news coverage of social problems.

The relation between CJ and traditional news could be examined by analyzing whether the former effectively steers the latter to cover social change news even when specific issues lack conventional news values. Can CJ influence major news organizations to cover social problems that lack obvious news pegs—official pronouncements, street demonstrations, newsy visuals, and predictable and routine news? This would mean, for example, covering poverty without statements from major political figures, gender-based violence without high-profile rapes, environmental degradation without weather emergencies, climate change without unusual heat waves, public safety without crimes in well-to-do neighborhoods, racial relations without minority-on-white homicides, and diarrheal diseases without cholera epidemics. The question, then, is whether CJ changes routine news on social problems (Waisbord 2010). As a measure of its impact on news conventions, this is also important to address the selectivity bias of news organizations that draw on CJ to report on conventional news events such as disasters, natural emergencies, wars, and violent protests. Countless citizen journalists routinely cover issues but, arguably, only a minuscule fraction gets attention from mainstream news.

We still lack answers to many questions. Does CJ contribute to expanding the coverage of topics and perspectives on social change and development in the legacy media? Are Western news organizations more likely to pay sustained attention to global social problems because they have easy access to CJ information? Do newsrooms bring out social issues by partnering with CJ or regularly checking news feeds produced by ordinary citizens or development/social change organizations? Does CJ successfully address traditional bottlenecks for social news, particularly as they affect citizens who rarely make news? Because these questions involve two large and fragmented actors, news organizations and citizen journalists, it is hard to produce categorical answers. We can find evidence of CJ's positive impact as well as minimal effects on legacy newsrooms.

If recent conclusions about the relationship between CJ and industrial journalism are applicable to social change news, it is doubtful that there have been major transformations. Old journalistic values have guided the "managed transition" into a new news ecology and the gradual opening of newsrooms to citizens' participation (Hermida & Thurman 2008; Meikle & Redden 2010; Paterson & Domingo 2008; Karlsson 2011). Although attitudes have been gradually changing, journalists initially didn't think bloggers and other forms of CJ have the credentials and expertise to produce quality reporting. To navigate the internet, they resort to standard news values—conventional definitions of newsworthiness (trust, facticity, timeliness and authority) to determine the value and the quality of information by citizen journalists (Pantti & Bakker 2009; Lewis, Kaufhold & Lasorsa 2010; Örnebring 2013; Reich 2008; Volkmer & Firdaus 2012).

Supporters of CJ have rightly criticized traditional journalism for ignoring a digital world brimming with democracy, half-heartedly accepting its inevitable ascent, and desperately seeking to reestablish a lost order. Unsure about the future, legacy newsrooms stick to conventional ideas to reaffirm their authority and insist on modern notions of "asymmetric" expertise ("we know, they don't") to survive in a postmodern environment. Some news organizations have promoted a "culture of collaboration" with CJ, but one shouldn't conclude that "participatory news" has triumphed. As demonstrated by reporters' use of Twitter for crowdsourcing and verifying information as well as experiments with "open source" reporting, "professional" journalism has been more receptive to citizen news and participation (Singer et al., 2011). These examples, however, aren't indications of the dawn of the "age of participatory journalism" and the passing of professional reporting. We lack solid evidence to conclude that CJ fill the gaps of legacy newsrooms.

The relationship between CJ and development/social change also needs to be examined by analyzing whether CJ enriches the news ecology. There is no question that this is true. A quick search about news/information about any social problem would likely produce long lists of sources. Consider HIV/AIDS, an issue that remains a critical global health challenge. Digital news includes massive amount of data produced by technical and scientific associations, information and research produced by activists' organizations and testimonies from bloggers living with the disease. There is an impressive amount of information about policies, drug treatments, vaccine development, and life experiences that rarely make it into industrial newsrooms. Even social problems that haven't received as much attention as HIV/AIDS in the past decades in global and domestic contexts, such as water access, poor sanitation and food security, are also widely covered by CJ. Citizens feed information about potential natural disasters, climate change, agricultural conditions, and disease surveillance to personal sites and open-source networks.

These examples show that the digital news ecology is more multileveled, open, complex, and diversified than the old news order. Information about social problems is abundant and easily available in a fragmented and chaotic news landscape. It is mistaken, however, to simply conclude that CJ's expansion of information about social problems settles the question about the diversity of news quality and perspectives. How sustainable is this information given that coverage of development/social problems, particularly their technical and political complexity, requires time and resources which, arguably, most individual CJ lacks? Certainly, bloggers who inform about their experiences and conduct original reporting make important contributions, but they are unlikely to have resources for producing regular, in-depth coverage or investigating the responsibilities and actions of governments and corporations. Sites dedicated to a range of "development/social change" news confront challenges to churn out quality information in a sustainable manner (Waisbord 2010).

Therefore, the crucial question shouldn't be whether CJ offers more opportunities for covering social problems, which undeniably it does. Although it expands news offerings, it is questionable whether CJ effectively matches the power of legacy newsrooms equipped with reporters, expertise and monetary resources. Nor is it obvious that CJ reaches "beyond the choir," that is, people who are already interested in certain social problems and find a wealth of citizen-produced information. What needs to be investigated are the connections between CJ and effective actions towards social change—its impact on public knowledge, participation, and policy making.

CITIZEN JOURNALISM AND COLLECTIVE ACTION FOR SOCIAL CHANGE

A second issue to consider is the relation between CJ and collective action. Just as CJ breaks down traditional distinctions between professionals and amateurs, it also erases conventional boundaries between journalism and activism, reporting and participation, witnessing and voicing demands. Not only "anyone can be reporter" (Bromley 2010), but anyone is potentially also an activist. Open-source technology makes it possible to link journalistic tasks such as reporting and data analysis to various dimensions of social activism such as organizing, petition, and voting. Citizens' reporting on social problems can be turned into evidence for advocacy. Crowdsourcing platforms may become springboards for digital activism. Unshackled by conventional notions of impartial, fact-based reporting, CJ can easily turn into mobilization.

To its defenders, here lies the enormous, innovative democratic potential of CJ—bringing out a wide spectrum of voices and engaging citizens in collective action (Shirky 2011). CJ attests to the dramatic reduction of obstacles for public organizing and acting without physical co-presence or conventional organizations. To its critics, the blurring of the distinctions between journalism and collective action is a major pitfall of CJ. It jettisons cardinal journalistic standards such as facts, verification and evenhandedness. For them, neither a "journalism without journalists" nor "journalism as collective action" are reasons for rejoice. Ideological agendas trump core journalistic ethics, a particularly worrisome trend at a time of the resurgence of partisan news.

Recent examples show that certain uses of CJ straddle reporting and collective action for social change. Citizens' digital postings sparked protest and other demands for attention and solutions to social ills (Agarwal, Lim & Wigand 2012; Antony & Thomas 2010; Harlow 2012). Other examples of effective CJ are the uses of open-source platforms in water access (www.nextdrop.org), climate change (climatecolab.org; iseechange.org) and disease surveillance (healthmap.org). Innovations are found worldwide: from oil spills in the Gulf of Mexico and conflict management in Atlanta in the United States to electoral monitoring in Kenya, from crime in the slums of Brazil to disaster management in Haiti. Citizen reporters produce and analyze data about local situations, discuss courses of actions, alert residents, and make evidence-based decisions (Zook, Graham, Shelton & Gorman 2010; Heinzelman, Brown & Meier 2011; Hirsch 2011; de Oliveira 2012).

While these cases illustrate the democratic uses of CJ by promoting participation and critical citizenship skills, we should not conclude that CJ necessarily leads to collective action. The problem here is similar to what I said previously regarding arguments about the impact of CJ on legacy newsrooms: it is difficult to generalize given multiple examples and contexts. Just as one finds examples of CJ enriching industrial journalism and digital news, there are also cases showing CJ's ability to organize and mobilize citizens. Yet we can't categorically affirm that these examples are representative of the unwieldy world of CJ. Not every citizen journalist is an activist who prefers to report about the humdrum of everyday life and personal interests rather than public issues. Also, accidental citizen-reporters document social realities yet their work doesn't necessarily materialize into activism. We should not cherry-pick cases to prove convictions, argument, and hopes. Random evidence shouldn't be interpreted as unquestionable proof that CJ is necessarily linked to collective action for social change. Instead, we need to explore under what conditions CJ is effectively linked to collective action and social change.

Enthusiastic views about CJ as collective action miss a crucial sociological dimension of social change: the role of organizations in collective action and the activation of networks of participation and policy change. Just as ordinary citizens posting family pictures or sharing digital diaries on social media aren't examples of participatory journalism, not all forms of CJ are similarly linked to organizations promoting social change. CJ needs to be analyzed in the context of how citizens mobilize (or fail to) for change.

Even if we consider it "organizing without organizations" (Shirky 2008), CJ doesn't make traditional organizations superfluous. Citizens using digital networks to organize and mitigate the impact of natural disasters and coordinate assistance attest to the blending of CJ with traditional humanitarian organizations. CJ has been intelligently used to mobilize citizens to report conditions, solicit input, define goals, and implement actions. Similar lessons come out of recent advances in global health, such as the expansion of AIDS and TB treatment worldwide, tremendous progress towards polio eradication in India, the remarkable reduction of malaria incidence and notable reduction in child and maternal mortality in sub-Saharan Africa. While these actions have incorporated old and digital forms of communication, none of these advances have been possible without collective actions involving government, international agencies, local organizations, and mobilization by technical experts and activists (Obregon & Waisbord 2012).

CJ may complement and facilitate the work of brick-and-mortar institutions such as government agencies and civic organizations. While CJ may lead and become part of collective action, sustainable social change demands institutions. Indeed, the contemporary professionalization of communication politics and social change attest to the renewed significance of institutions. Effective social change demands savvy, strategic organizations able to raise funding, implement data-driven actions, and navigate the complicated politics of social change. CJ may successfully articulate collective action in specific circumstances, and contribute to social change by documenting and bringing awareness to social problems. Sustainable change, however, particularly around highly political issues, demands institutional resources and appropriate conditions. The CJ of dissident movements has played important roles in political change, but it is not sufficient for social justice. Transforming policies and structures demands organizations. Long-term institutions and participation are needed for changing public priorities, social norms, policies, institutional performance, and/or power relations.

Nor is it clear that CJ is necessarily conducive to effective changes in governance and policy making. Recent experiments in e-activism render a more complex picture about its contributions to deliberative democracy that fall short of original optimism (Davis 2010; Moss & Coleman 2013; Loader & Mercea 2011). Instead of techno-romanticism and overblown predictions, caution is warranted to assess whether and how CJ promotes citizens' participation in policy making. CJ in support of crisis communication or protests is different from the kind of mobilization needed to foster social change. Mobilized citizens and institutions, not technologies, drive social change. Labels such as "the Facebook/Twitter" revolutions are appealing, particularly to Western media companies and audiences, as they reinforce apolitical visions of purely technology-driven "solutionism" for human improvement (Morozov 2013). Such labels, however, are deceiving. They neglect the constellation of factors that drive change, hide the laborious strategic and organizational work, and gloss over the messy political process that leads to social justice.

CITIZEN JOURNALISM AND NETWORKS OF SOCIAL CHANGE

Because social change comprises endless social problems as well as opportunities and obstacles to address them across the world, it is impossible to draw empirical conclusions about the lessons from recent CJ experiences.

CJ should not be narrowly seen as a form of reporting that either complements or challenges industrial journalism. Truly, as a set of innovative and empowering practices, CJ is a valuable source for legacy newsrooms and a welcome addition to digital news. CJ produces staggering amounts of information about social problems that are ignored or barely covered by the mainstream media, and it helps to report stories with more depth. Yet CJ is more than a complement to legacy newsrooms or an alternative supplier of information to the ever-expanding universe of digital news. Because it straddles reporting and collective action, it offers communication linkages during ordinary and extraordinary times. It is used to map out social conditions, identify problems, and discuss solutions. It contributes to organizing protests and coalescing critical discourse. It serves to connect citizens and organizations during natural disasters and other time-sensitive emergencies.

It is important, however, to examine CJ beyond examples of citizens' resistance and crisis management to assess its contributions to social change. Future studies are needed to analyze how different forms of CJ are used to mobilize opinion, pressure authorities, and promote issues for legislative debates and policy changes.

To understand the impact of CJ, the analysis needs to focus on the links between CJ and institutions and networks engaged in social change. News organizations aren't less significant for covering social problems because CJ contributes to diversifying coverage. Nor should we assume that trending Twitter topics, clicktivism, or digital organizing necessarily catalyze social change by overturning power inequalities. The analysis needs to be based on nuanced understanding of processes that contribute to solving social ills. We need to explore further its connections to conventional newsrooms as well as social and political organizations. Placing CJ within the politics of social change and examining how movements and networks harness CJ to advance social justice are necessary to refine conclusions and advance theory-building.

REFERENCES

Agarwal, N., Lim, M. & Wigand, R.T. (2012). "Online Collective Action and the Role of Social Media in Mobilizing Opinions: A Case Study on Women's Right-to-Drive Campaigns in Saudi Arabia" in *Web 2.0 Technologies and Democratic Governance*, eds C. G. Reddick and S. K. Aikins. New York: Springer pp. 99–123.

Allan, S. (2013). *Citizen Witnessing: Revisioning Journalism in Times of Crisis.* Cambridge: Polity Press.

Anderson, C. W., Bell, E., & Shirky, C. (2012). "Post-Industrial Journalism: Adapting to the Present." *Tow Centre for Digital Journalism, Columbia Journalism School, Centennial Report.*

Antony, M.G., & Thomas, R.J. (2010). "This Is Citizen Journalism at Its Finest: YouTube and the Public Sphere in the Oscar Grant Shooting Incident." *New Media & Society*, vol. 12, no. 8, pp. 1280–1296.

Bromley, M. (2010). "Anyone Can Be a Reporter: Citizen Journalism, Social Change and OhmyNews" in *An Introduction to Communication and Social Change*, eds P. Thomas & M. Bromley. St Lucia, Qld, Australia: UQP, pp. 187–201.

Cooke, Bill and Kothari, Uma (eds) (2001). *Participation: The New Tyranny?*, London and New York: Zed Books Ltd.

Davis, A. (2010). "New Media and Fat Democracy: The Paradox of Online Participation." *New Media & Society*, vol. 12, no. 5, pp. 745–761.

Earl, J., & Kimport, K. (2011). *Digitally Enabled Social Change: Activism in the internet Age.* Cambridge: MIT Press.

Escobar, A. (1995). *Encountering Development*. Princeton, NJ: Princeton University Press.

de Oliveira, R. G. (2012). "Citizen Journalism and Its Democratic Potential—Brazilian Case Studies: Viva Favela and Índios Online." *Selected Papers of internet Research*, (12.0).

Harlow, S. (2012). "Social Media and Social Movements: Facebook and an online Guatemalan Justice Movement That Moved Offline." *New Media & Society,* vol. 14, no. 2, pp. 225–243.

Heinzelman, J., Brown, R., & Meier, P. (2011). "Mobile Technology, Crowdsourcing and Peace Mapping: New Theory and Applications for Conflict Management" in *Mobile Technologies for Conflict Management*, Law, Governance and Technology Series 2, ed. M. Poblet. Dordrecht: Springer Netherlands, pp. 39–53.

Hermida, A., & Thurman, N. (2008). "A Clash of Cultures: The Integration of User-Generated Content within Professional Journalistic Frameworks at British Newspaper Websites." *Journalism Practice*, vol. 2, no. 3, pp. 343–356.

Hirsh, T. (2011). "More Than Friends: Social and Mobile Media for Activist Organizations" in *From Social Butterfly to Engaged Citizen: Urban Informatics, Social Media, Ubiquitous Computing, and Mobile Technology to Support Citizen Engagement*, eds M. Foth, L. Forlano, C. Satchell & M. Gibbs. Cambridge, MA: MIT Press.

Karlsson, M. (2011). 'The immediacy of online news, the visibility of journalistic processes and a restructuring of journalistic authority', *Journalism*, 12(3), pp. 279–295.

Lewis, S. C., Kaufhold, K., & Lasorsa, D. L. (2010). "Thinking About Citizen Journalism: The Philosophical and Practical Challenges of User-Generated Content for Community Newspapers." *Journalism Practice,* vol. 4, no. 2, pp. 163–179.

Loader, B. D., & Mercea, D. (2011). "Networking Democracy? Social Media Innovations and Participatory politics." *Information Communication & Society*, vol. 14, no. 6, pp. 757–769.

Meikle, G., & Redden, G. (eds.) (2010). *News Online: Transformations and Continuities*. London: Palgrave Macmillan.

Morozov, E. (2013). *To Save Everything, Click Here*. New York: Public Affairs.

Moss, G., & Coleman, S. (2013). "Deliberative Manoeuvres in the Digital Darkness: e-Democracy Policy in the UK." *The British Journal of Politics & International Relations*, pp. 1–18.

Obregon, R., & Waisbord, S. (eds) (2012). "Capacity Building (and Strengthening) in Health communication" in *The Handbook of Global Health Communication*, Oxford: Wiley-Blackwell.

Örnebring, H. (2013). "Anything You Can Do, I Can Do Better? Professional Journalists on Citizen Journalism in Six European Countries International." *Communication Gazette*, vol. 75, no. 1, pp. 35–53.

Pantti, M., & Bakker, P. (2009). 'Misfortunes, memories and sunsets Non-professional images in Dutch news media', *International Journal of Cultural Studies*, 12(5), pp. 471–489.

Paterson, C. & Domingo, D. (eds) (2008). *Making Online News: The Ethnography of New Media Production*. New York: Peter Lang.

Reich, Z. (2008). "How Citizens Create News Stories: The 'News Access' Problem Reversed." *Journalism Studies*, vol. 9, no. 5, pp. 739–758.

Rodriguez, C., Kidd, D., & Stein, L. (eds) (2009). *Making Our Media: Global Initiatives Toward a Democratic Public Sphere*. Cresskill, NJ: Hampton Press.

Servaes, J. (ed) (2008). *Communication for Development and Social Change*. London: Sage Publications.

Shirky, C. (2011). "The Political Power of Social Media *Foreign Affairs*, vol. 90, no. 1, pp. 28–41.

Shirky, C. (2008). *Here Comes Everybody: The Power of Organizing without Organizations*. New York Penguin.

Singer, Jane B., David Domingo, Ari Heinonen, Alfred Hermida, Steve Paulussen, Thorsten Quandt, Zvi Reich, Marina Vujnovic (2011). *Participatory Journalism: Guarding Open Gates at Online Newspapers*, Malden: Wiley.

Thakurta, P. G., & Chaturvedi, S. (2012). "Food and Nutrition Justice: How to Make It More Newsworthy?" *IDS Bulletin*, vol. 43, no. s1, pp. 58–64.

Volkmer, I., & Firdaus, A. (2012). "Between Networks and 'Hierarchies of Credibility'" in *Rethinking Journalism: Trust and Participation in a Transformed News Landscape*, ed. C. Peters & M. J. Broersma New york, Routledge, pp. 101–113.

Waisbord, S. (2008). Advocacy Journalism in a Global Context: The 'Journalist' and the 'Civic' Model in *Handbook of Journalism Studies*, Karin Wahl-Jorgensen and Thomas Hanitzsch eds., 371–385.

Waisbord, S. (2010). "Can NGOs Change the News?" *International Journal of Communication*, vol. 5, pp. 142–165.

Waisbord, S. (2013). *Reinventing Professionalism: Journalism and News in global perspective*. Cambridge: Polity Press.

Zook, M., Graham, M., Shelton, T., & Gorman, S. (2010). "Volunteered geographic information and crowdsourcing disaster relief: a case study of the Haitian earthquake." *World Medical & Health Policy*, vol. 2, no. 2, pp. 7–33.

CHAPTER FOURTEEN

A Latin American Approach TO Citizen Journalism[1]

CLEMENCIA RODRÍGUEZ

José Adolfo remembers how difficult it was to reach *Brisas del Oriente*, a neighborhood that is home to many recently displaced families. His citizen reporting team did not have any funding to hire moto-taxis or take the bus to the interview locations. The fares had to come out of their own pockets, but they did not give up. They remembered their journalism training and how crucial it is to find reliable sources, first-hand witnesses, people whose personal experiences can contribute to public conversations, and a diversity of perspectives to nourish the public sphere. Lesly, another citizen journalist and member of José Adolfo's team, recalls:

> The road was long and the sun did not take pity on us, and every minute the heat intensified. After talking with José Adolfo about how we were going to do this, I crossed the street and told him: "Come on! I'll take you!" His face sweating, with much effort, he tried to sit on the bar of my bicycle, and at the end he finally did it. I started pedaling with all I had, trying to joke about the whole thing. Five minutes later, sweat covered my face, and my legs started to shake; I stopped and told him to get off. I could not continue. Neither one of us had any money for minivan fare, so we decided to walk. I had to drag my bike all the way to the neighborhood (Ministerio de Cultura de Colombia, 2009:20).

José Adolfo, Lesly Adriana Cifuentes, and Manfry Gómez Ditta biked and walked to the refugee neighborhood where they conducted a series of interviews with men, women, and children recently displaced by Colombia's armed conflict. To complement their interviews, they gathered data about forced displacement from government documents, NGO's databases, and mainstream media. They confronted sources and included a variety of vantage points on the issue of forceful displacement. They treated personal narratives with respect and dignity, making sure the voices of the displaced men and women, their interviewees, played a central role in their report. A few months later, in February 2005, this citizen journalism team received the Golden Feather Award (*Pluma de Oro*) from a Colombian association of professional journalists for their series of radio reports titled "Memories of Forced Displacement in Barrancabermeja." The reports were broadcast via several local community radio stations.

This type of citizen journalism is very different from what has become known today as online citizen journalism. I believe it is important to analyze how these two styles of citizen journalism differ, in order to understand and do justice to a vast tradition of media activists and community/alternative/citizens' media makers. Current use of the term "citizen journalism" generally refers to ordinary people uploading content about current events to online platforms—known today as "user-generated content." What makes this practice "journalism" is the fact that the up-loader has unique access to images, sounds, or personal experiences about the event, or, in Stuart Allan's words, this type of action depends on: "the ordinary person's capacity to bear witness" (Allan, 2009:18). Non-professional reporters (citizen reporters) in the midst of Hurricane Katrina, the bombing of Baghdad during the war in Iraq, the tsunami on the Indian Ocean in 2004 (Liu et al., 2009) with access to digital technologies (cell phones with cameras and recording capabilities, tablets, PCs, etc.) provide a unique vantage point. In terms of quick access to the location of the event, personal experience, and online public spheres, digital technologies can clearly transform anyone into a citizen reporter. However, this type of citizen journalism differs greatly from they type of journalism José Adolfo and his team enacted, as I will explore in this chapter.

THE LATIN AMERICAN APPROACH

José Adolfo's team of citizen journalists received numerous hours of journalism training, attended several writing and reporting workshops, had access to conferences and discussions with professional journalists, read a variety of texts about journalism, journalistic genres, the basic principles of journalism, the role of journalism and journalists in processes of nation-building and democratic societies, and the importance of responsible and fair journalism in contexts of armed conflict and post-conflict. Their training was sponsored by *Radios Ciudadanas, Espacios para la Democracia* [Citizen Radio, Spaces for Democracy], an initiative led by Jeanine El Gazi at the Colombian Ministry of Culture.[2]

José Adolfo's team belongs to a deeply rooted branch of citizen journalism in Latin America. In this region, citizen journalism has been a practice traditionally connected to alternative media, social movements, and struggles for social justice. Latin American citizen journalism is typically carried out by people who are not professional journalists; instead, these citizen journalists make a living as farmers, office workers, teachers, etc. and practice journalism on the side. Community radio stations, televisions, or alternative print media broadcast their reports and news stories. In Colombia, community and alternative media have a strong presence in the national mediascape. Side by side with mainstream commercial media and government-controlled media, Colombian community and alternative media thrive, thanks in part to media regulation that guarantees community broadcasting licenses. Such regulation is the result of more than forty years of pressure on the part of social movements and media activists toward the democratization of media in Colombia. In 2008 the Colombian mediascape included 651 community radio stations, 553 community televisions, and 26 indigenous radio stations.

The Latin American approach to citizen journalism differs from the Global North's approach, where the term citizen journalism is used to refer to online platforms recently made accessible to any user for posting audience-generated content (Allan and Thorsen, 2009:4). In Latin America, citizen journalism is a practice of resistance that emerges as social movements, activists, and other social justice collectives refuse to embrace the notion that only professional news organizations can practice journalism and nourish the public sphere with information key to democratic processes. The Latin American approach to citizen journalism still maintains the basic foundational principles of traditional journalism, such as the importance of verifying facts, identifying, and securing specific sources, and maintaining autonomy (Guedes Bailey, 2009). This citizen journalism is understood as a practice driven by social responsibility and public interest. For example, most community radio and television stations in Colombia have their own journalistic code of ethics, in which community communicators specify their journalistic principles as they deal with issues ranging from responsibility and accountability to source selection and terminology (i.e., what terms will be used to refer to guerrilla combatants).

Despite the recent waves of enthusiasm and hype surrounding citizen journalism as a byproduct of the internet (Papandrea, 2007:523), citizens as journalists have been around since long before the rise of online public spheres. In the early twentieth century, social movements and media activists appropriated media technologies, divorced them from their corporate originators, and nurtured public spheres with their own counter-information. "For example, print technologies were appropriated by anarchist movements in the United States, and revolutionary groups in Mexico and Russia. By 1920, Bolivian tin miners were using radio technologies to mobilize their unions in their struggle for social justice against corporate and state oppressive forces (Downing, 2010; Huesca, 1996). Similarly, during the 1968 revolts, workers and students' movements benefited from technological progress and lower prices in the systems of reprography to publish their own newspapers, as shown with the *prensa marginal* in Mexico (Trejo, 1980), the 'parallel press' in France (Chadaigne, 2002), or the underground press in the United States (Lewes, 2000)" (Rodríguez, Ferron, and Shamas, in press).

In the following pages I would like to illustrate what I refer to as "the Latin American approach to citizen journalism." These examples emerge from fieldwork I conducted between 2004 and 2006 with citizens' media producers in regions of armed conflict in Colombia.

One of the main differences in this approach, also noted by Olga Guedes Bailey (2009) in the first volume of *Citizen Journalism: Global Perspectives*, is that in Latin America, much of citizen journalism exists in traditional media divorced from online platforms. Radio is still the number one source of information in the region, especially in poor urban and rural communities. In the following interview, Hernán Castellanos, a community radio producer from a marginalized neighborhood in Bogotá, describes the enormous effort it took to get a news program off the ground. Castellanos's words illustrate the depth of this radio station's commitment to training local reporters in order to address the community's information needs:

> Our station is in Soacha, a city of nine hundred thousand people, most of who are migrants from other parts of the country. We realized that, despite being so close to the capital city [Bogotá], Soacha is a completely uninformed community. We noticed that unless events in our community involve incidents of public unrest, they don't make it into the media, so we knew the main element of our programming had to be news. Currently we have a two and a half hour daily news program that runs from six to nine thirty in the morning. The program, which covers mostly local news, is produced by a team of twenty-two community reporters who comb Soacha inch-by-inch, neighborhood-by-neighborhood. These reporters don't receive a salary; they do this work out of their own commitment to their city. The learning process behind this news program required a lengthy process of trial and error. We began by making a public call for people with an interest in becoming community reporters; in response, people called to complain about stuff, but no one was willing to truly become a reporter. Next, we made arrangements with three local schools, so that one of the mandatory social service areas for students in their last year could be community reporting. This didn't work either, because participation was mandatory and these young people were not interested in journalism or the issues they had to cover. One day, assuming that the lack of willing participants meant that our news program was not what our people wanted, we discontinued it without any type of announcement. Complaints began pouring in; people wanted to know, "why had we stopped the news program?" That response prompted us to take things seriously: we raised funds, recruited a group of volunteers, and locked the whole for forty five days, Monday through Friday, in order to truly train a team. We offered basic journalism training and training on topics like the ethics of information management. Not to say that everything has been rose petals, but these community reporters have made tremendous improvements. Two weeks after the training ended, they independently

> decided to form a watchdog committee among themselves, to guarantee the project's success. (Unidad de Radio—Mesa Regional Cauca, Putumayo, Nariño, 2000, cited in Rodríguez, 2011:192–193).

In Colombia's community radio and television circles, and other incarnations of citizens' media, Gabriel García Márquez[3] is frequently mentioned, not as the only Colombian recipient of the Nobel Price of Literature, but as an icon of excellent journalism. Every community radio or television station wants to have its own "*franja informativa*," a Colombian localism for "news program." Since the rise of community media in the country in the 1980s, numerous governmental and non-governmental organizations and social movements have implemented journalism training initiatives for community media-makers and other non-professional reporters and journalists. After more than thirty years of continuous training and discussion about the role of citizen journalists, Colombia's citizen journalism goes far beyond "letters to the editor, man-on-the-street interviews, and call-in radio and television shows" (Allan, 2009:30; see also Chapter 4 in Rodríguez, 2011).

Citizen Journalism in Contexts of Armed Conflict

The accumulated wisdom of community media practitioners frequently plays a crucial role in countering the impact of armed conflict on civilian communities. One of the most significant examples Colombian citizen journalists have given us is their commitment to serving the information and communication needs of ***unarmed civilians***. While the reporting of mainstream media journalists is generally driven by the actions and words of those who wage war, Colombian citizen journalists shift their cameras and microphones toward those who did not take up arms. My fieldwork was conducted in regions of Colombia particularly impacted by armed conflict, so I found numerous instances in which community media's journalists re-directed all of their energy toward meeting the information and communication needs of civilians trapped in the crossfire between armed groups. The following examples are good illustrations of this use of citizen journalism:

> In 2000, during the three month long *paro armado*[4] protesting the fumigations in Putumayo, everything stopped between September and December; phone lines were cut, there was no transportation, and communication between families from the north and the south became impossible. Twenty days into the *paro*, the situation became truly dramatic; there was no food, because no trucks were allowed to move, and fear spread because the crisis was lasting so long;[5] people started traveling on foot, 'cause there was no transportation, and no gas. The community radio stations began asking, 'What do we do?' Gonzalo Portillas, one of the directors, told me "our stations have to do something because this is about our communities, the communities we belong to." So we found a way to link the stations from the south, middle, and north of Putumayo. There are ten community radio stations in the region, and five of these linked themselves together for daily broadcasts. So we started hearing messages

> like, "Mom, stay calm, I'm OK. I'm in Puerto Asis, but do not worry, I am fine." The daily shared broadcasts from the linked radio stations became crucial for communication between families separated throughout the region (Espitia, 2004, cited in Rodríguez, 2011:198).

In addition to facilitating communication between families, the linked broadcasts provided a public sphere in which unarmed civilians discussed daily issues. Javier Espitia, a long-time community radio leader who was living in Putumayo at the time, explained:

> Teachers began using the linked broadcasts to discuss what to do. Listeners might hear someone on air say, "The teachers from Orito have talked about what to do and we agree that we should start the holiday break now;" and the teachers from another municipality would reply, "in our case we will try to continue classes because we are not going to let them intimidate us." The entire debate happened on air, through the radio stations. The linked broadcasts were so successful that one of the stations in La Hormiga was the target of machine gun fire by the guerrilla because the guerilla interpreted these broadcasts as resistance. (Espitia, 2004, cited in Rodríguez, 2011:198–199).

Liberman Renjifo, director of Radio Ocaína Estéreo, a community radio station in Putumayo, described his station's actions during a guerrilla-mandated *paro armado*:

> ...after several weeks without any food, gas, or electricity in the region, the radio station joined forces with several other community organizations, especially women's collectives, to decide how to confront the situation; what emerged was a very beautiful action, because at the time, the station didn't have its own generator, so any time the electricity was cut, the station went off the air. At the meeting, it was decided that the community radio station was the most important medium to keep people informed, so the decision was made to pull together funds from all the different organizations and raise more funds in the community so we could buy a generator for the station. This would keep people informed. That was when we saw the community really begin to value their community radio station. At that time the guerrilla had a strong presence in town, and they convened meetings. Sometimes we had to attend those meetings, but when the guerilla saw that the community was getting really organized around the station, they realized they could not intimidate us any more. Before, people didn't leave the municipality to get food because everyone was afraid of the guerilla. Once people began to mobilize, we started going to other municipalities for food. Food came by plane to Puerto Asis, a municipality about an hour away, and we had to sneak out in a horse-cart to bring back food and gas. The station helped in all this, and we also helped by keeping people in far away rural areas, who were not allowed to come to town, informed about the food shipments. We monitored how much local stores were charging for food, because sometimes they tended to hike the prices, arguing that food was more expensive because it was coming by plane (Renjifo, 2004, cited in Rodríguez, 2011:199).

In what ways does this type of citizen journalism differ from typical online informational blogs or citizen reporting? First, these are collective endeavors, rarely driven by individual interests. Second, these citizen reporters operate within an institutional home—generally a community medium—that has a mission, policies, rules, and stated procedures. Third, these Colombian citizen reporters embrace journalism as a craft, an art to perfect; they perceive their reporting as something their communities need and value, making reporting for the community a source of empowerment and self-esteem. Most importantly, their reporting is driven by the information needs of their communities. If online citizen journalists report because they find themselves in the right place at the right time, Colombian citizen journalists report because they are deeply aware of their communities' information needs. While online citizen journalists live their lives and use technology to report from wherever life takes them, Colombian citizen journalists leave their everyday lives behind in search of the news and information the community needs. They are continuously trying to detect where the information that matters can be found; they leave their daily routines to access those places, find the sources, collect information, and spend hours and hours editing finished radio programs, reports, interviews, and so forth.

CITIZEN REPORTERS DIFFUSE VIOLENCE

In some cases, Colombian citizen journalism veers away from traditional journalism, as when citizen reporters detect exceptional situations in which specific information needs emerge in the community. In my fieldwork I was able to witness how Colombian citizen journalists design creative strategies to meet such needs. For example, in September 1996, twenty thousand demonstrating farmers arrived in Belén de los Andaquíes, a small town of six thousand villagers in the southern Colombian Amazon. The situation was volatile and potentially violent, because the local villagers perceived the demonstrators as radicalized puppets of the guerrilla, and the demonstrators perceived the villagers as "sold out" to a government known for neglecting rural, farming communities. Alirio González, director of the local community radio station remembers how radio reporters designed a strategy to avoid violent confrontation between these two hostile groups:

> We decided to open the mics. Instead of a narrow focus on the reasons behind the march, or whether the demonstrators were right or not, which at that point was quite confusing, we opened the focus, inviting anyone who wanted to sing, recite poems, etc. (González, 2004, cited in Rodríguez, 2011:72).

A few rules were agreed upon. The coverage would avoid any type of discourse for or against the demonstrators, maintaining instead dynamic programming that would meet the information and communication needs of the people marching and their hosts. Local reporters began a special broadcast that continued day and night, facilitating meetings, disseminating safety information (especially in regards to the well-being of the children camping on the shores of the fast-moving river), and transmitting messages from the demonstrators to their families and communities back home; in the words of González "the key is to align the medium with the needs of unarmed civilians" (González, 2004, cited in Rodríguez, 2011:72).

National and regional media covered these massive demonstrations in terms that aligned with armed groups. Mainstream media, aligned with the army and other Colombian armed forces, covered the march as proof that peasant organizations were manipulated by the guerrilla; other mainstream media covered the march as legitimizing the guerrilla's attempts to take over peasant social movements. By contrast, Radio Andaquí's agenda was driven by the information and communication needs of unarmed civilians. During the mobilization, Radio Andaquí served the needs of both local citizens and the demonstrators coming from distant communities. Soon, the station was filled with demonstrators transmitting messages to those who had stayed back home. Demonstrators sent messages such as: "sell the calf, because we ran out of money and nothing has been defined yet," "give corn to the chickens and sell a few hens because we have no money," "we are fine, in Belén, and God willing we will reach our goal," and "*m'hija* [my daughter], take good care of the animals, because we don't know how long this will take" (González and Rodríguez, 2008:114; Rodríguez, 2011:72).

Apart from meeting the information needs of demonstrators, Radio Andaquí invited both demonstrators and villagers to come to the station and share music, poetry, and songs. So many people came to play music, sing, and recite that the station had to improvise an outdoor stage in front of its studio. Soon, the new transmission site was baptized *La Tarima del Sol* [The Sun's Stage] because it sat under a canopy of trees that filtered the sunlight into beautiful shadows reflected on the artists. For eight days, The Sun's Stage was witness to what Alirio González calls, "the most significant festival of music, poetry, and song we have ever transmitted in ten years" (González and Rodríguez, 2008:115; Rodríguez, 2011, 72).

On The Sun's Stage, demonstrators and locals could learn about each other, and appreciate each other's creative abilities and identities as human beings, beyond any political agenda. Radio Andaquí provided demonstrators and their local hosts with a communication space for recognizing one another as multi-dimensional, complex subjects with a lot in common. Belén's radio audiences saw that the demonstrators were not the terrorists, delinquents, drug-traffickers, or guerrillas that the national media had portrayed. Radio Andaquí's communicators used their journalistic knowledge to facilitate information and communication among demonstrators and local villagers. They triggered new channels of information and new ways of speaking to each other. Information about the demonstrators reached the villagers and information about the villagers reached the demonstrators. This new information played key roles in de-stigmatizing both groups, de-escalating potential aggression and violence between demonstrators and villagers. What villagers saw and heard transformed their notion of demonstrators as puppets of the guerrilla and nurtured new perceptions of demonstrators as families of farmers, very much like themselves, struggling to have their rights recognized by the central government. Locals were able to see the demonstrators as members of loving families like their own, whose lives were disrupted by the need to join this effort to have their voices heard. González recounts:

> One afternoon don Pedro Vargas came to The Sun's Stage. He had walked all the way from Albania and once he was positioned in front of the microphone, he gave the following instructions: '*M'hija,* kneel down, you and the kids, because here goes dad's blessing.' With his right hand he made the sign of the cross over the microphone, recited the blessing, and finished by saying, 'Stop worrying. God is with us, I am fine, and we will make it to Florencia' (González and Rodríguez, 2008:115).

Radio Andaquí's role during the demonstration can be interpreted as what Chantal Mouffe calls the "democratic mobilization of affection in order to weave communal identities" (Mouffe, 2006). According to Mouffe, we need to reinstate the centrality of affection in politics. Mouffe states that "democratic politics need to have a real influence on people's desires and fantasies" so that, despite tremendous differences, people can see the "other" not as an enemy or antagonist to be eliminated, but as an "agonist" whose demands, although not shared, have a legitimate place in the public sphere (Mouffe, 2006). Radio Andaquí carefully designed an open communication space where locals and demonstrators could see each other as poets, singers, fathers, mothers, husbands, and farmers. With The Sun's Stage, Radio Andaquí helped deflate the potential antagonism between locals and demonstrators, and simultaneously fueled "agonist identities," as Mouffe conceives of the term.

The Sun's Stage helped defuse conflict between the demonstrators and the people of Belén, who may have felt threatened by the "invasion" of thousands of "terrorist criminals manipulated by the guerrilla." Although violence erupted in many places where the demonstrators stopped on their way to the capital city of Florencia, the people of Belén supported their guests until they left town a week later. The only violent incidents in Belén originated with the local armed forces, which attacked the demonstrators minutes before they left. The lone victim was a local man who, like many others, sided with the demonstrators and against the soldiers, and was injured by a soldier's gun shot as he was trying to shield his guests.

CONCLUSION

Based on my fieldwork and lessons learned from Colombian citizen journalists, I worry that the hype and enthusiasm around online citizen reporting will eclipse this other type of citizen journalism. I worry, because more and more, the resources available in the 1970s, 1980s, and 1990s to train citizen reporters in the Global

South, funding workshops, meetings, and so forth, are disappearing rapidly, replaced by the idea that access to online technologies will automatically trigger a wave of citizen journalists. I have no doubt increased access to technology will, in fact, multiply individual bloggers and citizen reporters—however, will this type of citizen journalism, based on being in the right place at the right time, be enough to meet all the information needs of communities in the Global South?

In the defense of online media enthusiasts, we may be entering a new world in which the logic of the bazaar is replacing the logic of the cathedral (Raymond, 2000), where the wisdom of the experts is replaced by "the wisdom of the crowds" (Surowiecki, 2004). According to these predictions, when everyone becomes a citizen reporter, all nooks and crannies in society will be reported, every issue we need to know about will have its own citizen reporter covering it, and we will all have access to enough technology to access such reporting, thus making news media obsolete. In this new world, the ability of citizen reporters to auto-correct will bypass the need for investigative journalists, editors, trained interviewers, and verifiers of facts. Crowds will produce as much expertise as trained experts.

However, in a town in the south of Colombia, such as Belén de los Andaquíes, the future appears bleak. We are still waiting for the crowds of citizen reporters, while funds to train community communicators vanish. Electricity is still precarious, let alone connectivity and access to personal computers, modems, and internet. Instead, away from public scrutiny, transnational mining and oil companies are striking deals to drill in the Amazon; government authorities are silently changing the policies that rule the Amazon, to allow mining, oil exploitation, and water use that previous, more environmentally sound policies forbade; corrupt mayors and governors are pocketing municipal budgets; and blatant violators of human rights (both leftist guerrillas and right-wing paramilitaries) run free with complete impunity. Ultimately, Belén de los Andaquíes will be left without trained citizen journalists, while it waits for the crowds of citizen reporters to show up, creating ideal conditions for mining companies, corrupt officials, and violators of human rights.

NOTES

1. Portions of this chapter were previously published in Chapters 1 and 4 of my book titled *Citizens' Media Against Armed Conflict. Disrupting Violence in Colombia* (Minneapolis: University of Minnesota Press, 2011).
2. *Radios Ciudadanas* was an ambitious and expensive project. El Gazi sought and secured most of the project's funding from national and international agencies, including Programa de las Naciones Unidas para el Desarrollo (PNUD), US-Agency for International Development, Organización Internacional de Migraciones, Agencia Rural para el Desarrollo y Parques Nacionales. More than 90 percent of the project's total cost of two thousand million pesos (about two million dollars), was funded by these agencies.

3. Gabriel García Márquez began his writing career as a journalist for different Colombian dailies, including *El Universal* (Cartagena), *El Heraldo* (Barranquilla) and *El Espectador* (Bogotá). In Colombia, García Márquez is as well known for his novels and short stories as he is for his chronicles and commitment to excellence in journalism. Colombians think of García Márquez not as a journalist ***and a*** novelist, but instead as an excellent chronicler; the writer himself has nurtured this idea, insisting that he is not a fiction writer, but a good observer and chronicler of the surrounding social and cultural reality—the basis of the literary school known as "Magic Realism." García Márquez's passion for journalism is so profound that in 1994 he founded the Fundación para el Nuevo Periodismo Iberoamericano [Foundation for Ibero America's New Journalism], a foundation with a mission "to work toward journalism's excellence, and toward the contributions of journalism to democracy and development in all Ibero American and Caribbean countries" (http://www.fnpi.org/fnpi/mision-y-valores/).
4. *Paro armado* is a type of demonstration in which guerrilla organizations order people to stay home, thus stopping all activities and bringing paralysis to local economies.
5. *Paros* typically last only a couple of days, not almost a month as in this case.

REFERENCES

Allan, Stuart. (2009). Histories of Citizen Journalism. In Allan, Stuart and Thorsen, Einar (eds.) *Citizen Journalism. Global Perspectives*, 17–31. New York: Peter Lang.

Allan, Stuart, and Thorsen, Einar (eds.). (2009). *Citizen Journalism. Global Perspectives*. New York: Peter Lang.

Downing, John D. H. (2010). *Encyclopedia of Social Movement Media*. Thousand Oaks, CA: Sage.

Espitia, Javier. 2004. Interview by author. Bogota, Colombia: Tape recording. October 6.

González, Alirio. 2004. Interview by author. Belén de los Andaquíes, Colombia. Tape recording. October 15.

González, Alirio, and Clemencia Rodríguez. (2008). "Alas para tu Voz. Ejercicios de Ciudadanía desde una Emisora Comunitaria del Piedemonte Amazónico." In Clemencia Rodríguez (ed.), *Lo que le Vamos Quitando a la Guerra. Medios Ciudadanos en Contextos de Conflicto Armado en Colombia*, 65–140. Bogotá: Centro de Competencias en Comunicacion—Fundación Friedrich Ebert Siebert.

Guedes Bailey, Olga. (2009). Citizen Journalism and Child Rights in Brazil. In Allan, Stuart, and Thorsen, Einar (eds.) *Citizen Journalism. Global Perspectives*, 133–142. New York: Peter Lang.

Huesca, Robert. (1996). Participation for Development in Radio: An Ethnography of the *Reporteros Populares* of Bolivia. *Gazette* 57(1): 29–52.

Lewes, James. (2000). The Underground Press in America (1964–1968): Outlining an Alternative, the Envisioning of and Underground. *Journal of Communication Inquiry*, 24(4): 379–400

Liu, Sophia B., et al. (2009). Citizen Photojournalism during Crisis Events. In Allan, Stuart, and Thorsen, Einar (eds.) *Citizen Journalism. Global Perspectives*, 43–63. New York: Peter Lang.

Ministerio de Cultura de Colombia. (2009). *Historias de Incidencia de Radios Ciudadanas*. Bogotá: Ministerio de Cultura. Available at http://www.mincultura.gov.co/?idcategoria=22319. Last accessed July 26, 2013.

Mouffe, Chantal. (2006). Las Identidades Colectivas Políticas en Juego. Paper presented at Conferencia de la Federación Latinoamericana de Facultades de Comunicación (FELAFACS), Bogotá, Colombia.

Papandrea, Mary-Rose. (2007). Citizen Journalism and the Reporter's Privilege. *Boston College of Law, Boston College of Law Faculty Papers*. Retrieved July 12, 2013 from http://lawdigitalcommons.bc.edu/cgi/viewcontent.cgi?article=1168&context=lsfp

Raymond, Eric Steven. (2000). *The Cathedral and the Bazaar*. Available at http://www.catb.org/esr/writings/homesteading/cathedral-bazaar/

Renjifo, Liberman. 2004. Interview by author. Bogotá, Colombia. Tape recording. October 7.

Rodríguez, Clemencia; Ferron, Benjamin; Shamas, Kristin. (In press) Four Challenges in the Field of Alternative, Radical and Citizens' Media Research. *Media, Culture, and Society*.

Rodríguez, Clemencia. (2011). *Citizens' Media Against Armed Conflict. Disrupting Violence in Colombia.* Minneapolis: The University of Minnesota Press.

Surowiecki, James. (2004). *The Wisdom of the Crowds*. New York: Doubleday. Available at http://teaching.p-design.ch/texts/Surowiecki2004_wisdom_of_crowds.pdf.

Trejo, R. (1980). *La prensa marginal*, México: El Caballito.

Unidad de Radio—Mesa Regional Cauca, Putumayo, Nariño. 2000. Recorded Session. Pasto, Colombia. Tape recording.

CHAPTER FIFTEEN

Getting INTO THE Mainstream: The Digital/Media Strategies OF A Feminist Coalition IN Puerto Rico

FIRUZEH SHOKOOH VALLE

My experiences as a reporter who covered human rights issues for five years at one of Puerto Rico's principal mainstream newspapers inform this research on the feminist movement and media. On many occasions, after interviewing mostly female survivors of atrocious acts of violence, I asked myself if I had accurately represented their experiences, feelings, and thoughts. How would they have narrated their own stories? How much did I interpret their experiences through my journalistic reporting? Was I depicting them solely as victims? As a feminist reporter at a mainstream newspaper I also asked myself: How could I be loyal to my interviewees' and my audience when, at the same time, I have to please my editors and, ultimately, contribute to newspaper sales and advertising? Similar questions also came up when I interviewed feminist advocates and activists. In what ways was I a feminist journalist? Was I interviewing women when I covered economic and political issues? Was a mainstream newspaper an adequate vehicle for my feminist journalistic goals and ethics?

These questions ignited my interest in the use of the internet as a form of alternative media and citizen journalism by the feminist movement in Puerto Rico, specifically the 18 organizations and groups in the feminist coalition *Movimiento Amplio de Mujeres de Puerto Rico* (MAMPR). The MAMPR has two Facebook accounts, with 1,289 and 267 members, a blog since 10th April 2008 that has received 33,152 visits[1], a Twitter account with 87 followers, and a listserv registered under Google since 2009 with approximately 100 participants.[2] Some MAMPR members also have their own blogs. My research goals are exploratory and descriptive. This chapter addresses how the MAMPR has employed the internet to advance a feminist agenda in the mainstream media. In order to answer this, I discuss issues around the internet as a medium of communication, exclusion, and access and participation in the public sphere.

Media scholars have used an array of concepts to describe media and journalistic practices "outside" mainstream media (see Fuchs, 2010; Atton, 2004; Couldry and Curran 2003; Downing et al., 2001; Rodríguez, 2003). Some of the concepts used to describe these are: alternative, citizen, autonomous, participatory, community, grassroots, independent, minority, popular, indigenous, emancipatory, critical, and radical. For the purpose of this chapter I use "alternative media" as a heuristic device to explore the virtual communication practices of the members of the MAMPR. As for "media" I adopt Downing's (2003) broader definition that includes radio, TV, print, theater, dance, songs, graffiti, video, and the internet. Mainstream media are described as large-scale and homogenous, state-owned or commercially driven, vertically and hierarchically organized, and "carriers of dominant discourses and representations" (Bailey, Cammaerts and Carpentier 2008:18). Alternative media is defined as being counter-hegemonic, non-commercial, small-scale, independent from the state and the market, and community oriented. Bailey, Cammaerts and Carpentier's (2008) definitions and theorizations are useful and provocative because they conceptualize "alternative media" as a hybrid, multidimensional, and changing form of expression that should not be encased in static or binary categories. For instance, there are alternative projects that emulate mainstream media structures and practices – in terms of how they generate capital, their internal organization, or editorial decisions. Bailey, Cammaerts, and Carpentier (2008) argue: "the definition of 'alternative' media should be amplified to include a wider spectrum of media generally working to democratize information/communication" (2008:xi) and "articulated as relational and contingent on the particularities of the contexts of production, distribution and consumption" (2008:xii). My research attempts to cross borders, blur boundaries, and reveal alternative media practices that defy traditional definitions.

In this chapter, I will explore how leaders of the MAMPRs member organizations, who participate in their listserv,[3] define their uses of the internet and their relationship with the mainstream media in Puerto Rico. Feminist literature on media has found that women are mostly misrepresented in, and lack access to, the mainstream media in their specific contexts (Byerly, 2012). This has contributed to women, and other underrepresented groups, creating their own media across the globe (Chambers, Steiner and Fleming, 2004.). In Puerto Rico, there is a rich history that dates back to the 19th century to feminist union leaders, reformists, and suffragists, creating alternative media—such as specialized magazines, newspapers, and leaflets – to have a space to represent themselves in their own terms and voice their concerns (Valle Ferrer, 1990, 2004; Rivera Lassén and Crespo Kebler, 2001; Dueñas Guzmán 1993; Bauzá 1987). Feminists in Puerto Rico have also had access to local mainstream media, due to a specific historical context that will be discussed later in the chapter.

I initially expected to find that the MAMPR's members were using the internet as a hybrid form of alternative media, but not necessarily as a space of contestation, because they have access to, and are mostly well represented by, the mainstream media, which spans newspapers, television and radio. Only the first premise of this initial hypothesis proved to be true, the one regarding access. I have found—through interviews with representatives of 14 of the 18 MAMPR organizations—that although almost all of the organizations have access to the mainstream media, they have mixed feelings about the coverage of gender issues, even though they are satisfied with the coverage of the MAMPRs activities. They opined that the mainstream media's coverage of gender issues in Puerto Rico—such as violence against women, reproductive rights, public policies, and sexual, racial and class discrimination—is contradictory, and definitely not a priority. But this has not strengthened their online voice or fomented a significant alternative narrative.

One of the possible explanations is that their access to mainstream media has been fairly guaranteed, which in itself is an important accomplishment. In Puerto Rico feminism and journalism have shared a close history for more than a century, since many feminist activists have also been journalists working in mainstream and alternative media alike. Therefore, critique of the mainstream media might threaten this access, and relationships with reporters and editors. There can be other explanations, such as lack of time, inadequate digital literacy and computer skills, or that in Puerto Rico mainstream media (mostly newspapers) still set the public agenda, and therefore remain the focus of many social movement's media strategies. Many of the participants were also concerned with issues of exclusion, in terms of how many women have access and the necessary digital skills to actually use the internet and engage in this relatively new space. This is one of the reasons they still continue to have face-to-face meetings. Indeed the MAMPR's internet practices are mainly focused on internal networking and communications, building consensus, decision making, formulating media strategies, distributing tasks, and voicing their opinions on a diverse range of issues, mostly through their internal listserv. The internal mailing group serves as a space of internal discussions that precede the more refined messages that will eventually go out to the mainstream media. Their internet practices are mostly reserved for internal communications, although their conversations in this space also "cross over" to face-to-face meetings and mainstream media, as will be discussed.

THEORETICAL FRAMEWORK

Little has been written about the connections between the feminist movement in Puerto Rico and its media practices (Valle Ferrer, 2004; Rivera Lassén, 2007). The focus of my research intersects with the literature of women and media, specifically the internet (Byerly, 2012; Byerly and Ross, 2006; Chambers, Steiner and Fleming, 2004). Alternative media research is also of interest since scholars are constantly reworking concepts of voice, participation, exclusion, the public sphere, and empowerment (Atkinson, 2010; Fuchs, 2010; Bailey, Cammaerts and Carpentier, 2008; Couldry and Curran, 2003; Atton, 2002; Downing et al., 2003; Rodríguez, 2001; Riaño, 1994). Moreover, I build on feminist analyses on the intersections between gender and the public sphere (Byerly and Ross, 2006, Chambers, Steiner and Fleming, 2004, Fraser, 1992). Manuel Castells' (2001, 2007, 2009) theories on the intersections of global corporate media and local grassroots networks, the rise of online "mass self-communication," and the internet as a potential space of public participation and communication power also frame this study.

Five themes run through the literature on alternative media, women and media, and women and the internet: (1) alternative media provides a space where groups can represent themselves when misrepresented by the mainstream media; (2) alternative media facilitates participation in the public sphere when groups have been excluded by the mainstream media; (3) groups that create alternative media are generally horizontally and non-hierarchically organized; (4) utopian and dystopian rhetoric of the internet as a space that permits diverse voices to express themselves, and (5) the internet holds the *potential* of transforming and redistributing unequal power relations.

Fraser (1992) describes Habermas's conceptualization of the public sphere as a "body of 'private persons' assembled to discuss matters of 'public concern' or 'common interest' in the context of absolutist regimes in early modern Europe (1992:112). Through the public sphere, society would hold the state accountable. But this idea of an open and accessible public sphere is based on exclusions, rather than inclusions. Using revisionist historiographies that have recorded how marginal groups such as women, workers, people of color, and gay and lesbians have created alternative publics, Fraser (1992) proposes the concept of "*subaltern counterpublics* in order to signal that they are parallel discursive arenas where members of subordinated social groups invent and circulate counterdiscourses to formulate oppositional interpretations of their identities, interests and needs" (1992:123). As an example she mentions the alternative media outlets and projects of feminists in the United States, such as bookstores, journals, video networks, academic programs, and festivals. Through these counterpublics, feminists have been able to insert issues such as "domestic violence," "sexual harassment," and "the double shift" into the official public sphere.

Journalism and feminism have shared a long history in Puerto Rico. Well-known journalists have openly identified with feminist causes and participated in the mainstream public discourse through their journalistic work and political activism since the late 19th century.[4] Feminist journalists also contributed significantly to the 1970s brewing feminist activism (Valle Ferrer, 2004). Therefore, the feminist movement in Puerto Rico has participated significantly in the public sphere maintained by mainstream media. It has inserted its messages into the mainstream media and had the opportunity to influence public discourse. Feminist organizations have also created alternative publications in which they have exposed, discussed, and debated a wide range of issues. In other words, having access to the mainstream media did not conflict with the need to create alternative spaces of expression. They used leaflets, bulletins, flyers, banners, magazines, and newspapers. During the second wave hundreds of feminist art pieces, such as paintings, murals, drawings, silkscreen posters, and caricatures were created (Fernández Zavala, 2007). Although most of the publications were short-lived, mainly due to financial limitations, they were important platforms for alternative communication.

Increasingly, the internet is playing a crucial role in the analysis of public discourse and the public sphere (Papacharassi, 2010, 2002). Many social movements are using the internet for networking, communicating, sharing information, coordination actions, creating transnational alliances, and for participating in the mainstream public realm in some contexts (Gerbaudo, 2012; Earl and Kimport, 2011; Juris, 2008; Rucht, 2004). Although the importance of the internet in the 2011 Arab Spring has been contested, it is undeniable that it played a role in mobilizing protests in some of the countries (Castells, 2012; Tufecki and Wilson, 2012; Lynch, 2011). Castells (2007) states that the appearance of mass self-communication, through SMS, blogs, podcasts, wikis, and social media such as Facebook and YouTube, and Twitter, "offer an extraordinary medium for social movements and rebellious individuals to build their autonomy and confront the institutions of society in their own terms and around their own projects" (2007:249). There is a long and rich history of citizens' media around the world, especially in Latin America, that has mainly taken the form of community radio, guerilla TV, and video production (Allan and Thorsen, 2009; Riaño, 1994; Rodríguez, 2001; Martín-Barbero, 1987).

Although social movements have been using media for a long time to participate in public debates, Castells (2007) believes that for new social movements "the internet provides the essential platform for debate, their means of acting on people's minds, and ultimately serves as their most potent political weapon" (2007:250). Castells (2009) also envisions the possibility of social grassroots movements using the internet horizontally while they simultaneously insert their messages in the mainstream media. In other words, one strategy does not exclude the other. Listservs and email groups have also been valuable platforms for social movements' participation in the public sphere. Bailey, Cammaerts and Carpentier (2008) explain that: "For civil society organizations, activists and social movements, e-mail, mailing lists, and forums increasingly represent a (cost-) efficient means to distribute alternative-counter-hegemonic-information, to mobilize online as well as offline direct action, to debate issues, and even at times to become a tool for internal decision-making. Because of this, it is often claimed that the internet is increasingly important for strengthening the public sphere or public spaces" (2008:97). This understanding of listservs resonates with the findings of this study. The MAMPRs listserv can be considered a form of citizen media in the sense that it provides a critical space for alternative content production that eventually—although not always—can be transferred to the mainstream media through press releases, interviews, and other events.

Some feminist media scholars have argued that the internet may have the potential to provide a more democratic, open, and participative space for women and gender issues (Byerly, 2012; Youngs, 2004; Harcourt, 1999). Analyzing certain women groups' internet use, Chambers, Steiner and Fleming (2004) state that "independent information distribution media projects on the internet are quickly growing in both number and sophistication, spurred on both by dissatisfaction with the mainstream news media and increasingly easier access to new media technology [...] Importantly, several of these internet news and information networks challenge mainstream definitions of news by overtly advancing feminist initiatives" (2004:190). It is precisely the networking aspect of the internet that has made it a potentially significant medium for many social movements. But despite the internet's potential for networking, sharing information, transnational communication, blurring of consumer/producer binaries, and challenge to mainstream media's authority, there are important issues of skills, exploitation, violence, and unequal power relations (Choi et al., 2006; Atton, 2004). Women's participation in the public sphere, or rather the mobilization of gender and feminist issues and voices in the public sphere and the possible overlapping of counter-public spheres, is a crucial consideration when studying feminist online alternative media and communications.

Feminist media and social movement scholars are also skeptical about the democratic potential of the internet (Friedman, 2005). Rodríguez and Kidd (2010) remain critical of web 2.0 projects because of the possibility of government and corporate surveillance and data mining of participants by corporations. Media activists, for example, debate whether corporations such as Microsoft, Google, and Yahoo, just to name a few, should provide and sustain the services and platforms used by grassroots and radical activists. Rodríguez, Kidd and Stein (2009), therefore, argue that community radio is the "world's most significant medium, especially for marginalized groups, in both rural and metropolitan areas" (2009:6). Radio broadcasting can function without electric current, it can be extremely mobile, low-powered, and it does not require reading or writing skills. But, in terms of internal organization, radio projects can be as problematic as any other medium in terms of power relationships. As an example, the Asia Pacific Women's sector of AMARC (the World Association of Community Radio Broadcasters) studied the situation of women working in community radio. The study was prompted by concerns of gender inequality within community radio's internal structures. The result was the publication of a "Gender Policy for Community Radio" (AMARC-WIN, 2008) to ensure women's full participation in these grassroots organizations.

This leads us to the following questions that are debated incessantly in the literature: How democratic is the internet? How plural is it? The internet is not a panacea for marginalized people and communities (Atton, 2004). There are issues of power that must be taken into account. Leslie Regan Shade (2002) describes the exclusion of Zapatista women's voices in the widely celebrated internet strategies used by the *Ejército Zapatista de Liberación Nacional* after its insurrection in Chiapas, México, in 1994. And in their case study of the Independent Media Centers (IMCs), Lisa Brooten and Gabriele Hadl (2009) reveal the subtle network media practices that reinforce hierarchies that exclude women. In "Women@Internet," Laura Agustín (1999) problematizes the question of voice: "For, with all the rhetoric about the need to liberate 'unheard voices', we miss an essential point: those voices have been talking all along. The question is who is listening" (1999:155). These are some of the issues concerning the possibilities of the internet as a terrain for transcending traditional communicative and participative barriers.

My research question is located in this literature in multiple ways. Through their digital practices, specifically their internal listserv, the MAMPR engages in a hybrid form of media that blurs traditional boundaries between alternative and mainstream, internal and external communications, and online and offline. Activists have established a space in which they can build consensus, network, dissent, consult, plan meetings and events, and design media strategies aimed at positioning themselves within the mainstream mediascape. In this way, the listserv also supports a *subaltern counterpublic*, which at times intersects with the official mainstream public sphere in Puerto Rico. One the other hand, there are also problems involving access, media literacy, exclusion, participation, and ability to truly transform the mainstream mediascape.

FINDINGS

The MAMPR, founded in 2007, is a horizontally organized coalition comprised of 18 member organizations, 30 individual members, and nine affiliated organizations, mobilized across issues of gender, race, sexuality and class. It does not work with a budget or have physical headquarters. Some of the member organizations are embedded within formal academic institutions, some are informally organized, and most are established non-governmental/non-profit organizations. Some depend completely on voluntary work, and some have full-time, part-time, and freelance employees.

This chapter is based on semi-structured interview with representatives from 14 of the 18 member organizations of the MAMPR, all of who participate in the listserv (see also Shokooh Valle, 2010). Evidence on the nature and content of the messages rely on the interviews with the activists. MAMPR's documents and multimedia presentations were also analysed for context (to date there is no scholarly research on the MAMPR). The MAMPR started its listserv on Google in February 2009.[5] The listserv had 101 members, as of July 14, 2013. The listserv is administered by a member of the coalition, and is only open to members and friends of the movement. Between August 2012 and July 2013, 3,662 emails were sent within the group, for an average of 305 messages sent monthly.

The internet (i.e., listserv, blogs, websites, social media) seems to be an important part of these organizations' work. In some way or another, in the form of websites, blogs, email, or as social media users, the internet plays a fundamental role in the lives of these organizations. Nine out of 14 have their own websites or professional blogs. Interestingly, though, only three of the interviewees had "personal" blogs. The majority of the participants said that the internet, mainly through the use of the listserv, has been a crucial tool of internal communication, networking, strategizing, reaching consensus, and source of information. The responses suggest that the MAMPR's use of the listserv blurs traditional alternative media concepts because it provides an arena for alternative communication that is not media-centric, though it may intersect with mainstream media at some point. It is also a space that allows some form of self-presentation, participation, and inclusion. Some of the participants argued that face-to-face meetings are still crucial for the feminist movement. The "difficult" and more confidential debates are discussed in meetings and reunions.

The issue of exclusion surfaced repeatedly in the interviews: exclusion from the internet in general, and from the MAMPR's listserv in particular. Although some participants, believe that the internet has provided a much needed space for traditionally marginalized groups, such as young women, many respondents noted that the internet establishes boundaries along lines of access, computer literacy, and time. There is a general sense that face-to-face meetings are still necessary, and that traditional methods of mobilization and communication should converge with online strategies.

Eleven of the participants said that their organizations have had access to the mainstream media in Puerto Rico. The other three participants did not explain whether they had been denied access, or simply had made no attempt.

One participant said: "the media has identified the MAMPR as the most updated feminist network in Puerto Rico." This presents a unique case within the literature reviewed for this study, because women generally do not have access to the mainstream media in their countries. This, as has been explained, is mostly due to the close relationship between feminism and journalism on the Island. Another interesting finding is that these eleven participants said that journalists contact them for stories, which means that reporters are actually seeking out their perspectives. The participants also contact reporters, who respond. Also, the MAMPR's activities have been widely covered in the media, specifically by the three major mainstream newspapers in Puerto Rico, *El Nuevo Día, Primera Hora* and *El Vocero*. Many participants said that the MAMPR's success in obtaining access is because of the presence of renowned feminists, such as Josefina Pantoja and Ana Irma Rivera Lassén, who have well-established relationships with the media. Another finding is that participants report that there are specific feminist reporters who regularly cover feminist events and include a gender perspective in their articles.

Whether the MAMPR and its organizations' use of the internet play a meaningful role in gaining access to the media is still being debated. Mostly, participants believe that it is still indispensable to mobilize and work offline. The listserv is used to discuss media strategies and design content that will eventually be exposed in the public realm through the mainstream media. But participants still have to call reporters, send press releases, and mobilize in different ways for those messages to get out. Of course, the MAMPR's listserv is closed, although also porous because content frequently makes a "cross-over" to mainstream media. Reporters have rarely used the organizations' websites, the participants' blogs, or the MAMPR's blog, as sources of information for their articles. Two respondents recalled specific moments in which the media published stories that cited their blogs. Yet these examples are hardly representative.

Fraser's (1992) theories on the intersections of gender and the public sphere and "subaltern counterpublics" is useful in understanding the implications of the MAMPR's communication strategies. In my reading the MAMPR has created a form of subaltern or counter-public sphere through their listserv. At *prima facie* this analysis seems contradictory and even impossible. How can a closed listserv be described as public? It is public in the sense that many of the internal discussions, debates, and dialogues are at some point transformed into public messages. It is subaltern because it is a space for the discussion of ideas, thoughts, opinions, and feelings that are generally excluded from official mainstream discourse, notwithstanding the movement's relationship with the media on the Island. There is also a conundrum here. Is it still "subaltern" when included in the mainstream media? To what extent is it still subaltern when it is also exclusionary?

In my view mainstream media may effectively co-opt certain subaltern discourses. It appropriates ideas that at some point were considered different, underground, rebellious, anti-establishment, dangerous, and even revolutionary. For example, mainstream media in Puerto Rico has "discovered" that domestic and sexual violence against women are worth covering. Once a marginal issue, news on domestic violence now fills the main pages of the newspapers. Journalists interview relatives, friends and neighbors of the victims. They interview activists and experts on domestic violence. They publish graphs and tables. Domestic violence has become a mainstream issue in Puerto Rico thanks to the feminist movement. Violence against women is part of the mainstream public discourse. This is a very important achievement, but at the same time activists should reflect on the consequences of this "mainstreaming" process. That is, to consider how mainstream media is shaping public discourse on domestic violence, and its impact on the situation of battered women on the Island.

Interestingly, the participants' responses on how the mainstream media in Puerto Rico covers gender issues—such as, but not limited to, domestic violence, unequal wages, sexual harassment, racism, and homophobia—contradicts my initial hypothesis: that the MAMPR's members were not using the internet as a space of contestation, because they were mostly well represented by the mainstream media. As a former reporter who mainly covered gender issues at the second largest newspaper in Puerto Rico, I thought the members of the MAMPR not only had access, but were also pleased with the coverage. The general sense is that the mainstream media does a fair, balanced, and accurate job of covering the MAMPR's events. But respondents criticized the media's coverage of women's and gender issues in general. Coverage of women's issues was described as contradictory, superficial, and sensationalist. Many participants said that mainstream media disproportionately covers domestic violence, because it sells newspapers. Many noted the good work of specific reporters, but gave no credit to the media as institutions. In other words, individual reporters have been able to make changes, yet the media institutions are still commercially driven and exclusive.

These findings demonstrate that the listserv has offered a space in which MAMPR activists feel they can express themselves. But, this space is also crisscrossed with a range of exclusions. The internet has yet to provide a powerful public counter-narrative vis-à-vis the mainstream media in Puerto Rico, and offline mobilization is still a fundamental part of the feminist movement's agenda. This research has also found that the listserv, although closed, is a space in which different configurations of power and alternative forms of communication are emerging. Many-to-many communication through the internet facilitates networking, consensus building and strategizing as a collective. Also, members of the MAMPR have successfully accessed and inserted their messages into the local mainstream media, although they are critical of the coverage of women's and gender issues. The MAMPR could take more advantage of the internet, especially blogs and social media, to attempt to influence the media's coverage of gender issues.

CONCLUSION

The internet is not a power equalizer. It is a platform that can facilitate certain forms of power redistribution in specific historical and cultural contexts. This does not necessarily mean that the "marginal" will gain more power through the internet. The internet, however, may provide a space in which more people can distribute their messages and participate in public debate. I do not want to idealize the internet as a perfect space for marginalized groups. What is undeniable, though, is that the internet is being used to communicate a wide range of information. Mainstream media still is an important actor in many contexts, and it's far from its demise, as some have proclaimed. However, mainstream media must learn how to share the communicative space. The flow of the conversation has changed from one-to-many to many-to-many. These transformations can open a world of opportunities for new social movements and marginalized communities.

The feminist movement in Puerto Rico, specifically the member organizations of the MAMPR, is using the internet in diverse ways in order to advance a feminist agenda in the media. The urgency and immediacy of internet communications has certainly facilitated the discussion and dissemination of issues crucial to the feminist movement. As a coalition the organizations mostly use the listserv, although the MAMPR has its own blog and many individual organizations have websites. The listserv is a space for alternative communications and a form of citizen journalism because it provides a stage for representation and counter-discourses; it is horizontally structured; and it is non-commercial, small-scale, independent from the state and the market, and community oriented (the feminist community). It also serves as an important tool to participate in public debate sustained by the mainstream media.

The listserv is a "private" space for discussions, deliberation, networking, strategizing, consensus building, decision making, information, task distribution among members, yet these efforts can produce public messages that will circulate in the mainstream/corporate/commercial media. It is alternative and mainstream, private and public, inclusive and exclusive, small-scale and large-scale, horizontal and vertical, and independent and dependent. The mainstream is still considered an important and valuable space. The MAMPR's uses of the internet are multidimensional, hybrid and unsettled. Its digital practices cross, transgress, and merge across borders and boundaries. This intermixing and border crossing enhances the MAMPR's media practices because it defies traditional, static, and binary assumptions and definitions that generally only serve to enclose, suppress, and circumscribe creative, transgressive, and innovative media manifestations and relationships. My research introduces an interesting case in the literature on gender and media because the MAMPR, through diverse online and offline strategies, has been able to participate in public discourse and gain access to mainstream media in Puerto

Rico. But it is important to point out that this success did not occur out of the blue. These achievements must be situated within the context of the long and fruitful relationship journalism and feminism have shared in Puerto Rico.

The access that the MAMPR's organizations have gained is unique in the literature on gender and media. This accomplishment must be situated in its context and history. The women's movement in Puerto Rico, dating from the suffragists, has had a long and rich relationship with the media (alternative and mainstream). Throughout decades, prominent feminist leaders and activists have founded their own alternative publications, inserted their political messages into the mainstream media, and been journalists themselves. In Puerto Rico, feminism and journalism have walked together, although this does not mean the path has been free from conflicts and tensions. As demonstrated in this investigation, this long-standing relationship has not necessarily resulted in satisfactory representations in the media.

In this research, participants divided the issue of representation into, on one hand, coverage of the MAMPR's activities, and on the other, of gender and women's issues in general. Participants said that the MAMPR receives adequate coverage, but that the representation of gender issues could be improved. Participants recognized the work of individual feminist reporters, and they faulted media institutions for the poor coverage of gender issues. Although there are instances of resistance within these powerful institutions, there have not been profound structural transformations. The fact that many women, including feminist women, in Puerto Rico have become hard-news reporters and editors in the mainstream media has not, according to these findings, resulted in deep-rooted changes regarding perceptions of gender representations. For further research it would be interesting to conduct in-depth interviews with reporters, editors, and owners of mainstream media outlets in Puerto Rico, specifically of newspapers, which continue to be the main agenda setters and shapers of public discourse in the country.

In my view the MAMPR's access to mainstream media, and its participation in mainstream public discourse, has inhibited their online presence. Participants' responded that their organizations have access to mainstream media in Puerto Rico. They also argued that mainstream media generally distort gender issues, but this problem is not one of the MAMPR's priorities at this moment. Participants' were more focused on designing strategies to attract the mainstream media (to cover their activities) than to transform the coverage of gender issues in Puerto Rico. The literature shows that alternative narratives generally flourish when people assess that they are being excluded from, and misrepresented by, the mainstream media. Perhaps members of the MAMPR would have a stronger online presence if mainstream coverage of gender issues were one of their top priorities, and if they thought that their use of the internet could effectively influence and transform public discourse. Mainstream media, especially newspapers, still shape daily public discussions on the Island.

NOTES

1. Data provided through personal communication with MAMPR activist who maintains the blog.
2. Statistics are as of July 16, 2013.
3. Listserv is defined as an electronic mailing list distributed among a group of people.
4. For an excellent investigation on how women journalists made their entrance to the news sections of the mainstream newspapers in Puerto Rico please see Maricelis Rivera Santos' MA unpublished thesis (2008): "Las periodistas al control del cuarto poder en Puerto Rico: Desde la llegada de la imprenta hasta su incursión en las páginas frontales de los diarios." Department of History, University of Puerto Rico.
5. This information was provided by an individual member of the MAMPR, and one of the administrators of the listserv, through personal communication.

REFERENCES

Agustín, Laura. 1999. "They Speak, but Who Listens?" pp. 149–155 in *Women@Internet: Creating New Cultures in Cyberspace*, edited by Wendy Harcourt. New York: Zed Books.

Allan, Stuart, and Einar Thorsen. eds. 2009. *Citizen Journalism: Global Perspectives.* New York: Peter Lang Publishing.

Atkinson, Joshua. 2010. *Alternative Media and the Politics of Resistance: A Communication Perspective.* New York: Peter Lang Publishing.

Atton, Chris. 2002. *Alternative Media.* Thousand Oaks, CA: Sage Publications.

———. 2004. *An Alternative internet.* Edinburgh, Scotland: Edinburgh University Press.

Bailey, Olga Guedes, Bart Cammaerts, and Nico Carpentier. 2008. *Understanding Alternative Media.* Maidenhead: Open University Press.

Bauzá, Nydia. 1987. "El feminismo y *El tacón de la chancleta*." Masters Thesis. University of Puerto Rico, San Juan, PR.

Brooten, Lisa and Gabriele Hadl. 2009. "Gender and Hierarchy: A Case Study of the Independent Media Center Network." pp. 203–222 in *Making Our Media: Global Initiatives Toward a Democratic Public Sphere*, edited by Clemencia Rodríguez, Dorothy Kidd and Laura Stein. Cresskill, NJ: Hampton Press.

Byerly, Carolyn. 2012. "The Geography of Women and Media Scholarship." pp. 3–19 in *The Handbook of Gender, Sex, and Media.* Hoboken, NJ: John Wiley and Sons.

Byerly, Carolyn M., and Karen Ross. 2006. *Women and Media: A Critical Introduction.* Malden, MA: Blackwell Publishing.

Castells, Manuel. 2012. *Networks of Outrage and Hope: Social Movements in the internet Age.* Cambridge, UK: Polity.

———. 2009. *Communication Power.* Oxford; New York: Oxford University Press.

———. 2007. "Communication, Power and Counter-Power in the Network Society." *International Journal of Communication* 1: 238–266.

———. 2001. *The internet Galaxy: Reflections on the internet, Business, and*

Society. New York: Oxford University Press.

Chambers, Deborah, Linda Steiner, and Carole Fleming. 2004. *Women and Journalism.* London, England: Routledge.

Choi, Yisook, Linda Steiner, and Sooah Kim. 2006. "Claiming Feminist Space in Korean Cyberterritory." *Javnost-The Public* 13(2): 65–84.

Couldry, Nick, and James Curran, eds. 2003. *Contesting Media Power, Alternative Media in a Networked World.* Lanham, MD: Rowman & Littlefield Publishers.

Downing, John, Tamara Villarreal. Ford Genève Gil, and Laura Stein. 2003. *Radical Media: Rebellious Communication and Social Movements.* Thousand Oaks, CA: Sage.

Earl, Jennifer, and Katrina Kimport. 2011. *Digitally Enabled Social Change: Activism in the internet Age.* Cambridge, MA: MIT Press.

Fernández Zavala, Margarita. 2007. "La huella gráfica de los feminismos en Puerto Rico: Una aproximación a la memoria de la desmemoria." *Identidades* 6: 27–70.

Fraser, Nancy. 1992. "Rethinking the Public Sphere: A Contribution to the Critique of Actually Existing Democracy." pp. 109–142 in *Habermas and the Public Sphere*, edited by Craig Calhoun. Cambridge, MA: MIT Press.

Friedman, Elisabeth J. 2005. "The Reality of Virtual Reality: The internet and Gender Equality Advocacy in Latin America." *Latin American Politics & Society* 47(3): 1–34

Fuchs, Christian. 2010. "Alternative Media as Critical Media." *European Journal of Social Theory* 13(2): 173–192.

Gerbaudo, Paulo. 2012. *Tweets and the Street: Social Media and Contemporary Activism.* London, United Kingdom: Pluto Press.

Guzmán, Maximiliano Dueñas (1993). "La prensa alternativa en Puerto Rico: diecisiete experiencias" PhD dissertation, Universidad de Puerto Rico.

Harcourt, Wendy, ed. 1999. *Women@ Internet: Creating New Cultures in Cyberspace.* New York: Zed Books.

Juris, Jeffrey. 2008. *Networking Futures: the Movements Against Corporate Globalization.* Durham, NC: Duke University Press.

Lynch, Marc. 2011. "After Egypt: The Limits and Promise of Online Challenges to the Authoritarian Arab State." *Perspectives on Politics* 9(2): 301–310.

Martín-Barbero, Jesús. 1987. *Procesos de comunicación y matrices de la cultura: Itinerario para salir de la razón dualista.* Mexico City, Mexico: Ediciones G. Gili and Federación Latinoamericana de Facultades de Comunicación Social (FELAFACS).

Papacharissi, Zizi A. 2010. *A Private Sphere: Democracy in a Digital Age.* Cambridge, UK: Polity Press.

———. 2002. "The Virtual Sphere: The internet as Public Sphere." *New Media & Society* 4(1): 9–27.

Riaño, Pilar, ed. 1994. *Women in Grassroots Communication: Furthering Social Change.* Thousand Oaks, CA: Sage Publications.

Rivera Lassén, Ana Irma. 2007. "Las organizaciones feministas en Puerto Rico o el holograma del poder." *Identidades* 5: 117–137.

Rivera Lassén, Ana Irma and Elizabeth Crespo Kebler. 2001. *Documentos del feminismo en Puerto Rico: Facsímiles de la historia Volumen 1 (1970–1979).* San Juan, PR: Editorial de la Universidad de Puerto Rico.

Rodríguez, Clemencia, Dorothy Kidd, and Laura Stein. 2009. *Making our Media : Global Initiatives Toward a Democratic Public Sphere.* Cresskill, NJ: Hampton Press.

Rodríguez, Clemencia. 2001. *Fissures in the Mediascape: An International Study of Citizens' Media.* Cresskill, NJ: Hampton Press.

Rucht, Dieter. 2004. "The Quadruple 'A': Media Strategies of Protest Movements Since the 1960s" pp. 29–56 in *Cyberprotest: New Media, Citizens, and Social Movements.* London, England: Routledge.

Shade, Leslie Regan. 2002. *Gender and Community in the Social Construction of the Internet.* New York: Peter Lang Publishing.
Shokooh Valle, Firuzeh. 2010. "Getting into the Mainstream: The Virtual Strategies of the Feminist Movement" MA Thesis, Department of Journalism, Northeastern University.
Tufekci, Zeynep, and Christopher Wilson. 2012. "Social Media and the Decision to Participate in Political Protest: Observations From Tahrir Square." *Journal of Communication* 62(2): 363–379.
Valle Ferrer, Norma. 1990. *Luisa Capetillo, Historia de una mujer proscrita.* San Juan, PR: Editorial Cultural.
———. 2004. "Voces de fuego: Un estudio de radio recepción." Ph.D. Dissertation, Center for Advanced Studies of Puerto Rico and the Caribbean, San Juan, PR.
———. 2006. *Las mujeres en Puerto Rico.* San Juan, PR: Instituto de Cultura Puertorriqueña.
World Association of Community Radio Broadcasters Women's International Network (AMARC-WIN) (2008). *Gender Policy for Community Radio.* Montreal, Canada: AMARC-WIN.
Youngs, Gillian. 2004. "Cyberspace: The New Feminist Frontier?" pp. 185–205 in *Women and Media: International Perspectives*, edited by Karen Ross and Carolyn M. Byerly. Malden, MA: Blackwell Publishing.

CHAPTER SIXTEEN

Reporting a Revolution and Its Aftermath: When Activists Drive the News Coverage

YOMNA KAMEL

Afflicted with problems of unemployment, injustice, corruption and decaying political systems, the Arab region has been witnessing a wave of unprecedented uprisings since December 2010, sweeping away ruling regimes that had been in place for several decades. The Arab uprisings, also known as the Arab Spring, were arguably sparked by images of Mohammed Bouazizi, a young Tunisian man who immolated himself in protest of injustice and corruption. Tunisian activists blogged his story and quickly made it known across Tunisia and the world. Bouazizi's saga along with images of Tunisians protesting Bin Ali's regime were then picked up by the international news media, which amplified the Tunisian voices to the world and inspired more suppressed Arabs to revolt against their dictators. In taking down Bin Ali's authoritarian regime, the revolution became contagious, triggering massive demonstrations and protests across other parts of the Arab World. Egyptians followed Tunisians and revolted against Mubarak's regime.

Similar to Bouazizi's story, the Egyptian Revolution was sparked by police torture and murder of Khaled Said, a young man from Alexandria who posted a video clip online showing police corruption. A Facebook page has been named after him titled 'We are all Khaled Said' called for a massive protest against Mubarak's regime on January 25th 2011, marking the annual Egyptian Police Day. Thousands of Egyptians, led by anti-regime groups and individual activists, joined the online call and took to the streets in massive demonstrations across Egypt calling for the collapse of Mubarak's regime. These anti-regime movements are social media–equipped groups and individual activists. They have been using blogs, Facebook and Twitter not only to organize and mobilize, but also to record, in text, images and footage, human rights violations committed by these dictatorial regimes and make them known to the international community, particularly news organizations.

From the image of the Tunisian Bouazizi to the image of the Egyptian blue bra-girl who was dragged, stripped and brutally beaten by security forces in Cairo's Tahrir Square, huge amounts of visual and textual content have been posted and circulated on social media platforms by citizen journalists, who are also activists after a cause. Citizen journalists are often described as 'activist reporters' as it is difficult to distinguish between activists and citizen journalists because both perform similar online activities using the same social media tools to disseminate information and promote their causes (Kempa, 2012). Activist reporters have been collaborating with professional journalists, networking with them via social media, to amplify their voices and make their stories heard. For several news organizations, social media-equipped activists have become indispensable sources of news, particularly under intense circumstances like the Arab uprisings. Thus, it is argued that their collaboration with news organizations have brought changes to journalism practices, particularly journalists-source relationship, and raised questions about the role of social media in reshaping this relationship. This chapter offers an attempt to understand the role of social media–equipped activists in driving the news coverage of the Egyptian Revolution and its associated events through a case study that looks at the reporting of five selected news organizations—Al-Jazeera, BBC, CNN, Russia Today and XINHUA—which represent a geopolitical spectrum of media ecologies.

Al-Jazeera	The Doha-based station is owned by the state of Qatar. It was launched in 1996, first in Arabic and in 2006 it added the English service. Al-Jazeera offers 24-hour news services and programs on controversial issues.
BBC	It is one of the oldest news media in the world, founded in 1922 (Radio) and in 1936 (TV). It launched its first foreign service (Arabic) in 1938. Now, BBC has 32 other languages beside English.
CNN	Launched in 1980, CNN is part of a US media empire, Time Warner. Its coverage of the first Gulf War made it known internationally and it now reaches to millions of viewers worldwide.
RT	Launched in 2005, Russia Today is funded by the Russian government. It has TV and online services in Arabic, English, Spanish and Russian, reaching to 550 million in 100 countries.
XINHUA	It is the news agency of the Chinese government, founded in 1931. With almost all Chinese media, electronic and print, counting on its services, XINHUA is considered the news gateway to China. It has services in Chinese, English, Spanish, French, Russian and Arabic.

AN EVOLVING JOURNALISTS-SOURCE RELATIONSHIP

Media scholars studying the relationship between journalists and sources and how it manifests itself in news reporting have repeatedly criticized what they describe as 'elite-sourcing routine'. It refers to news organizations' practice of over-relying on sources of authority (elite-sources) while marginalizing others. The issue becomes more noticeable when analyzing news coverage of violent political conflicts to find out that such sourcing practices might let human rights violations go unreported and leave activists' and victims' voices often unheard (Chomsky, 1982). For example, although anti-regime protests and demonstrations are not new to the Arab countries, prior to the 1990s, they were almost absent from the news agenda. The 1970s and the 1980s witnessed waves of demonstrations and strikes in protest of cutbacks in subsidies and price rise (in Egypt and Morocco), and in other cases for political reasons (in Syria and Jordan). Protesters and anti-regime activists were met by security forces' severe brutality, which killed and injured many of them, but had none or limited media coverage, which was dominated by a state-perspective. Official (elite) sources were predominant, giving their own versions of the stories. For example, the state-controlled Egyptian media during President Anwar Sadat's time (1970–1981), repeatedly labeled university students' protests against food price rise in 1977, 'The uprising of thieves' (Brownlee, 2011). Similarly, media reports on the 1982's Syrian uprising were limited, and the voices of the rebels were absent. In February 1982, the *Guardian* reported heavy fighting in the city of Hama between the Syrian government forces and rebels and stated that the Syrian army sealed off the city, and there were no eyewitness accounts (Rodrigues, 2011). "Before there was Facebook, Twitter, YouTube, or even Al Jazeera, there was Hama, Syria", reflected former CBS News Middle East correspondent, Lawrence Pintak.

> It was 1982 and an anti-regime protest was put down with ferocious violence. The Syrian government simply destroyed whole sections of the city, leaving at least ten thousand people dead. But the slaughter went unreported in that closed society. Those of us trying to cover the story from nearby Beirut had little more to work with than hearsay, and certainly no pictures. (Pintak, 2011)

Voices of anti-regime individual and group activists in the Arab region were absent from the local media and had limited access to the international media. The Arab public in general had narrow access to foreign radio services like BBC World Service, Voice of America or Monte Carlo, which provided limited coverage of what local media ignored (Wheeler, 2006). As a result, news coverage of the last century's violent political conflicts, particularly in authoritarian regimes, was dominated by state elite-sources and lacked voices of anti-regime activists and victims of human rights violations.

As of the early 1990s, the development of the internet has given power to anti-regime groups and individuals. Empowered by the tools the internet provides, activists started to get more attention from mainstream media. Media scholars started to notice possible shifts in news sourcing practices as voices of activists, who were usually marginalized by mainstream media, have become more visible (Walejko & Ksiazek, 2008; De Keyser & et al., 2008; Kenix, 2011; Aouragh and Alexander, 2011). Also, several studies highlighted the new practice of using blogger-activists as sources in news stories (Walejko & Ksiazek, 2008; Kenix, 2011; Oriella, 2011). According to Kafi (2010):

> The support the bloggers lend to popular causes has brought some of them exceptional visibility. This impact, remarkably, extends to traditional media. Here their comments often make an impact unseen in countries with highly refined systems of news generation and distribution. For instance, many Egyptian media have begun to regularly reprint entries from well-known blogs of well-known bloggers such as Kareem Amer or Wael Abbas thereby tapping into news sources normally inaccessible to them and increasing their circulation. (Kafi, 2010)

Yet shifts in news sourcing practices remained relatively slow and limited[1] until the development of the internet's latest novelty: social media. The tools of social media have 'communicationally' empowered marginalized individuals and groups; giving them means of communicating and networking with stakeholders, including journalists. Anti-regime activists in countries ruled by authoritarian regimes, like most Arab countries, who were voiceless and marginalized for years, have become visible to the international media, and their voices are now heard by a wider global audience (Kaplan, 2008; Heinrich, 2008; Hermida, 2010; Hermida, et al., 2012; Oriella, 2011; Wallsten, 2011; Aouragh & Alexander, 2011; Harlow & Johnson, 2011). This was evident in the news coverage of the Egyptian uprising where news organizations relied on content posted on Twitter, Facebook and other social media platforms from a variety of (official and unofficial) sources including citizen journalists, activists and NGOs (Bossio & Bebawi, 2012). Even a few years prior to the Arab uprisings when social media platforms were introduced to the region, Egypt, for example, witnessed the rise of numerous anti-regime groups and individual activists who were equipped by social media tools. They reported intensively on human rights violations and collaborated with journalists to amplify their voices to the international community. *Kefaya,*[2] The April 6th Movement, the National Association for Change and 'We are all Khaled Said' are just few of many social media-equipped anti-regime movements established a few years before the Egyptian uprising (Lim, 2012; Sanders, 2012).

As the Arab region, since the end of 2010, has been witnessing a wave of anti-regime uprisings led by social media–equipped individual and group activists, the uprisings coincided with a remarkable increase in internet and social media users in the region. According to the International Telecommunication Union, the number of internet users in the Arab region in 2010–2011 has increased by 23%, which is higher than the world average growth of 13%. Internet users in Egypt, for example, in June 2012 reached 31.2 million, with a growth of 21% from June 2011 (Al-Ahram, 2013). Also, social media users in the Arab region have also increased by 20–30% (ITU, 2012). Facebook, the most popular website in Egypt, had 12.17 million users in December 2012, up from 9.4 million in 2011 and 4.2 million in 2010 (Al-Ahram, 2013).[3]

Also, the Arab Social Media Report 2011 recorded growth and shift in social media usage from being social to becoming primarily political. The shift was noticeable with the uprisings in Tunisia in December 2010 and in Egypt in January 2011 (Dubai School of Government, 2011).[4] Arab activists have become more visible on the social media sites, mainly Facebook and Twitter, not only to network and mobilize, but also to report on the events they are part of. They have been recording human rights violations committed by parties involved in the Arab uprisings and uploading, on social media platforms, text, images and footage, which are often picked up by journalists and remediated to a wider audience. Sami Ben Gharbia, a leading Tunisian blogger, believes that much of the content from the Tunisian revolution that appeared in traditional media originated on Facebook, re-posted on Twitter for journalists and others (Ghannam, 2011).

MEDIA–ACTIVISTS COLLABORATION

Several scholars argue that social media have contributed in creating strong ties between activists and journalists to the benefits of both sides, in most cases. Iskander (2011) argues that activists on social media will always need the traditional national media to reach out to those with no internet access. Idle and Nunns (2011) also noted "The activists on Twitter were not only talking to their fellow Egyptians, but to the international media and the world [...] Professional journalists also used the site as did more opinion-orientation bloggers. The result was like a company of artists painting constantly updated pictures of events" (Idle & Nunns, 2011). Iskandar's view agrees with Idle and Nunns that "it was the relationship and interaction between social media and traditional media" that created "an environment for renewed political activism." (Iskander, 2011) Also, Radsch (2012) argues "Twitter became a real-time newsfeed, connecting journalists directly with activists and becoming a key tool in the battle to frame the protests and set the news agenda, particularly in the international media like Al Jazeera and elite Western outlets."[5]

Such collaboration between news organizations and activists in reporting on the Egyptian Revolution and its associated events has developed into what seems like an interdependent relationship. This relationship is visible on social media platforms, where news organizations network with citizen journalists and activists. As news organizations might not be equally embracing this interdependent relationship, its impact on the coverage varies. For example, Al-Jazeera, more than any news channel, has been continually under fire from the ruling authorities in Egypt both pre and post January 25th Revolution. The interim military regime, post January 25th, accused Al-Jazeera of inciting the Egyptian public against them. The tension between the military regime and Al-Jazeera escalated on September 12th, 2011 when the Egyptian security forces raided its office in Cairo, disrupted the transmission of Al-Jazeera Mubasher Misr,[6] confiscated equipment and arrested its staff. Such harsh measures against Al-Jazeera came out of the accusation that the news organization was not only reporting, but also igniting tension through its collaboration with and privileging of anti-regime activists. It was not unusual to see protesters in Tahrir Square carrying signs saying "Thank you Al Jazeera Channel," next to signs thanking Facebook and Twitter.

As political cyber-activists need news channels to amplify their voices, news organizations need their feed, particularly from areas of the world that are not accessible by their own journalists. Prior to the collapse of Bin Ali's regime in Tunisia, Al-Jazeera had to rely on a network of bloggers in Tunisia and considered them, according to senior correspondent for Al-Jazeera English, Alan Fisher, "a vital link in a country where Al-Jazeera was banned" (Fisher, 2011). Similarly, Al-Jazeera was keen to build a network with key bloggers and activists in Egypt, according to Riyaad Minty, who leads the news organization's social media initiatives (Mir, 2011). International news organizations have also worked on enhancing audience engagement and introducing interactivity features to their online platforms, inviting more citizens to participate in the coverage through posting text, audio or video materials (Duffy, 2011).

Al-Jazeera, for example, launched Mubasher Misr, a channel and a website dedicated to covering Egypt's news and providing online space and airtime for citizens to participate with content. Several news organizations have also increased the interactivity features of their online platforms and launched their own pages and accounts on social media. Through their Facebook pages and Twitter accounts, news organizations have enabled citizens and activists to connect with their editorial teams, or simply post their entries online and allow the channels to use their contributions in news services.

In both revolutions, the Tunisian and the Egyptian, online content provided by citizen journalists and activists played a central role not only in calling for an action, but also in reporting on the action and putting their revolutions' stories on the news agendas. Several studies on cyber-activism and citizen journalism, done before and after the Arab Spring, have highlighted the role that citizen journalists and cyber-activists play in putting certain issues on the news agendas and arguably in mobilizing people and bringing change on the grounds.

> The internet enables people to not only set their own media agendas but to influence others' issue agenda by helping them locate and contact people who care about similar issues. It also gives more power to people who have an agenda item that is not normally reported in the major mass media. (Rostovtseva, 2009)

As Arab citizen journalists and activists were after their cause, telling the world their own stories via various new media tools, dozens of international news organizations have intensively covered the Arab Spring. Several international news organizations which offer multilingual news services (including Arabic) do not only provide news service to their international audiences, but also to the Arab public (whose majority does not speak English). In Egypt, at a time when the local Egyptian Television, which is a state-run organization, lacked credibility and sided with the regime against the January 25th protesters, Egyptian spectators turned to several international news organizations which offer Arabic service, such as Al-Jazeera, BBC and CNN. "This credibility crisis was attributable to coverage from transnational satellite channels such as Al Jazeera, as well as to the reporting of citizen journalists, who provided minute-by-minute unedited accounts of actions on the ground" (Khamis, 2011).

REPORTING SCAF AND THE EGYPTIAN REVOLUTION

Post January 25th Revolution, which ousted Mubarak, international news organizations have continued their coverage of news coming from Egypt under its new ruling authority; the Supreme Council of Armed Forces (SCAF). Media focus on SCAF-associated events was not only because they, collectively, represented a crucial transitional stage to a new democratic Egypt, but also because of the dramatic nature of bloody clashes and violence between the January 25th Revolutionaries and SCAF. The military council, which heads the Egyptian Army, is one of the strongest organizations of the Egyptian state. It stems its powerful and respected status from a history of siding with the people against a corrupt royal regime and the British occupation of the country, leading a coup d'état in July 1952 that ousted King Farouk and established the first Egyptian Republic. Since then and until the January 25th Revolution, all presidents of Egypt, Mohamed Naguib (1953–1954), Gamal Abdel Nasser (1954–1970), Anwar El Sadat (1970–1981) and Hosny Mubarak (1981–2011), came from the military organization (Al-Ahram, 2012). The president of the Republic of Egypt is also the head of the armed forces and the head of the Supreme Council of Armed Forces (SCAF), which is a group of senior officers governing the Egyptian Army.

In response to the uprising on January 25th 2011, SCAF, headed by Mubarak, held an emergency meeting and decided to deploy army soldiers to bring order and security to the streets. Throughout the first 18 days of the uprising, the army neither used violence against the January 25th Revolutionaries nor tried to dismantle their protests. Army soldiers were stationed around the government's strategic buildings to protect them while remaining neutral and avoiding clashing with protesters. The public sentiment was positive towards the army, and many protesters chanted slogans like 'the army and people are one hand' (El-Khalili, 2013). As the situation escalated in 18 days, Mubarak was forced to step down on February 11th 2011, and transfer his authorities to SCAF and its new head, Field Marshall Mohamed Hussein Tantawi. Upon taking over, SCAF suspended the constitution, dissolved Parliament and issued a number of constitutional declarations as steps towards reform. Initially, SCAF promised to fulfill the revolutionaries' demands and announced it would remain in power for a six-month transitional period, begin reforms and prepare the country for elections.

However, it was a short truce between the interim military regime and activists as army soldiers clashed with hundreds of protesters who remained in Tahrir square and violently evicted them. As the January 25th revolutionaries were not satisfied with SCAF's pace of reform, they took to the streets several times organizing million-man marches to put pressure on the military council to speed up the transitional process and meet their demands. Scores of protesters were killed, injured and arrested in clashes with the army. The public sentiment changed to anti-SCAF—particularly as it extended the transition for over a year and did not call for elections after six months as it had promised earlier. As SCAF remained in power for 16 months between February 2011 and June 2012, the country witnessed bloody clashes that killed and injured tens of protesters. Also, more than 12,000 civilians were tried before military courts, and gross violations of human rights, including systematic harassments of female protesters, were committed (Amnesty, 2012). Similar to the beginning of the uprising, activists used social media to make SCAF practices known and record, in text, image and footage, violations of human rights committed under the interim military rule. They launched anti-SCAF campaigns on social media, using hashtags like #NoSCAF and #NoMilTrials on Twitter (El-Khalili, 2013). Also, several anti-SCAF pages in Arabic and English were launched on Facebook. Text, images, footage and live streaming were provided by activists on their social media platforms to tell the world about SCAF's practices and to catch the attention of the international media (Bengtsson, 2013; El-Khalili, 2013).[7] Such intense circumstances provide a rich case study to examine the role of social media–equipped activists in driving the news coverage of the Egyptian revolution and its associated events.

Based on a quantitative study of SCAF-driven coverage by the Arabic and English news sites of the five selected news organizations, a sample of 194 news articles and social media entries was extracted from the weeks characterized by media coverage intensity of SCAF (55+ stories per week) between January 2011 and January 2012. The sample included 72 Arabic and 68 English news articles published on the websites of the five selected news organizations, Al Jazeera, BBC, CNN, RT and XINHUA, and 54 posts by six political activists/movements on their social media platforms: the April 6th Movement (Facebook), We are All Khalid Said (Facebook), Hossam Al-Hamalawy (blog), Asmaa Mahfouz (Twitter), Zeinobia (Twitter) and Nawara Negm (Twitter). The study used both quantitative content analysis and qualitative (textual) analysis of the sampled news stories, examining: dominant voices in headlines, main players in leads and sources used in stories. Under each category of sources, a record of references to and inclusion of social media sources was kept. The study also analyzed the lexical choices (non-attributed key words and phrases) made by the selected news sites and compared them to the lexical choices made by political activists on their social media platforms. The interpretations of numerical and textual data aggregated served in understanding how the relationship between each of the selected news organizations and political activists impacted the coverage of the Egyptian Revolution and its aftermath.

The study findings show that activists were the most frequent voices in Al-Jazeera's headlines with **62%** recorded for them against **20%** for SCAF. The BBC came second with **52%** of dominant voices in headlines being activists. Russia Today and CNN followed with **47%** and **48%** for activists. The least frequency of activists' voices the study recorded was in the XINHUA's headlines with a slice of **18%** against **61%** going to SCAF. Also, the study findings show that activists were the main player in 63% of Al-Jazeera leads against **23%** recorded for SCAF. Activists in BBC, CNN and RT came in close occurrences **(46%, 47%** and **48%).** XINHUA recorded the least mention of activists in leads with a share of **18%** against **68%** recorded of SCAF, which was the highest among all.

For attribution (selection of sources) Al-Jazeera topped the selected news sites in privileging activists as sources with **42%** against **15%** for SCAF. XINHUA, on the contrary, gave activists a thin slice of **9%** of all sources against **27%** given to SCAF. BBC came second in privileging activists as sources with **38%** followed by CNN **(30%)** and RT **(29%).** Graph (1) shows sources distribution in the selected news sites.

Similarly, Al-Jazeera came first among the selected news organizations in including content from social media, mostly from activists. In Al-Jazeera sampled news stories, there were at least **17** references to and inclusion of social media content; **11** of them were from political activists' entries on their Facebook pages, Tweets and blogs, and there were **6** references to the SCAF Facebook page. BBC came second in including content from social media, and similar to Al-Jazeera, it used more content from social media entries by activists **(9)** more than SCAF (2). Graph (2) shows references to social media in the selected news stories.

For the non-attributed key words/phrases, Al-Jazeera's lexical choices appeared to be the most matched with activists' lexical choices, followed by CNN. BBC and RT came under equal influence while XINHUA was the least influenced, as table (1) shows.

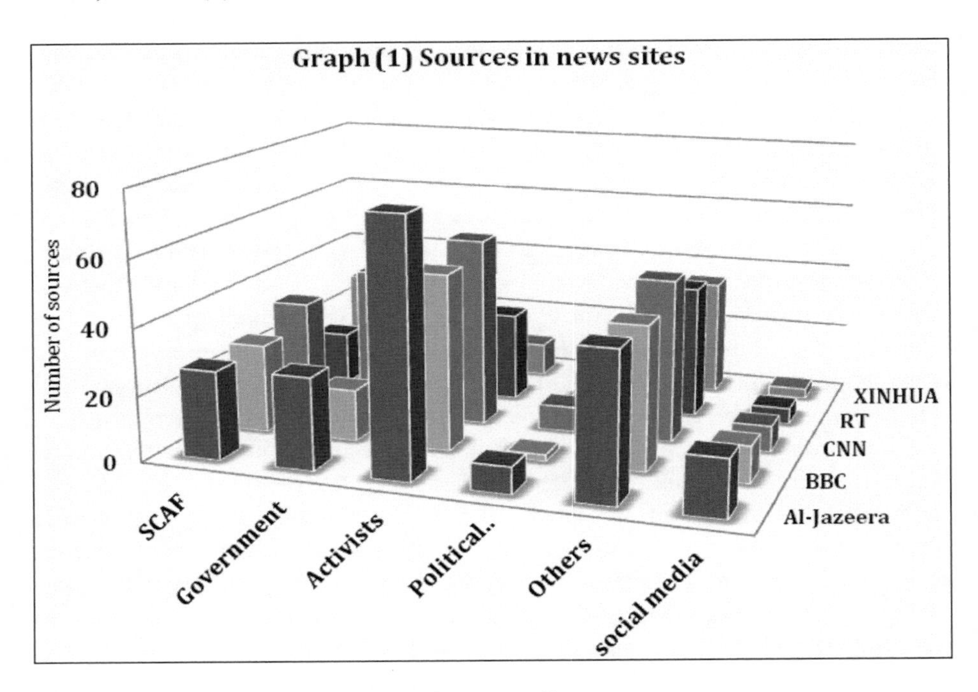

WHO IS DRIVING THE COVERAGE?

Numerous studies looking at news media performance criticize journalists' heavy reliance on elite—official, authoritative and institutional—sources, to the extent that they (journalists) allow the elite to define the news agenda. They argue that when elite sources receive privileged access to media, they become the "primary definers" of news agendas (Reese, 1990; Glasgow Media Group, 1993; Yoon, 2005; Freedman, 2005; Stromback & Nord, 2006; Mason, 2007; Macdonald, 2008). Journalists' reliance on elite sources as a media practice and the impact of such practice on news coverage are strongly tied to the concept of "indexing" proposed by Bennett (1990), who argues that voices included in news stories or expressed in editorials reflect the range of views expressed in mainstream government debate. Bennett (ibid.) predicts that the news media "indexing" system, which "compresses public opinion to fit into the range of debate between decisive institutional power blocs," might change if triggered by outside events that would reshape the relationships among the systems' actors. Studies researching media development in the 1990s onwards demonstrate that Bennett's predictions have been partially fulfilled with the development of the internet, the rise of alternative media, citizen journalism and more recently social media. Mass-driven, social media–equipped events, like the Arab uprisings, have provided circumstances for the reshaping of the relationship between journalists and ordinary citizens, activists and other "usually" marginalized groups. Also, media coverage of the Arab uprisings has made it clear that the news media "indexing" system is changing, and news organizations are embracing new practices, at varying degrees.

My findings indicate there were disparities in the coverage among the five selected news organizations (Al Jazeera, BBC, CNN, RT and XINHUA). Such disparities were the result of their different sourcing practices, their non-attributed lexical choices and their inclusion of materials from social media. As Al-Jazeera's coverage recorded the heaviest visibility of activists, comparing to other parties involved in the Egyptian Revolution, this indicates a strong relationship between the news organization and activists. The other selected news organizations have not been equally embracing this relationship, and again this is evident when comparing activists' visibility in their coverage to other parties involved. As the study recorded the heaviest visibility of activists in Al-Jazeera's coverage, its non-attributed lexical choices were also the most agreed with activists' lexical choices. Additionally, Al-Jazeera had the heaviest inclusion of social media materials in its news stories. This feature was seen relatively less in the coverage by BBC and CNN, and were almost absent from RT and XINHUA, as table (1) below shows:

Table 1. Prominence of activists as sources and their social media related practices

Elements of comparison	Activists' visibility in coverage	Including activists' social media entries	Influence of activists' lexical choices
Al-Jazeera	47%	65%	57%
BBC	41%	81%	33%
CNN	34%	50%	48%
RT	35%	0%	33%
XINHUA	12%	0%	28%

At least four of the five news organizations included in the study were found to be practicing "counter-elite sourcing," which is commonly associated with alternative media outlets when "they oppose the conventions and representations of the mainstream media" by privileging activists and other marginalized voices over elite and official sources (Atton and Wickenden, 2005). The study suggests that social media–equipped activists have played a significant role in pushing news media to change their practices; departing from their conventional elite-sourcing routine towards more source diversity and non-elite sourcing practices. This comes in agreement with Hermida (2010), who argues that the traditional news model, which defines news as information and quotes from official sources is in a period of transition. It is getting re-shaped by social media tools that "facilitate the immediate dissemination of digital fragments of news and information from official and unofficial sources over a variety of systems and devices." (ibid.) Based on the case study findings and scholars' views, there is an indication that the interaction between news organizations and social media–equipped activists has altered sourcing practices. In reporting on the Egyptian Revolution and its aftermath, the more a news organization was connected with activists on social media, and the more content it used from their social media entries, the more likely it privileged them against other elite sources allowing activists to drive the coverage.

As the unprecedented Arab uprisings have sent shock waves across the world, these volatile circumstances have reassured a fact: social media–equipped activists, who could also be described as "activist reporters" have become an essential part of today's media scene. Besides their role as citizen journalists covering their own communities and reporting on events that news organizations might miss or ignore, they also help bridge the gap between news organizations and the non-elite marginalized public, and drive journalists' attention to the voices of ordinary citizens, activists and other marginalized groups.

NOTES

1. Carpenter (2008) noted that while online citizen journalists were more likely to cite unofficial sources, journalists working for online newspaper remained more likely to cite official sources.
2. *Kefaya*, founded in 2004, did not have a physical headquarters or permanent meeting place. "It spread news, hosted online forums, and coordinated activities through its main website, Haraka- Masria.org, and through MisrDigitial.com, which hosted "Egyptian Awareness," the country's first independent digital newspaper." (Lim, 2012)
3. In the Arab region, the total number of internet users exceeded 125 million, and the number of active users of social media exceeded 53 million in 2013 (DSG, 2013).
4. According to the Arab Social Media Report 2011, 94% of people in Tunisia and 88% of people in Egypt said they got their news from social media (DSG, 2011).
5. "A 2009 survey of Twitter demographics found that nearly 60 percent of respondents said they interacted most often with media and journalists, coming in just after friends at 70 percent." (Radsch, 2012).
6. Al-Jazeera Mubasher Misr is a news channel and a website dedicated to cover Egypt's news 24/7 and provides an online space and an airtime for citizens to participate with content.
7. Also, activists launched an off-line street campaign against SCAF titled "*Askar Khazeboun*" or (Lying Officers) where they went to low-income neighborhoods around the country, used a projector connected to a laptop to show images and footages posted on social media recording SCAF's violations (El-Khalili, 2013).

REFERENCES

Abdulla, Rasha (2012). *The Revolution Will Be Tweeted.* The Cairo Review of Global Affair, School of Global Affairs and Public Policy, the American University in Cairo.

Al-Ahram (2012, June 24). Muhammad Ali Pasha to Mohamed Morsi: Egypt's Modern Rulers. Retrieved from http://english.ahram.org.eg/NewsContentPrint/1/0/46033/Egypt/0/Muhammad-Ali-Pasha-to-Mohamed-Morsi-Egypts-modern-.aspx

Al-Ahram (2013, February 4). Internet users in Egypt reach 31 million. Retrieved from http://english.ahram.org.eg/NewsContent/3/12/63999/Business/Economy/Internet-users-in-Egypt-reach--million.aspx

Allan, Stuart, & Thorsen, Einar (2009). *Citizen Journalism: Global Perspectives*, New York: Peter Lang.

Amnesty International (2012).The Annual Report 2012: the State of the World's Human Rights. Retrieved from https://www.amnesty.org/en/region/egypt/report-2012

Aouragh, Miriyam and Alexander, Anne (2011). The Egyptian Experience: Sense and Nonsense of the *internet Revolution. International Journal of Communication.*

Atton, Chris, & Wickenden, Emma (2005). Sourcing Routines and Representation in Alternative Journalism: A Case Study Approach. *Journalism Studies*, Napier University.

Bengtsson, Rebecca (2013). 'Even if it is not your fault, it is your responsibility': Live streaming as means of civic engagement, a case study of citizen journalism in Egypt and Syria. Master Thesis, Malmö University, Sweden.

Bennett, W. Lance (1990). Toward a Theory of Press-Station Relations in the United States. *Journal of Communication*, 40 (2) spring.

Bossio, Diana and Bebawi, Saba (2012). Reaping and sowing the news from an Arab Spring: the politicised interaction between traditional and alternative journalistic practitioners. *Global Media Journal, Australian Edition.* Retrieved from http://www.commarts.uws.edu.au/gmjau/v6_2012_2/bossio_bebawi_RA.html

Brownlee, Jason (2011). Peace before Freedom: Diplomacy and Repression in Sadat's Egypt, https://webspace.utexas.edu/jmb334/www/documents/brownlee.2011–2012.psq.pdf

Carpenter, Serena (2008). Source Diversity in U.S. Online Citizen Journalism and Online Newspaper Articles. The 9th International Symposium on Online Journalism. http://online.journalism.utexas.edu/2008/papers/OnlineCitizenJournalism_Carpenter.pdf

Chomsky, Noam (1982). 'Towards a New Cold War: US Foreign Policy from Vietnam to Reagan', in *The Essential Chomsky* [2008], Bodley Head: London.

Cottle, Simon. (2011). Media and the Arab uprisings of 2011: Research notes. Sage Publications.

Dubai School of Government (2011). Civil Movements: the Impact of Facebook and Twitter. Retrieved from http://www.dsg.ae/portals/0/DSG_Arab_Social_Media_Report_No_2.pdf

Duffy, M. J. (2012). Networked journalism and Al-Jazeera English: How the Middle East network engages the audience to help produce news. *Journal of Middle East Media*, 7(1), 1-23.

El–Khalili, Sara (2013). Social media as a government propaganda tool in post–revolutionary Egypt. First Monday, Volume 18, Number 3, 4 March. Retrieved from http://journals.uic.edu/ojs/index.php/fm/article/view/4620/3423

Fisher, Alan. (2011). The Arab Spring, Social Media and Al Jazeera. In: Mair, John et al. (ed.) *Mirage in the Desert? Reporting the "Arab Spring"*. Suffolk, UK: Arima Publishing.

Freedman, E. (2005). Coverage of the Central Asian Political, Press, and Speech Rights Issues by Independent News Websites. *Asia Pacific Media Educator*. Issue 16.

Ghannam, Jerfrry (2011). *Social Media in the Arab World: Leading Up to the Uprisings of 2011.* Washington, DC: Center for International Media Assistance.

Glasgow University Media Group & John Eric Thomas Eldridge (1993). *Getting the Message: News, Truth and Power.* London: Routledge.

Harlow, Summer and Johnson, Thomas (2011). Overthrowing the Protest Paradigm? How The New York Times, Global Voices and Twitter Covered the Egyptian Revolution. *International Journal of Communication.*

Heinrich, Ansgard (2008). Network Journalism: Moving towards a Global Journalism Culture. Paper delivered to the RIPE conference in Mainz, October 09–11, 2008 'Public Service Media for Communication and Partnership'.

Hermida, Alfred (2010). Twittering the News: The Emergence of Ambient Journalism. *journalism practice*, Vol. 4, No. 3, pp. 297–308, July 2010.

Hermida, Alfred, & et al. (2012). Sourcing the Arab Spring: A Case Study of Andy Carvin's Sources During the Tunisian and Egyptian Revolutions. Paper presented at the International Symposium on Online Journalism in Austin.

Hermida, Alfred, Lewis, Seth C. and Zamith, Rodrigo (2012). Sourcing the Arab Spring: A Case Study of Andy Carvin's Sources During the Tunisian and Egyptian Revolutions. Paper presented at the International Symposium on Online Journalism in Austin.

Idle, N., & Nunns, A. (eds) (2011). *Tweets from Tahrir: Egypt's Revolution as It Unfolded, in the Words of the People Who Made It.* New York: OR Books.

Iskander, E. (2011). Connecting the National and the Virtual: Can Facebook Activism Remain Relevant After Egypt's January 25 Uprising? *International Journal of Communication*

Kaplan, Joel (2008). Expectations for Objectivity and Balance in Multi-Platform Distribution: Traditional and New Media. Retrieved: http://www.cpb.org/aboutcpb/goals/objectivity/whitepapers/

Kempa, Dave (2012). Documenting Death—Syria's Information Pipeline. *Social Media Chimps.* https://socialmediachimps.com/news/documenting-death-syrias-information-pipeline/

Khamis, Sahar (2011). The Transformative Egyptian Media Landscape: Changes, Challenges and Comparative Perspectives. *International Journal of Communication.*

Khamis, Sahar, and Vaughn, Katherine (2011). Cyberactivism in the Egyptian Revolution: How Civic Engagement and Citizen Journalism Tilted the Balance. *Arab Media and Society,* Issue 13.

Lim, M. (2012). 'Clicks, Cabs, and Coffee Houses: Social Media and Oppositional Movements in Egypt, 2004–2011', *Journal of Communication*, 62(2012), pp. 231–248.

Macdonald, Isabel (2008). "Parachute Journalism" in Haiti: Media Sourcing in the 2003–2004 Political Crisis. *Canadian Journal of Communication,* Vol 33, 213–232.

Macnamara, J. (2005). Media content analysis: Its uses, benefits and Best Practice.

Methodology. *Asia Pacific Public Relations Journal*, 6(1), 1– 34.

Mason, Anthony (2007). Reporting the Fiji Coup: Elite Sources, Journalistic Practice and the Status Quo. *Pacific Journalism Review.*

Mir, Mashaal (2011). Was Al-Jazeera English's Coverage of the 2011 Egyptian Revolution "Campaigning Journalism"? In: Mair, John et al. (ed.) *Mirage in the Desert? Reporting the "Arab Spring"*. Suffolk, UK: Arima Publishing.

Oriella (2011). The State of Journalism in 2011. Oriella PR Network Digital Journalism Study. www.oriellaprnetwork.com

Pintak, L. (2011). 'Breathing room toward a new Arab media', *Columbia Journalism Review*, http://www.cjr.org/cover_story/breathing_room.php, 5 May.

Reese, Stephen (1990). Setting the Media's Agenda: A Power Balance Perspective. Paper presented to the Association for Education in Journalism and Mass Communication, Minneapolis, Minnesota.

Radsch, Courtney C. (2012). *Unveiling the Revolutionaries: Cyberactivism and the Role of Women in the Arab Uprisings*. Houston, TX: The James A. Baker III Institute for Public Policy, Rice University.

Rodrigues, Jason (2011). '1982: Syria's President Hafez al-Assad crushes rebellion in Hama', *The Guardian*, http://www.guardian.co.uk/theguardian/from-the-archive-blog/2011/aug/01/hama-syria-massacre-1982-archive, 1 August.

Rostovtseva, Nataliya (2009). Inter-Media Agenda Setting Role of the Blogosphere: A Content Analysis of the Reuters Photo Controversy Coverage During the Israel-Lebanon Conflict in 2006, University of North Carolina at Chapel Hill.

Russell, Adrienne (2011). Extra-National Information Flows, Social Media, and the 2011 Egyptian Uprising, *International Journal of Communication*.

Sanders, Christoph (2012). 'Deconstructing the "Facebook Revolution" Specifying the Contribution of the Youth to the Radicalization of Contentious Politics in Egypt between 2000 and 2011', *State of Peace Conference & Peace Report 2012 "Democracy in Crisis: The Dynamics of Civic Protest and Civic Resistance."* Schlaining, Austria: Freidensburg Peace Castle.

Soueif, Ahdaf (2011). Image of Unknown Woman Beaten by Egypt's Military Echoes Around World. *The Guardian*, Sunday 18 December 2011.

Stromback, Jesper, & Nord, Lars W. (2006). Do Politicians Lead the Tango? A Study of the Relationship between Swedish Journalists and their Political Sources in the Context of Election Campaigns. *European Journal of Communication*, vol 21(2): 147–164.

Wallsten, Kevin (2011). Microblogging and the News: Political Elites and the Ultimate Retweet. In *Politics and Policy in the Information Age*.

Wheeler, Deborah (2006). *Empowering Publics: Information Technology and Democratization in the Arab World—Lessons from internet Cafes and Beyond*. Oxford: Oxford internet Institute.

Wimmer, Roger D. and Dominick, Joseph R. (2006). *Mass Media Research: An Introduction*. Boston: Thomson Wadsworth.

Yoon, Youngmin (2005). Legitimacy, Public Relations, and Media Access: Proposing and Testing a Media Access Model. Communication Research, SAGA publications, http://crx.sagepub.com/cgi/content/abstract/32/6/762.

CHAPTER SEVENTEEN

Citizen Journalism in Indonesia's Disputed Territories: Life on the New Media Frontline

KAYT DAVIES

To paraphrase an old truism: some things change while others do not. Sadly in Indonesia it appears that what has not changed much since the 1970s is its iron fist attitude towards its disputed territories and the lawless brutality practiced by some members of its armed forces.

In Balibo, back in 1975, the bullet-riddled bodies of five Australian-based newsmen were burnt (Allard, 2009). Not only that, but their rolls of 16mm film were also destroyed, denying the world the opportunity to see what their eyes had witnessed. This begs the questions: Could such a thing happen today? If Greg Shackleton had carried an iPhone would he have been safer, or killed sooner? Would we all have seen the invasion as it happened? And would the Timorese have needed Australians to tell their stories for them?

Between the invasion in the 1970s and the internationally supported conflict of the 1990s, East Timor suffered under the pall of occupation, supported by a few dedicated activists and largely ignored by mainstream media. As Hill documented in a research paper about the role of the internet in the East Timor campaign:

> For the first fifteen years, before the internet, such civil society organizations used "snail mail," telephone, and fax to sustain their campaign for East Timor's self-determination. Internet and e-mail were articulated almost unconsciously into the activists' armory, so surreptitiously that some of the most active in the campaign have difficulty recalling precisely when they adopted the evolving technologies (Hill, 2002:50).

Hill argued that the pioneering of online organisation by campaigns, such as the pro-separatist movements in Timor and Aceh, proved: "how these new technological possibilities could exert international political leverage" (2002: p. 26). This chapter will explore the way emerging media technologies are being used by citizen journalists reporting on conflict in Indonesia's disputed territories and the problems they are encountering. It will also look at how the work they are doing fits with models and definitions of journalism practice and the role of media in democracies.

WHY TRUTH TELLING MATTERS

While personal horror stories about atrocities committed in East Timor between 1974 and 1999 continue to emerge (Soares, 2010), the big picture has been painted. Initially the story was told by mainstream media journalists. Tony Maniarty, the Balibo five and Roger East were among the first to start drafting the history of the conflict, as it happened. But the long official version is the 2,800-page report entitled "Chega," created through the Commissao de Acolhimento, Verdade e Reconciliacao de Timor-Leste (CAVR). Rendered in English, CAVR was the Timor-Leste Commission for Reception, Truth and Reconciliation, set up in 2001 and dissolved in 2005. It was an independent statutory authority led by seven East Timorese Commissioners and mandated by United Nations Transitional Administration in East Timor to undertake truth seeking, facilitate community reconciliation, and report on its work (CAVR, 2013). The report is based on interviews with 8,000 witnesses and blames the Indonesian government for the deaths of as many as 183,000 civilians (Powell, 2006).

Since Chega was released, political wrangling has confused and delayed moves to transform CAVR's recommendations into legislation in Timor Leste (Walsh, 2011) and very little has come of the efforts to have members of the Indonesian military —the Tantari Nasional Indonesia (TNI)—face criminal charges for gross human rights abuses in East Timor (Moore, 2004; Byrne & Gibson, 2007), despite prosecutions being one of the CAVR recommendations.

The effort that was invested in CAVR was justified, however, because Chega now exists. As Naqvi (2006) pointed out, by exposing the truth, societies are better able to prevent the recurrence of similar events; because knowing the truth has been deemed essential to heal rifts in communities; to foster accountability; and to support reconstructing of national identities by unifying countries through dialogue about a shared history. According to Naqvi, it was all these "slightly amorphous considerations" about the value of truth telling en-masse that led to the UN Commission on Human Rights adopting, in its 61st session, Resolution 2005/66, which "recognizes the importance of respecting and ensuring the right to the truth so as to contribute to ending impunity and to promote and protect human rights."

So while Timor Leste continues to rebuild, Chega is now on record and work continues on disseminating its findings via the post-CAVR Secretariat based in Dili. However, not all of the victims of relatively recent Indonesian military action have had the opportunity to so cathartically and authoritatively speak their truth. The similarity of the experiences shared by people in Indonesia's disputed territories is due in part to redeployment of key TNI personnel, who were active in East Timor in the 1970s, to other provinces after the Timor conflict. It is also partly due to the dynastic nature of the TNI that has some 3rd generation generals—and partly to the fact that some people who were junior officers in East Timor, have matured into being senior military personnel. In the past two decades threats to the "territorial integrity of Indonesia" have been raised by separatist groups in, among other places, Maluku province (the Moluccan Islands), Aceh and West Papua.

WHOSE TRUTH IS TOLD?

The Republik Maluku Selatan (RMS) proclaimed its independence from Indonesia on April 25, 1950 (RMS, 2013). This led to a violent struggle that officially ended with the execution by Indonesian forces of the RMS leader Chris Soumokil in 1966 (Amnesty, 2009). Since then the RMS has described itself as a "government in exile" and Maluku Province is recognised by most of the world as being part of Indonesia, which has for decades been supporting transmigration to the islands diluting the mostly Christian, ethnic Moluccan population. In 1999 violent clashes between the TNI and the Moluccan separatists broke out, and while the RMS claims these clashes were over the push for independence, much of the world media reported them as sectarian clashes between Christians and Muslims. While the conflict officially ended again in 2002, with the Malino Peace agreement, violence broke out again in 2004 (Algionby, 2004). Then, in 2007, 22 men were arrested for performing a traditional Moluccan dance and displaying the independence flag in front of Indonesian President Susulio Bambang Yudhoyono (Amnesty, 2009). According to the RMS website, "Current events that show Moluccans who use their natural right to freedom of speech are arrested, beaten, tortured and given long prison terms, only because they exercise their rights and stand out for their political beliefs upon a legitimate endeavour" (RMS, 2013). While the latest news on the RMS website is from 2011, the website of the Moluccas International Campaign for Human Rights also provides background information about the conflicts and features a current news stream comprising links to relevant articles in other media such as *The Jakarta Post* and Reuters. It claims the "the Moluccan people still remain under foreign occupation" and describes the struggle as ongoing (MICHR, 2013).

In 2003, Indonesia had 50,000 troops on the group in Aceh, fighting 5,000 pro-separatist "guerillas" from the Free Aceh Movement (GAM). The conflict dragged on and the Indonesian military was again on the front foot with "image management"—this time by "keeping its domestic press under control and virtually barring foreign correspondents from covering the military offensive" (Neumann, 2003).

In 2005 the Indonesian Government and GAM signed a peace accord in Helsinki aimed at ending the conflict, which had raged at varying levels of intensity since 1976. Agreements struck in 2000 and 2002 had failed to maintain peace and while the current agreement is ostensibly working, there is still contention about the rights of the Acehnese to fly a flag that resembles the GAM flag (Citrawan, 2013; Bachelard, 2013), and concerns have been expressed about the human rights implications of Sharia Law and its penalties of stoning and death (Garg, 2012). In addition the 2005 agreement mandated the establishment of a Truth and Reconciliation Commission and this has yet to happen (Harsono, 2013).

THE NEW MEDIA TIPPING POINT

Despite the ongoing unrest and the abundance of graphic images of the Moluccan and Aceh conflicts shared online through human rights networks—it was a natural disaster that put Aceh onto the front pages of world newspapers in late 2004. Marveling that the news they watched during the 2004 holiday season "did not come from and was not wholly controlled by the 'media giants,'" Robinson and Robinson (2006) wrote:

> they gathered and shared news through mobile gear, and sent information on the go (for example, from shops with internet or wireless access). Survivors' tales were related via email, mobile texting and digital cameras. Vacationers phoned in from cellphones and transmitted images through cellphone cameras and attached email files. Locals and amateurs captured stories and remarkable video with camcorders. Bloggers and rescue workers published online on a regular basis, establishing internet-based missing person's networks. (Robinson and Robinson, 2006:85)

They concluded the coverage of the tsumani of 12/26 was a "tipping point for citizen journalism" (2006:87) that provided challenges complicating the relationship between citizen journalists and the mass media.

Richard Sambrook (2009) also noted what he calls the "Asian tsumani," along with the London bombings, in a list of events that changed the way mainstream media thought about citizen journalism over the course of 2005 and 2006. Sambrook, who was a senior BBC journalist and news executive until 2010, described a number of issues coalescing around that time, changing the scale of the collective citizen media enterprise. He particularly notes the publication of Dan Gillmor's "seminal" (2006) book *We The Media*, which documented shifts up to that point in time in the way internet connectivity was being used. Meanwhile in Indonesia's disputed territories, a group of former East Timor activists, who had not read Gillmor's book, were using the new tools they had available to maximize their impact and efficiency.

Describing the gradual uptake leading to the tipping point, founder and editor of West Papua Media, Nick Chesterfield, said: "In Timor bulletin boards and basic list servers were replacing the old telephone tree as a fundamental tool for organizing people, we used group mailers and to a large extent these things are still the basis of our work." (Chesterfield, Personal communication, May 16, 2013).

In 2006, Chesterfield and a crew of activists, turned human rights monitors and media fixers, who had followed the TNI to West Papua, paused to consider their options and the best way to proceed. The Indonesian Government had imposed a foreign media ban in West Papua (IFJ, 2006) similar to the restrictions it had put in place in Aceh in 2004 (Neumann, 2003). This situation has not improved, with Perrottet and Robie, reporting in late 2011 that: "Jakarta still upholds its prohibition on all foreign journalists and media workers from entering either province in West Papua, unless pre-approved under a slow and bureaucratic process from the Ministry of Information. [And] Even after approval, journalists are always accompanied by a minder from the Badan Intelijen Nasional (National Intelligence Body)" (Perrottet and Robie, 2011:178).

According to Chesterfield, the media ban led to the conclusion that more news would get out if they trained local journalists in credible news gathering techniques than they would shepherding occasional illegal foreigners through the jungles, towns and villages. This led to the establishment of West Papua Media (WPM) one of the most sustained and organized citizen media outlets in the Asia Pacific region (Davies, 2012).[1]

In its early days in 2007, WPM simply used an email list—a makeshift news group through which articles and news tips were sent to journalists they had met through their fixing work. The crew founding WPM were aware that there were already some activist voices online, in the form of the civil resistance online organizing networks that they had been supporters of and participants in since the Timor days in the 1990s, but the point of difference they were seeking in order to cut through and attract main stream media attention was credibility, in a non-partisan way. The ambition of WPM members/participants was to build an organization which had processes that could that be trusted to meet journalistic standards of accuracy and veracity, in accordance with what Ward (2010) describes as the journalistic requirement of "pragmatic objectivity."
In Chesterfield's words:

> When activists were trying to get West Papua into the media there were always complaints from mainstream journos that the information wasn't credible and certainly there have been some cases of stories where claims couldn't be verified. And there were very, very powerful

> voices that were rubbishing all of the claims that came out of West Papua. So from a human rights data perspective we were looking at a situation where the best way to ensure high quality factual reportage coming out of Papua was to make sure there was some sort of basic training and methodologies that were agreed upon by local people. And in reality as well, the more we, as media fixers, were seeking to organise and get factual stories out of Papua, the more we realised that we were doing the work of journalism and journalism training. (cited in Davies, 2012:72–73).

This determination to provide an alternative to the limited "official" version of events was in many ways proving up Gillmor's (2006) prophetic comment about the citizen reporting after September 11, 2001:

> Journalists did some of their finest work and made me proud to be one of them. But … Another kind of reporting emerged during those appalling hours and days. Via emails, mailing lists, chat groups, personal web journals—**all nonstandard news sources**—we received valuable context that the major American media couldn't, or wouldn't, provide. We were witnessing—and in many cases were part of—the future of news.

QUALITY CONTROL AND CREDIBILITY

Having worked as human rights monitors, the WPM founders considered that official sources—especially the sort that had been behind the decision to shoot the Balibo five to prevent them reporting what they had seen (Allard, 2009), and had burdened reporters in Aceh with minders (Neumann, 2003) and put in place a total foreign media ban in WP (Perrottet & Robie, 2011)—were unlikely to be completely honest and transparent about their activities when speaking to the media.

The WPM approach was in line with Gillmor's (2006) call for collectively developed new standards of verification in citizen journalism. Through their previous work experience with human rights monitoring organizations the crew had already been trained in methodologies and protocols for in-the-field data gathering. They adapted these methodologies to suit the needs of discrete undercover journalism work (required because of the risk of persecution/prosecution) and incorporated the classic question set "Who, What, Where, When, Why and How". Lamble (2004) described this question set as being sufficiently core to journalism practice that it can be used as one of the features that identifies journalism as a distinct research methodology. Other defining characteristics of journalism, according to Lamble (2004) include a preference for empirical data, a neutral point of view approach to reporting, and a combination of deliberate and network sampling to recruit subjects for interview. The WPM protocol incorporates all of these features. It also encourages the use of tables for recording the names, ages, addresses, injuries, and other specific details about people involved in violent incidents and a layer of protocol about recording when and where photos were taken and how information should be transmitted and archived. Once this protocol had been established, they started recruiting Papuans in Papua to become their network of stringers. Since 2008 West Papua Media has trained over 40 citizen journalists/stringers and the training work has taken the form of one-on-one, small group work, the development and use of training materials, scenario-based workshops, and assessments to ensure comprehension, retention and proficiency (Chesterfield, Personal communication, January 16, 2012).

This training model is similar in some ways to training being offered by the British Government funded Syrian Commission for Justice and Accountability, which in 2012 started "training Syrian activists to professionalise their amateur investigations" (Analysis: The beginnings of transitional justice in Syria, 2012:1).

THE PROS AND CONS OF WEB 2.0

WPM is still going and today it operates as a website (westpapuamedia.info) that garnered 182,372 direct page views in 2011 and 231,828 in 2012. It has 218 subscribers to the website's news feed and an additional 144 journalists on a media-only list. It also sends out links to content to 1,347 Facebook followers and 3,027 Twitter followers, with potential for much wider viral reach. Asked about the use of proprietary social media platforms Chesterfield said: "We use them to promote the content on our website and in that sense they are great." He added that the group struggled with dissemination of news via Facebook because of the platform's automated processes which allow links to specific sites to be blocked, if they are reported by other Facebook users, who can use this function to suppress the sharing of information. Chesterfield said that this type of malicious reporting was an ongoing problem with Facebook that the team worked around, by not "putting all its eggs in one basket" and not relying on Facebook to be a trusted or essential part of its dissemination strategy. Attempts to post links to WPM content have sometimes been blocked by Facebook, which presents a pop-up explaining that the site has been reported for hosting "spammy or offensive content." Chesterfield said another concern about Facebook and the role it plays in information sharing is the growing realization that it is a filtered medium, and that its responsive algorithms over time only give people information that accords with things they have liked in the past and therefore does not challenge their base assumptions.

WPM content is also distributed by several other e-lists reaching a combined total 30,000–40,000, through content-sharing arrangements. In addition, content is mirrored on other sites including Pacific Scoop and Scoop New Zealand, Pacific Media Centre Online, AK Rockefeller and a host of other civil resistance and media sites, some of which also publish the WPM Twitter feed. WPM also continues to liaise with media companies, offering both fixing services and access to text, images, audio and footage when big news breaks or investigative pieces are completed. While this range of services sounds professional, it is worth noting that the organization is entirely funded by donations and no one is paid a salary. When media companies pay for content or services, that revenue goes to the individual or team responsible, but the money is usually donated back to WPM to pay the bills.

CITIZEN JOURNALISM AND THE ROLE OF THE MEDIA

One way of analyzing what citizen journalists in Indonesia's disputed provinces have achieved is to consider Sambrook's (2009) distinction of four kinds of citizen journalism. These are:

- eyewitness reporting;
- opinion sharing;
- investigative reporting; and
- knowledge sharing.

He argues that they have "different motivations, different purposes and different effects" (2009:227):

- If both "eyewitness" and "investigative" reporting are considered to be forms of "watchdog journalism";
- if "knowledge sharing" is the same as the "provision of information";
- and "opinion sharing" is "participation in public debate,"

Then Sambrook's categories align with the three functions of media in a liberal democracy identified by Errington and Mirragliotta (2011:8). This suggests that mainstream professional journalism no longer has the monopoly hold on this role and that where it is failing to fulfill this role, citizen journalism is filling the vacuum.

Eyewitness reporting, that "gets around the media ban," has been the key priority for WPM. In addition to ensuring that newsrooms can be reassured about the veracity of the reports, attention has been paid to establishing protocols suitable for hostile environment work, in order to protect both reporters and footage. Inspired by the Hostile Environment First Aid Training that major news organizations now require foreign correspondents to undergo (Cramer, 2009), WPM developed a training package and a set of detailed guidelines, in both English and Bahasa. It is available at http://westpapuamedia.info/safe-witness-broadcasting-tips/

The following is an example of the kind of immediate eye-witness reporting by WPM, and disseminated as a breaking news report through its list and promoted via Twitter and Facebook and picked up by a number of major news outlets around the world:

> Indonesian police in Jayapura have this morning violently dispersed a pro-independence rally being held by the West Papua National Committee (KNPB), arresting its leader Victor Yeimo, media worker Marthen Manggaprouw and two KNPB activists, according to early reports.
>
> The rally was being held to commemorate the shootings and violent crackdown by Indonesian security forces on peaceful demonstrations across Papua on May 1, which left four people dead and drew international condemnation up to the UN Human Rights Commissioner, Navi Pillay.
>
> Reports from witnesses at the scene have confirmed that police conducted several rounds of baton charges against rally participants who arrived on motorbikes, and then joined by over 1000 other participants who continued to resist the police charges outside the gates of Cenderawasih University in Abepura. Injuries have been reported by but no particulars are yet available. More arrests are expected according to witnesses.
>
> (West Papua Media, May 13, 2013).

Given the events taking place, it is reasonable to assume that had the citizen journalists involved been caught in the act of reporting to an international audience there may have been adverse consequences. According to Chesterfield, this is why WPM reporters do not use the mobile devices they have with them to post news directly to the readership, even though that is technologically within the organization's capability. Chesterfield said:

> There are really two reasons why we don't broadcast live from the field, instead the field reporters send to the editors and the editors broadcast to the readership. The first is because not all reporters in the field have access to all of the information that the editors have, and so this process allows for fact checking. The increase in accuracy justifies the time delay, and the instantaneous capability of the technology does speed up that process a lot. The other fundamental danger with all mobile technology is the embedded phone-home capability that without user input sends back the GPS data from the last transmission. The good reasons for this are that it allows for the provision of better signal strength and bandwidth but the downside is that it can be used by whoever has access to the metadata, not just at a network level, but also at an embedded social media level as well. So if you are broadcasting stuff that those with access to the network don't like, they know exactly where you are to shut you up. This is why we generally suggest that people broadcast from a different location so the actual transmission location of a source can't be identified.

Detailing the process that WPM believes the TNI uses, he continued:

> The Indonesian military use car-based systems to triangulate locations. Stage one is deep packet inspection, a process used to sniff out key words in real time in tweets, emails and sms messages. They can then go to the next level and match IP addresses to phone numbers, IMEI [International Mobile Station Equipment Identity] and IMSI [internet

> Mobile Subscriber Identity] to geolocate the actual phone. They fine tune this using triangulation devices small enough to be carried in police cars to get down to the person level and they can do this even when an activist is at a crowded demonstration or in a marketplace. This GSM cellular interception equipment is supplied by Australia, Germany, the UK and the US and it has been used in the arrest of many media activists, citizen journalists and human rights workers worldwide.

The claim that TNI uses this equipment is supported by court records from the trial of Buchtar Tabuni in Jayapura in 2009 (available from the author on request). Logs from GSM Cellular Interception Equipment were presented to the court as evidence that Tabuni had been involved in organizing a pro-independence protest.

WPM's release of footage of a West Papuan man being tortured by TNI personnel in 2010 is another example of eye-witness citizen reporting that gained international media attention (Vaswani, 2010). At the time WPM was using an open-source structured website on a privately hosted site. The release of the video prompted a retaliatory DDOS attack on WPM and all other sites that were co-hosting the footage, such as the Asian Human Rights Commission, Friends of People Close to Nature, Free West Papua Campaign and Survival International (Press Association, 28 October, 2010; West Papua Media, 5 November, 2010). Chesterfield argued that:

> The attack was large volume and high intensity but not very sophisticated. An investigation that we conducted and a much appreciated pro-bono investigation by Operation Aurora, found that they had used Russian and Chinese botnets. We can't prove who did it but on the balance of probabilities it was probably whoever wanted to shut us up, and who had the funds to do that. We did find out that the source IP addresses for the attack came from computers associated with Indonesian academic, military and government institutions and the investigation also found IP addresses for the Cybersecurity Centre in Canberra, one associated with the US military and one from GCHQ in the UK. (Chesterfield, Personal communication, May 16, 2013).

After these attacks WPM deduced which hosting platform the intelligence-run pro-Jakarta West Papua websites were using, and migrated to it, so that "if they took us down they'd take themselves down too—and since then there's been not a skerrick of trouble from DDOS attacks". He added that:

> The only threat has been in the form of a Tweet storm in July 2012 when a group purporting to be part of Anonymous, but who in reality were an off-shoot of a pro-military Indonesian nationalist group said it would take us down because we were threatening to break-up Indonesia. We dared them to try and nothing happened. (Chesterfield, Personal communication, May 16, 2013).

As Sambrook (2009) pointed out there has been an increase in the willingness of mainstream media to broadcast material such as this from non-mainstream sources, however the process of sourcing this material from local eye-witness citizen journalists raises many issues. Not only is the death toll among local reporters higher than it is among other foreign correspondents (INSI, 2012), the issue of payment is also less clear cut, with some news organizations asking citizen media operations to provide for no charge "exclusive" footage—requiring extra journalism work and extra bandwidth to transmit. This places the financial burden onto philanthropists funding citizen media and raises ongoing global questions about the business model underpinning the provision of fourth-estate journalism (Rosen, 2008).

Knowledge sharing and/or the provision of information is the second category of citizen journalism idenitifed by Sambrook (2009) and the second role of the media identified by Errington and Mirragliotta (2009). This is achieved on West Papua Media, and other citizen media sites via timelines, maps, glossaries and other documentation of historic and current events.

The third category/role is participating in, and/or facilitating public debate by sharing opinions. WPM's contribution on this front is to share, via its dissemination network, opinion pieces produced by commentators embedded in other organizations. The sources of these pieces are clearly identified, to distinguish them from eyewitness and investigative journalism and efforts are made to share pieces that are both supportive and critical of the pro-separatist movement, specifically its armed and militant branches.

As Bräuchler (2004) pointed out in her study of social media and the Moluccan diaspora, online conversations about internal conflict events include participants who are not in-country, on the ground. Just as Ramos Horta, described himself as East Timor's "foreign minister in exile" for many years while working for the movement, West Papua has a number of pro-democracy activists who are not in-country. These include Benny Wenda, now based in the UK, and Jacob Rumbiak in Australia. According to Chesterfield, while WPM posts opinion pieces written by diaspora members, such as Wenda and Rumbiak, he has some reservations about the effects of online conversations about specific events:

> While it's important to have online conversations that the diaspora can take part in, in order to socialise the issue and keep people engaged, they can also sometimes add inaccuracies to a story by incorporating details that later on turn out not to be true, or to be based on heresay. This means we need to distinguish these kinds of conversations from legitimate exercises in crowd-sourcing information—and this means paying attention to how much information has to be qualified in order to get through the rumours and second-and third-hand stories, and the habit people have of including older more graphic images to sets of fresh images of abuses in order to increase the impact, but not realising the damage that does to the credibility of the whole story. (Chesterfield, Personal communication, May 16, 2013).

NO SITE IS AN ISLAND

WPM is not standing alone in taking advantage of the potential of citizen-generated media and social networking to do the work that mainstream media could be doing better, the victims of violence in Aceh, Moluccas and Timor Leste have also, to varying extents, adapted to the times and technologies available. All have Wikipedia pages, web pages, Facebook and Twitter presences. There are also groups, such as the East Timor and Indonesia Action Network (www.etan.org) and Tapol (tapol.org), documenting human rights abuses. While WPM is unique in some of the work it is doing in the region, because of its journalistic focus, what it does is complemented by other nodes in the network of organizations and groups in the region and internationally. "We are embedded in the movement that we serve, and we serve by having a credible and independent and factual voice," Chesterfield noted. "The movement is free to use our information, which is based on fact, and it should do, because the movement itself should be based on fact." (Chesterfield, Personal communication, May 16, 2013).

Other citizen media projects active in the area include EngageMedia's Papuan Voices Project that is the work of a group of independent filmmakers, Papuans Behind Bars (a project documenting the incarceration of Papuan activists) and WestPapuaBackground.wordpress.com (which provides easily accessible background information and links for journalists new to covering West Papua). There are also several activist and human rights organizations that are online, sharing information and perspectives. In addition, the network of voices includes creative and musical elements such as filmmaker Charlie Hill-Smith, whose 2009 film *Strange Birds* in Paradise continues to evoke discussion, and *Rize of the Morning Star* that uses music to engage people worldwide in the Papuan cause. Collectively they constitute an online community that not only discusses events in Indonesia's disputed territories, but seeks to create change. The mechanism through which they expect to achieve this change has been attributed to Gene Sharp, a scholar in the field of nonviolent resistance—and reduced to the words "push, pull, squeeze." The three-pronged strategy involves civil disobedience that pushes out the occupying power; while spreading awareness of the issue pulls away the consent of the population of the occupying power; and international media attention creates embarrassment that squeezes the occupying power, further motivating their withdrawal (Davies, 2012).

CONCLUSION

Whether the WPM strategy will succeed in securing freedom for West Papua, Aceh or the Moluccas is yet to be seen, and it is not difficult to find arguments that it is doomed to fail because of Indonesia's interest in West Papua's mineral wealth and its success in getting other nations to sign treaties with clauses about not supporting threats to its "territorial integrity." The citizen journalism is still valuable however, as it is subverting attempts to simply silence the pro-independence movements through heavy-handed military action under the cover of a media ban. It is also providing a dynamic and internationally available version of "truth" that is valuable in the ways that Naqvi (2006) pointed out in her discussion of the emergence of the right to truth as a legal concept at national, regional and international levels. As stated earlier, her paper details the evolution of acknowledgement in international law that by exposing the truth societies are better able to prevent the recurrence of similar events; to heal rifts in communities; to foster accountability; and to support reconstruction of national identities by unifying countries through dialogue about a shared history. While the political situation in West Papua is not at the point where a Truth and Reconciliation Commission on the scale of CAVR is likely to be resourced, citizen journalists in West Papua are documenting, curating and publishing their own history in a way that the victims of the abuses in East Timor were not able to between the 1970s and 1990s.

In seeing this transformation in action, however, it is important to remember that it has not only occurred because the technology has improved and there are now more pixels and bandwidth than ever before. As Dixit (2011:12) pointed out "technology alone is never the answer. Information technology is not value-free either and is not, by itself, going to provide answers to deep-seated structural problems of governance, social justice and equity." The only thing that has ever changed the world is a small group of dedicated people, and it is the evolution of their understanding about the potential and hazards of citizen journalism over the past two decades that is now creating the new new new media.

NOTES

1. The other members of the team cannot be named here because they work in-country, and naming them would endanger them and their associates.

REFERENCES

Algionby, J. (2004, April 26). UN office burnt as upsurge of violence in Moluccas kills 10. *The Guardian*. Retrieved from: http://www.guardian.co.uk/world/2004/apr/26/indonesia.johnaglionby

Allard, T. (2009). Balibo Five executed, soldier admits. *Sydney Morning Herald*. Retrieved from: http://www.smh.com.au/national/balibo-five-executed-soldier-admits-20091207-kff6.html

Amnesty International (2009). Indonesia: Jailed for waving a flag. Prisoners of conscience in Maluku. Amnesty International Publications. Retrieved from: https://www.amnesty.org/en/library/asset/ASA21/008/2009/en/83bb8344-a4d3-425d-95d2-d36eb886e307/asa210082009en.pdf

Analysis: The beginnings of transitional justice in Syria. December 14, 2012, IRIN Humanitarian News and Analysis: UN Office for the Co-ordination of Humanitarian Affairs. Retrieved from http://www.irinnews.org/Report/97045/Analysis-The-beginnings-of-transitional-justice-in-Syria

Bachelard, M. (2013, April 16). Aceh wants to raise separatist flag. *Sydney Morning Herald*. Retrieved from: http://www.smh.com.au/world/aceh-wants-to-raise-separatist-flag-20130415-2hvvr.html

Bräuchler, B. (2004). Public Sphere and Identity Politics in the Moluccan Cyberspace. *The Electronic Journal of Communication*. 14, 3–4. Retrieved from: http://www.cios.org/EJCPUBLIC/014/3/01438.html

Byrne, M., and Gibson, K. (2007). Indonesian war criminals and Australian law. Uniya Occasional Paper no. 13, May 2007. Retrieved from http://www.uniya.org/talks/byrne_may07-op2.html

CAVR (2013). Comissão de Acolhimento, Verdade e Reconciliação de Timor Leste. Retrieved from: http://www.cavr-timorleste.org/

Citrawan, H. (2013, April 18). Aceh's Flag: A human rights approach. *The Jakarta Post*. Retrieved from: http://www.thejakartapost.com/news/2013/04/18/aceh-s-flag-a-human-rights-approach.html

Cramer, C. (2009). Taking the right risk. In *International news reporting: Frontlines and deadlines* (J. Owen & H. Purdey eds). pp. 163–190. London: Wiley-Blackwell.

Davies, K. (2012). Safety vs credibility: West Papua media and the challenge of protecting sources in dangerous places, *Pacific Journalism Review*, 18(1), 69–82.

Dixit, K. (2011). Real investigative journalism in a virtual world. *Pacific Journalism Review*, 17 (1), 12–19.

Errington, W. & Mirragliotta, N. (2011). *Media and politics: An introduction* (2nd edition). Melbourne: Oxford University Press.

Garg, A. (2012, November 13). Indonesia's Aceh Province adopts Sharia law in conflict with human rights standards. The Human Rights Brief. Retrieved from: http://hrbrief.org/2012/11/indonesia%E2%80%99s-aceh-province-adopts-sharia-law-in-conflict-with-human-rights-standards/

Gillmor, D. (2006). We the media. Retrieved from: http://www.hypergene.net/wemedia/download/we_media.pdf

Harsono, A. (2013, April 17). Human Rights Watch statement to Aceh House of Representatives: Hearing on proposed Truth and Reconciliation Commission. Retrieved from: http://www.andreasharsono.net/2013/04/human-rights-watch-statement-to-aceh.html

Hill, D. T. (2002). East Timor and the internet: Global political leverage in/on Indonesia. *Indonesia*, 73, 25–51.

International News Safety Institute (INSI) (2012). Killing the messenger (January-June) Update 2012. Retrieved from: http://www.newssafety.org/images/pdf/pdf/KTM2012_Jan-June(No%20CJ).pdf

International Federation of Journalists (IFJ) (2006, February 17). IFJ says foreign media ban in West Papua continues to obstruct press freedom. Retrieved on February 12, 2012, from: http://www.ifj.org/en/articles/ifj-says-foreign-media-ban-in-west-papua-continues-to-obstruct-press-freedom-

Lamble, S. (2004). Documenting the methodology of journalism. *Australian Journalism Review*, 26(I), 85–106.

MICHR (2013). Moluccas International Campaign for Human Rights. Retrieved from: http://www.michr.net/

Moore, M. (2004, August 7). Last Indonesians cleared of Timor killings. *The Age*. Retrieved from: http://www.theage.com.au/articles/2004/08/06/1091732085250.html

Naqvi, Y. (2006). The right to truth in international law: Fact of fiction? *International Review of the Red Cross*, 88 (862), 245–273.

Neumann, A.L. (2003, July 22). Indonesian military hems in press on Aceh citizens. *Asian Times* Online. Retrieved from http://cpj.org/2004/12/indonesian-military-hems-in-press-on-aceh-citizens.php

Perrottet, A., & Robie, D. (2011). Pacific media freedom 2011: A status report. *Pacific Journalism Review*, 17 (2), 148–156.

Powell, S. (2006). UN Verdict on East Timor. Global Policy Forum. Retrieved from: http://www.globalpolicy.org/component/content/article/163/29216.html

RMS (2013). Republik Maluku Selatan. Retrieved from: http://www.republikmalukuselatan.nl/en/content/home.html

Robinson, W. & Robinson, D.J. (2006). Tsunami mobilisations: Considering the role of mobile and digital communication devices, citizen journalism and the mass media. In Kavoori, A. and Arceneaux, N. (eds) *The Cellphone Reader: Essays in social transformation*, (pp. 85–104). New York: Peter Lang Publishing.

Rosen, J. (2008, April 22). Where's the business model for news, people? *Pressthink: Ghost of democracy in the media machine*. Retrieved from: http://archive.pressthink.org/2008/04/22/business_model.html

Sambrook, R. (2009). Citizen journalism. In International news reporting: Frontlines and deadlines (J. Owen & H. Purdey eds). pp. 220–242. London: Wiley-Blackwell.

Soares, A.W. (2010). A sojourn in journalism: Building trust, meeting challenges. Honours thesis. Murdoch University, Australia. February 2011.

Press Association (2010, October 28). Survival international website attacked over torture video. *The Guardian*.Retrievedfrom:http://www.guardian.co.uk/society/2010/oct/28/survival-international-website-torture-video

Walsh, P. (2011). Where to now after the CAVR and CTF? Summary of session with comments. University of Timor-Lorosae/Victoria University Joint Conference, Dili 4–5 July 2011. Retrieved from: http://www.dtp.unsw.edu.au/documents/ReportonVUUNTLfutureofCAVRCTF2.pdf

Vaswani, K. (2010, October 22). Indonesia confirms Papua torture. BBC World-Asia-Pacific. Retrieved from: http://www.bbc.co.uk/news/world-asia-pacific-11604361

Ward, S. (2010). *Global journalism ethics*. Montreal: McGill-Queens University Press.

West Papua Media (2013, May 13). Breaking News: Beatings, arrests as KNPB rally forcibly broken up by police. West Papua Media. Retrieved from: http://westpapuamedia.info/2013/05/13/breaking-news-beatings-arrests-as-knpb-rally-forcibly-broken-up-by-police/

West Papua Media (2010, November 5). Indonesian authorities suspected of launching cyberattacks on NGO websites. Retrieved from: http://westpapuamedia.info/2010/11/05/indonesian-authorities-suspected-of-launching-cyber-attacks-on-ngo-websites/

CHAPTER EIGHTEEN

Civic Responsibility and Empowerment: Citizen Journalism in Russia

KARINA ALEXANYAN

Russian citizen journalism made international front-page news in February 2013, when a meteor entered the earth's atmosphere and crashed in Chelyabinsk Oblast. Dramatic eyewitness videos emerged on social networking sites showing the fireball darting across the sky, followed by a loud explosion. Ordinary citizens driving near the area captured much of the available footage, often accidently so because of the widespread use of so-called "dashcams" (video camera mounted on dashboard in car) to secure eyewitness material against police corruption or attempted insurance fraud. Eyewitness footage of the meteor helped verify the nature of the asteroid, and also helped to quell initial scepticism about vague official reports emerging in the hours immediately after the incident.

Whilst the Chelyabinsk meteor helped to bring global attention to dashcam reporting in Russia, citizen journalism has long been recognised within the country as a means to empower individuals intent on countering the dominance of state-controlled media. The internet is widely seen as having transformed the relationship between individuals and all types of information, and in particular news and other public affairs related content. User-generated content and interactive features of mobile and online communications have fractured the traditional mass media dichotomy of broadcasters vs receivers and producers vs. audiences, empowering "audiences" to participate in content creation and dissemination. Citizen journalism, public journalism, grassroots journalism, participatory journalism—all of these terms are used to describe the public's new role in the production and diffusion of information, and of news. It is in this context that NYU journalism professor Jay Rosen's definition of citizen journalism is most applicable—"when the people formerly known as the audience employ the press tools they have in their possession to inform one another" (Rosen, 2008). Today, the spread of information is no longer one directional, top down, from mainstream media to the masses. Internet technologies empower individuals as members of a many-to-many network of "self-communication," enabling conversation and coordination using a variety of mobile and online media, including blogs as well as social networking sites, blurring traditional boundaries between producers and consumers of information, reducing the power of information gatekeepers, and building networks of communication and organization that span the globe.

The ramifications of technological change, however, are determined not merely by what the technologies enable, but how and by whom they are used. The impact of online and mobile media on journalism and society, I argue, is culturally and nationally specific. The significance of "citizen journalism" depends on the socio-political context, and a nation's understanding of the role of the citizen and of journalism itself. By examining specific local contexts, scholars can go beyond universal assumptions to consider the socially and politically situated relationship between online media, information and individuals. Thus developing a more nuanced assessment of the implications of online and mobile media on the role of public affairs reporting, and on social and political action. This approach underscores the importance of expanding theories of new media beyond US-centric models, or models that universalize US-centric assumptions about the interplay between online media, journalism and the public.

To that end, this chapter places examples of citizen journalism and engagement in Russia from 2009–2012 within the context of Russia's contemporary media landscape, illustrating the socially and politically situated relationship between online media and offline action in Russia. In semi-authoritarian environments like Russia, I will argue, the participatory journalism and activism enabled by online and mobile technologies fosters a novel and essential quality—a sense of civic society, civic responsibility and civic empowerment. In Russia, citizen journalism is inextricably tied to socio-political activism, calling for an expanded understanding of the concept of participatory journalism, going beyond the standard definition of "citizens informing one another" to include not only the dissemination and broadcast of information, but also interpersonal communication and organization.

RUSSIAN MEDIA LANDSCAPE

Russia's media landscape presents a spectrum of freedom and independence across the various Russian news media outlets, traditional to web native. Universally available top-down, government-controlled federal television lies on the most restricted end of the spectrum, a vulnerable and disappearing printed press teeters in the middle, and a liberal, growing, but still marginal internet media lies on the most independent end. Despite their diversity and vibrancy, independent and liberal media (online, print and radio) are sidelined by television in Russia. According to a number of leading public opinion surveys, television serves as the primary source of information, trust and entertainment for the majority of the Russian electoral base (FOM Public Opinion Foundation, 2011; Levada Analytical Center, 2009). While the population of other countries, such as the US for example, access multiple media channels and platforms to get news (PEW internet and American Life Project, 2010), only 30% of Russians turn to a medium in addition to TV (FOM Public Opinion Foundation, 2011). As a result, the information diet of the majority of the Russian populace is particularly narrow and monolithic. The 2010 Freedom House *Freedom of the Press* report describes the contemporary scenario as one in which "government-controlled television was the primary source of news for most Russians, while lively but cautious political debate was increasingly limited to glossy weekly magazines and news websites that were accessible mostly to urban, educated, and affluent audiences." (Freedom House, 2010).

In Russia, the internet remains a relatively elite medium with an as yet modest nationwide reach—internet users making up approximately 52% of the population in 2012. On a global scale, this places Russia between Brazil (39%) and Italy (59%), but still significantly behind the US (79%) (FOM Public Opinion Foundation, 2012). In Russia's two major cities, St. Petersburg and Moscow, the rate is consistently higher, matching that of the US (FOM Public Opinion Foundation, 2012). Those Russians who do go online, however, do so frequently, actively using the internet as a source for information, news and social networking (FOM Public Opinion Foundation, 2009). In particular, recent studies of global social media use find that the percentage of internet users who are also active social media users is consistently higher in Russia than it is in the US, or even globally (Universal McCann, 2009).

As a result, the Russian media ecology exhibits a particularly stark distinction between a television-watching majority exposed to a restricted diet of government-regulated and -approved news content, and a (growing) minority of active internet users, who access and trust a more independent array of news and information. This distinction has led a variety of social commentators to divide Russian in two—the "TV audience" and the "internet audience." In 2011, Radio Free Europe described these as "the 'internet Russia' of mostly young, increasingly globalized readers who can access any information they want; and the 'television Russia' that listens to what the authorities want it to hear" (Feifer, 2011). Similarly, in a July 2011 article in *The Nation*, Russian analyst and Foreign Affairs blogger, Vadim Nikitin, describes Russia's "two nations" as "an offline mass of older, poorer, disaffected but largely inert and atomized consumers of state-controlled TV" and "an incestuous city-state of upwardly mobile, internet-savvy young urbanites organizing and networking on [the popular social media platform] LiveJournal" (Nikitin, 2011a).

Media analysts have traditionally drawn a direct, causal line between information diet and socio-political activism. From this perspective, independent information can stimulate an informed citizenry into action. At the very least, social and political activism is virtually impossible without access to uncensored information and news. So, for instance, in a June 2011 American Enterprise Institute report, analyst Leon Aron describes Russia's "two nations"—the "television nation" and the "internet nation" in terms of information diet and politics. "Although most Russians still get their daily news from television," he argues, "the minority who rely on the internet are more politically engaged" (Aron, 2011). The connection between media diet and politics, however, is not as direct as it may appear. While the politics of the television-watching majority reflect a relatively straight-forward support for the ruling party, the politics of the "internet minority" are diffuse, and often contradictory. A 2010 editorial on the Russian news website Gazeta.ru highlights this distinction, explaining that, while a majority of Russia's television viewers vote for Putin, Medvedev and "any other United Russia member," the political views of active internet users are more complex. While they are "by and large sharply critical of the government" they also take "differing positions that are at times diametrically opposed" (Gazeta.ru, 2010).

In my view the relationship between independent information and political engagement (including, but not limited to, citizen journalism) is not unidirectional, from information to engagement. In Russia, as in other "wired" semi-authoritarian states, analysis of the influence of internet technology on discourse, journalism and political action should be expanded to highlight communication and coordination, in addition to information. Russia's relatively elite, young, urban and globalized minority, I contend, is not more socially and politically engaged merely because of the independent information they receive online. Rather, the minority that are involved in public affairs turn to the internet not only for information, but also, and more significantly, leverage online media for interaction, publication and organization. In a *Foreign Affairs* blog post from October, 2011, Nikitin affirms this connection between the social elite and the politically active: "It is precisely the small group of wealthier and more educated people who live in the largest cities and have the strongest real-life social capital," he writes, "that also dominates the country's virtual civil society: the most likely to get their news online, read and write blogs, and participate in social activism" (Nikitin, 2011b).

FROM ONLINE INTERACTION TO OFFLINE ACTION

As the internet and social media diffuse throughout Russia, its users are going online, not only for information, but also for communication, coordination and community.

Since 2009, when Russia's nationwide internet penetration reached one-third of its citizens, there has been growing evidence of the internet's role in helping individuals coordinate, organize and mobilize offline socio-political action. Over the last few years, Russia's vibrant online sphere (dubbed the "RuNet"), has responded to the lack of civic and political options by producing a number of internet-enabled grassroots civic campaigns, volunteer and relief efforts and other non-ideological "bottom-up" movements. Russia's grassroots movements illustrate the internet's role as a medium of interpersonal communication and coordination, above and beyond its "liberating" power as an information medium. In semi-authoritarian environments like Russia, I argue, the participatory journalism and activism enabled by online and mobile technologies fosters a novel and essential quality—a sense of civic society, civic responsibility and civic empowerment. Supported by a strong network of active social media users, Russia's citizen journalism fosters an emerging sense of civic consciousness, and, ultimately, political activism.

The uses of the internet for offline activism before the dramatic events surrounding the 2011–2012 election cycle provide compelling examples of Russians mobilizing their active online media networks to address specific issues that have a concrete effect on everyday life. These pre-2012 actions are united by their non-ideological nature, sharing an overall concern with corruption, abuse of power and privilege, and a lack of official accountability. This understanding is supported by a 2010 study published by the Berkman Center for internet and Society at Harvard University, which found that the leading politically oriented YouTube videos posted by individuals in 2009–2010 consistently addressed issues of corruption, accountability and lack of transparency, "pushing back against abuse of power by business and finances elites, the government and the police" (Etling et al., 2010).

Russian citizen journalism and internet-enabled activism of 2009–2011 can be grouped into three main categories—civil rights–oriented social movements, such as those surrounding motorist rights, environmental/preservationist issues and the right of assembly; individually sponsored, targeted, grassroots civil action campaigns; and volunteer, aid and crisis intervention efforts. These include successful, targeted grassroots actions instigated by individuals—mostly popular bloggers who were able to leverage and mobilize their social networks for specific change. For example, in 2010, 26-year-old Ilya Varlamov (known as "zyalt") founded the blogging campaign "A Country without Idiocy" to address a specific instance of abuse of vehicular and official privilege—the brazen violation of handicapped-parking-spot use near the headquarters of Moscow's traffic police. Mr. Varlamov led a letter-writing campaign to the prosecutor's office that successfully ended the practice, and his actions have, according to an October, 2010 Radio Free Europe article, earned him "the devotion of tens of thousands of microbloggers and turned him into a cult phenomenon on the Russian internet" (Balmforth, 2011). "I wouldn't put it as glamorously as doing my 'civic duty,'" Varlamov claimed,

"I just do what I like doing and what I think is right because I would really like to change things [...] I simply try to bring these little things to people's attention so that we can change the situation together" (ibid.).

Similarly, in 2010, prominent travel blogger Sergey Dolya launched the "Country without Garbage" or "Bloggers against Garbage" campaign, leveraging his blog's popularity to raise awareness about littering. The campaign went on to organize cleanup operations in 100 locations nationwide, in one case removing 200 tons of trash from Russia's streets in a single day (Dolya, 2011). The most prominent example of an individually sponsored, targeted and successful grass-roots effort is the anti-corruption site RosPil.net, launched in 2010 by well-known activist Alexei Navalny. The site allows people to anonymously report suspicious government tenders, and claims to have prevented more than $10 million worth of misappropriations (Nikitin, 2011a).

In addition to undergirding popular movements, online media have been used to coordinate collective legal action, as well as crisis intervention, volunteer and aid efforts. The most successful and well known of these is the Wildfires Help Map (http://russian-fires.ru/), initiated as a result of the state's inadequate response to widespread wildfires near Moscow during the summer of 2010. The Help Map was created using the open-source Ushadi platform. Members of the Russian team at Global Voices Online (globalvoicesonline.org, an international citizen media website based in the US) worked together with Russian blogger and activist Marina Litvinovich on the project. The site allowed users throughout the area to coordinate volunteer and relief efforts, and went on to win a prestigious RuNet Award in 2010 (Sigal, 2010). In addition to the Help Map, there were many other cases of people from different regions organizing online to deliver humanitarian assistance to the affected areas. Other examples of collaborative action enabled by online media include: the collective lawsuit organized against Moscow airports and airline officials in response to violations that occurred in December 2010, when foul weather caused 1,200 flights to be delayed or cancelled and 20,000 people stranded (RIA Novosti, 2011); and the response to the tragic July 2011 sinking of a Russian cruise ship, when volunteers used *Vkontakte* (Russia's version of Facebook) to coordinate aid and support (Sindelar, 2011). Strategies including LiveJournal, Vkontakte, Facebook, Twitter, and YouTube bulletins have also been used to organize searches and successfully locate missing persons (Sindelar, 2011).

By the summer of 2011, a few months before the dramatic protests surrounding the 2011–2012 election cycle, this surge in grassroots activity fostered by online networks was being interpreted by Russia analysts as evidence of a nascent civil society, notable for its non-ideological bent, more concerned with specific abuses of power and privilege, than with abstract "authoritarianism." In October 2011, Radio Free Europe reported on the way that the "internet is changing the rules of engagement" in Russia, echoing the key issues—spectrum of media freedom and independence, monolithic information environment, two audiences—discussed above. The article also highlighted the non-ideological nature of the Russian language online sphere, bridging diverse political perspectives by fostering civic movements that target specific, non-ideological everyday concerns:

> In an otherwise bleak media landscape, the rapid rise of the internet is shifting the Russian information industry from a top-down operation to a looser, more pluralistic affair where regular Russians are no longer expected to be passive consumers of traditional news [...] Television, which still reigns supreme as a source of news for 85 percent of Russians, may provide a glossy and orchestrated image of the world that suits the Kremlin's needs. The dwindling newspaper trade may deliver tailored bulletins to niche audiences. But the Russian internet, or RuNet, is the first medium in the country to come without a built-in ideological bent [...] And along the way, it's fueling a new wave of civic activism—one that may not bring sweeping political change or find common cause with the traditional opposition, but which is rapidly giving regular Russians power to bear on issues that affect them most, from car inspections to community safety to bureaucracy and corruption (Sindelar, 2011).

Leon Aron similarly argues that the internet is the "backbone of civil society in Russia – giving people both a voice and the tools to self organize", in his report for the American Enterprise Institute entitled "Nyetizdat – How the internet is Building Civil Society in Russia (Aron, 2011). Where others emphasize the targeted and non-ideological nature of these grassroots movements, Leon views the active Russian internet sphere not only as the "main alternative public platform and the engine of grassroots self-organization," but also as "a growing force against authoritarianism [...] at once a national 'town hall' and party headquarters, vital to the emergence and maintenance of thousands of social and political movements" (Aron, 2011).

CITIZEN JOURNALISM DURING THE 2011–2012 ELECTORAL UNREST

The election period of 2011–2012 was a turning point for socio-political action in Russia. The relationships, energy, experience and civic confidence fostered by the non-ideological movements, participatory journalism and grassroots campaigns of the preceding years came to a fore, triggered by a series of government moves that were interpreted as blatant political affronts. These began with Vladimir Putin's unilateral announcement in September 2011, that he would be running for President, in effect swapping places with Dmitry Medvedev. "The Russian public saw the two leaders' trading of places as evidence that they held their citizens in full contempt," according to Maria Lipman and Nikolay Petrov, "especially when

Medvedev, lamely, added that their decision to switch offices had been made long ago" (Lipman and Petroy, 2012). This "affront" was followed by Parliamentary (Duma) elections in December 2011, that were riddled with perceived fraud, including a wave of DDOS (Distributed Denial-of-Service) attacks targeting independent and watchdog online media that were understood as a flagrant attempt to block election monitoring (Barry, 2011). The result was a year of massive street protests, and an unprecedented wave of citizen journalism and political engagement, supported by Russia's strong social networks and the creative application of internet technologies.

In Russia, 2012 was full of examples affirming the internet's role as a medium of interpersonal communication and coordination, above and beyond its "liberating" power as an information medium, illustrating the extent to which citizen journalism is intertwined with activism. These include the resilience of online networks in face of DDOS attacks; internet-enabled innovations in election monitoring and other accountability efforts; crowd-sourced volunteer and aid efforts; the online production and distribution of grassroots symbols and publicity materials; the flood of documentation, self communication and citizen journalism related to protests and election violations on social media and YouTube; and the creative "virtual" organization of protests and demonstrations that work around official restrictions.

The DDOS attacks on 4th December 2011 were timed to overlap precisely with the Duma elections, beginning as soon as polls opened and ending once the polls were closed. The attacks were a focused effort, unprecedented in scale, directed at the popular LiveJournal blogging and social media platform—the online home of the majority of Russia's independent journalism, the election monitoring group Golos (www.golos.org), and a number of other leading independent journalism sites, such as *Echo of Moscow*, *Novaya Gazeta*, *NewTimes*, *Bolshoi Gorod*, slon.ru, ikso.org, ridus.ru, zaks.ru and pryaniki.org (Asmolov, 2011). The timing and targets of the attack, and especially the fact that many of the targeted sites were publicizing submissions on election violations, led observers to interpret the attacks as a clear "attempt to inhibit publication of information about violations" (Roberts and Etling, 2011). Despite the unprecedented scale of the 4th December 2011 DDOS attacks, these were not the first, and many media organizations were prepared, swiftly transitioning to alternative platforms such as Facebook, Vkontakte (Russian social network similar to Facebook), Twitter, YouTube and Google Docs to continue distributing information. The large scale, targeted attack was unable to stifle the flow of information, and RuNet was full of exit polls, reports and videos about voting violations (Asmolov, 2011). *Russia Profile* analysts argue that, in fact, the DDOS attacks served as an additional mobilization call, triggering a massive reaction in the Russian blogging community: "One blogger after another, from such popular figures as anti-corruption crusader Alexei Navalny to rank-and-file civic journalists, offered their blogs as vehicles for publishing reports of violations" (Zolotov and Roth, 2011). Social networks, especially Twitter, disseminated exit poll data from Moscow and the regions, as well as evidence on alleged vote rigging. The persistence of these reports attests to the power and significance of Russia's interpersonal networks, above and beyond the individual platforms or channels used. It demonstrates that attacking or shuttering a specific channel or online resource is not effective when the interpersonal network is strong and flexible enough to migrate elsewhere.

Of the many internet-enabled election monitoring and accountability tools that emerged from the election cycle of 2011–2012, one of the most well known was an online election monitoring platform created by Golos, (golos.org) an independent Russian election monitoring organization active since 2000. Golos' *Map of Violations* (KartaNarusheniy.ru), supplemented with a text messaging service (sms.golos.org), served as an interactive, crowd-sourcing platform with real-time data showing the location of perceived election violations throughout the country (Asmolov, 2012). Another election monitoring application, developed especially for smartphones, "Web Observer" (webnablyudatel.org), enabled users to instantly share video, photos and reports of violations. Introduced one week prior to the presidential election, Grakon (grakon.org) is a social networking platform designed to coordinate and share election monitoring information, enabling members to register as voters, observers, members of voting commissions, lawyers, and official representatives.

After the contested Parliamentary elections of 4th December, a number of leading journalists and public figures united to create the "League of Voters" (ligaizbirateley.ru), an online hub for various opposition activities, aggregating information on various activist groups around Russia and enabling users to register and search for groups by topic or location. The League sponsored and supported protest activity, as well as initiatives related to election monitoring and accountability, such as the website *Svodny Protocol* (svodnyprotocol.ru), which verified election monitoring reports, and the website *The Black List* (chernyspisok.info), which crowd-sourced information relating to election fraud. Additional platforms were developed to help citizens register as official observers on behalf of any party or candidate taking part in the election. These included the website *Citizen-observer* (nablyudatel.org) and Alexey Navalny's initiative Rosvybory.org. Regional websites, such as *Saint Petersburg Observers* (spbelect.org) also emerged, launched by local activists with the goal of increasing the number of local observers (Asmolov, 2012). Ultimately, these efforts enabled 28,000 volunteers throughout Russia to serve as official observers for the 2012 presidential election—an unprecedented upsurge in civic responsibility and grassroots activism, which, as Aron argues, was "all the more remarkable because the result was widely believed to be predetermined" (Aron, 2012).

Further citizen journalism efforts include the special election page created by the online service Kuda-Komu ("Where-To Whom"), which helped users generate letters of complaint. The special election page assisted users with letters that focused on particular types of electoral fraud and suggested where they should be sent. An online petition was also started, on churovu.net, calling for the resignation of the Head of the Russian Central Election Committee, Vladimir Churov—a figure who was seen by many as responsible for significant election fraud. Last, but not least, the official initiative to install webcams in nearly all of Russia's 95,000 polling stations, enabling visitors to the presidential election monitoring website webvybory2012.ru to observe the voting process. Prime Minister Putin announced this initiative in December 2011, in response to agitation regarding voter fraud. The webcam installations, cost an estimated 10–15 billion rubles ($320–$480 million) and equipped every polling station with two cameras: one focused on the ballot box and the other providing a broad view of the polling station. Once voting was over, one of the cameras also broadcast the vote counting process. Around 600,000 Russians registered on the site to observe the presidential elections (RIA Novosti, 2012).

The frequent demonstrations of 2012 also sparked a number of creative approaches to public protest and assembly, designed to circumvent restrictions on mass demonstrations and other authoritarian measures. These solutions were empowered by internet technology, which fostered a decentralized, collective style of organization that made the enforcement of restrictions difficult. One example was the car-based protest, such as the ones held around Russia in January and February 2012, in which cars marked with white ribbons (the movement's symbol) circulate in a particular area. These protests were organized via social networks and blogs, and have led to a large-scale turnout of thousands of vehicles (Bulay, 2012). Another "flashmob" type protest organized, facilitated and publicized online was the Big White Circle action of 26th February 2012, in which protestors wearing white formed a nine-mile human chain around Moscow's city center, surrounding the Kremlin. This demonstration did not require a permit because it was, in actuality, a multitude of individual, one-person protests. In a few cases, the demonstrations and protest activities were organized and coordinated via "pop-up" websites created specifically for the event (GlobalVoicesOnline, 2012). Protest agenda and demonstration speakers were selected via online voting, using dedicated pages on LiveJournal, Facebook or SurveyMonkey (for example: http://golosovalka.livejournal.com/95354.html or https://www.surveymonkey.com/s/CSC899Z or https://www.facebook.com/questions/237935562943756/).

The demonstrations and activism of 2012 were augmented by a plethora of citizen journalism efforts in the form of YouTube videos, blog posts and mobile camera reports that documented the experiences of rally participants. Mobile phones were used to live-stream protest coverage, enabling protesters to broadcast live from such unusual locations as police vans and police stations. Mobile-based, real-time broadcasting platforms such as ustream.com and bambuser.com were used to provide live coverage of election protests from the heart of the crowd. Streams included images of clashes between protesters and police, and in some cases, live broadcasts of personal arrests. Some of the streams had an audience of more than 40,000 people at one time (Asmolov, 2013). The detained participants of the rallies also actively used Twitter to update of their arrests, as well as to share information about the location of the police car taking them to the station. At the peak of the arrests, Twitter feeds were full of dozens of reports from those detained (Asmolov, 2013). The live broadcasting and tweeting of arrests increased transparency and accountability around police action. When an individual broadcast news that he or she had been detained, friends and volunteer lawyers could follow them to the police station and demanded their release. Another website, ovdinfo.org, aggregated information from different sources about arrests, and another site, created by activist Maxim Katz, coordinating information and dispatching volunteer lawyers (Asmolov, 2013). Technology and citizen journalism combined to generate images and figures on the size and number of rally participants, challenging official figures. Innovative photo-journalists used remote-controlled aerial cameras to produce "birds-eye" views and 360 degree images, documenting the size and scope of the rallies. Applications such as "The White Counter" created by programmer Anatoliy Katz, analyzed a large number of images taken every second to generate counts of protestors. The White Counter, used for the first time at the 12th June 2012 protest, calculated 54,000 protestors, as opposed the official figure of 18,000 (Asmolov, 2013).

Cable and online channels, broadcast coverage of the demonstrations, as well as online debates and dialogue between various oppositional leaders. The leaders in this category are Networked Public TV (SOTV) (rusotv.org), created as a public television channel by a number of liberal public figures just prior to the Duma elections in November 2011, and Dozhd TV (TVRain.ru), which launched in 2010. The vibrant protest coverage provided by professional and citizen journalists, as well as the unprecedented magnitude of the demonstrations themselves, forced a transformation in the content of state-controlled television. After ignoring the initial post-election protests on 5–6th December 2011, state television channels began to provide national coverage of the large protests that swept Moscow and other cities on 10th December. In some cases, journalists and editors on the state channels allegedly refused to broadcast unless the demonstrations were covered (Asmolov, 2012). On the state-run Rossiya 1 television channel, Presidential election night coverage featured a variety of opinions and topics unheard of on national television in the past decade, including coverage of the discussions of fraud publicized in social media and the new phenomenon of online election monitoring (Zolotov and Roth, 2011).

CONCLUSION

While overtly triggered by a series of explicit government "affronts," 2012's momentous wave of citizen journalism and political engagement was incubated in the relationships, energy, experience and civic confidence fostered by the non-ideological movements and grassroots campaigns of the preceding years. These earlier movements leveraged Russia's vibrant, interconnected and flexible online networks to address the same fundamental concerns with corruption, abuse of privilege and lack of accountability that ultimately fueled the protests of 2012. Taken together, they illustrate the internet's role as a medium of interpersonal communication and coordination, above and beyond its "liberating" power as an information medium. In Russia, the ability to take targeted action and create successful change is a new social characteristic, fostered, at a fundamental level, by the interpersonal communication networks of online media. What's more, it is not just simply the ability, but the symbolic power of a belief in the necessity and possibility of grassroots change, that contain the seeds of civil society. Online and mobile technologies have nurtured an emergent sense of civic society, civic responsibility and civic empowerment, fostering participatory journalism, socio-political activism, and a novel appreciation for the role and power of the citizen.

REFERENCES

Aron, Leon (2011). *Nyetizdat: How the Internet Is Building Civil Society in Russia*. Russian Outlook. American Enterprise Institute, June 28, 2011. http://www.aei.org/outlook/society-and-culture/nyetizdat-how-the-internet-is-building-civil-society-in-russia.

Aron. Leon (2012). *Russia's Protesters: The People, Ideals and Prospects*. AEI, August 9, 2012. http://www.aei.org/outlook/foreign-and-defense-policy/regional/europe/russias-protesters-the-people-ideals-and-prospects/

Asmolov, G. (2011). "Russia: Massive DDoS Attacks Against Independent Websites on the Election Day" *Global Voices Online*, December 4, 2011. http://globalvoicesonline.org/2011/12/04/russia-massive-ddos-attacks-against-independent-websites-on-the-election-day/.

Asmolov, G. (2012). "Russia: 11 Areas of Election-Related ICT Innovation." *Global Voices Online*, March 6, 2012. http://globalvoicesonline.org/2012/03/06/russia-11-areas-of-election-related-ict-innovation/

Asmolov G. (2013). "Dynamics of Innovation and the Balance of Power in Russia" in Muzammil M. Hussain and Philip N. Howard (eds) *State Power 2.0 Authoritarian Entrenchment and Political Engagement Worldwide*, Survery, UK: Ashgate.

Balmforth, T. (2011). "As Russian Bloggers Gain Prominence, The Kremlin Takes Notice." *RadioFree Europe/ RadioLiberty*, October 12, 2011. http://www.rferl.org/content/russian_bloggers_gain_prominence_kremlin_takes_notice/24357352.html

Barry, Ellen (2011). "Tens of Thousands Protest in Moscow, Russia, in Defiance of Putin." *The New York Times*, December 10, 2011, sec. World/Europe. http://www.nytimes.com/2011/12/11/world/europe/thousands-protest-in-moscow-russia-in-defiance-of-putin.html.

Bulay, Andrey (2012). "Russia's Putin Protests: Cars Circle Central Moscow." *Huffington Post*, January 29, 2012. http://www.huffingtonpost.com/2012/01/29/russia-protests-putin-car-circle_n_1240113.html.

Dolya, Sergey (2011). "Popular Blogger Prompts Thousands of Russians to Clean Their City", *The Observers France24.com*, November 7, 2011. http://observers.france24.com/content/20110711-popular-Sergey-Dolya-prompts-thousands-russians-clean-hometown-streets-bloggers-against-garbage

Etling, B., Alexanyan, K., Kelly, J., Faris, R., Palfrey, J. and Gasser, U. (2010). *Public Discourse in the Russian Blogosphere: Mapping RuNet Politics and Mobilization*. Berkman Center for internet and Society: Harvard University. http://cyber.law.harvard.edu/sites/cyber.law.harvard.edu/files/Public_Discourse_in_the_Russian_Blogosphere_2010.pdf

Feifer, Gregory (2011). "Russian Media Landscape Varied Despite Heavy State Control." *RadioFree Europe/ RadioLiberty*, October 7, 2011 http://www.rferl.org/content/russian_media_landscape_varied_despite_heavy_state_control/24352432.html

FOM Public Opinion Foundation (2009). *Media Preferences: Internet Approaches Television* April 23, 2009 http://bd.fom.ru/pdf/d16lp.pdf.

FOM Public Opinion Foundation (2011). *Media Preferences—News and Information*. February 10, 2011. http://bd.fom.ru/pdf/d06niip11.pdf.

FOM Public Opinion Foundation (2012). *Internet in Russia, Fall 2012*. http://runet.fom.ru/Proniknovenie-interneta/10738.

Freedom House *Freedom of the Press* (2010). http://www.freedomhouse.org/report/freedom-press/freedom-press-2010

Gazeta.ru (2010). "Internet vs the Idiot Box." December 10, 2010. http://www.gazeta.ru/comments/2010/12/21_e_3473009.shtml

Levada Analytical Center (2009). *Public Opinion* 2009. http://www.levada.ru/books/obshchestvennoe-mnenie-2009.

Lipman, Maria, and Nikolay Petrov (2012). "What the Russian Protests Can—and Can't—Do." *Foreign Affairs*, February 9, 2012. http://www.foreignaffairs.com/articles/137091/maria-lipman-and-nikolay-petrov/what-the-russian-protests-can-and-cant-do.

Nikitin, V. (2011a) "The Rebirth of Russian Civil Society." *The Nation*, July 18, 2011. http://www.thenation.com/article/162108/rebirth-russian-civil-society#.

Nikitin, V. (2011b) "Russia's Internet Polyarchy." *Foreign Policy Blogs*, October 10, 2011. http://foreignpolicyblogs.com/2011/10/10/44626/.

Pew Internet and American Life Project (2010). *Understanding the Participatory News Consumer*, March 1, 2010. http://www.pewinternet.org/Reports/2010/Online-News/Summary-of-Findings.aspx.

RIA Novosti (2011). "Angered Passengers to Sue Moscow Airports." December 28, 2011. http://en.rian.ru/russia/20101228/161971086.html

RIA Novosti (2012). "About 600,000 Russians to Monitor Presidential Elections Online." March 3, 2012. http://en.rian.ru/russia/20120303/171703448.html.

Roberts, H., and Etling, B. (2011). "Coordinated DDoS Attack During Russian Duma Elections." *Internet & Democracy Blog, Berkman Center for Internet and Society*, December 8, 2011.

Rosen, Jay (2008). *PressThink Archive*, http://archive.pressthink.org/2008/07/14/a_most_useful_d_p.html

Sindelar, D. (2011). "As Other Media Stagnate in Russia, Internet Changing Rules of Engagement." *RadioFreeEurope/ RadioLiberty*, October 9, 2011. http://www.rferl.org/content/as_other_media_stagnate_in_russia_internet_changing_rules_of_engagement/24354068.html https://blogs.law.harvard.edu/idblog/2011/12/08/coordinated-ddos-attack-during-russian-duma-elections/

Sigal, Ivan (2010). "Russia: 'Help Map' Wins Runet Award." *Global Voices Online*, December 3, 2010. http://globalvoicesonline.org/2010/12/03/russia-help-map-wins-runet-award/

Universal McCann (2009). *Social Media Study Wave 4*. July 2009 http://www.slideshare.net/IN2marcom/power-to-the-people-wave-4-social-media-study-by-universal-mccann

Zolotov and Roth (2011). "The Internet's Watching." *Russia Proflie.org*, December 5, 2011. http://russiaprofile.org/politics/50537.html

CHAPTER NINETEEN

Beyond THE Newsroom Monopolies: Citizen Journalism AS THE Practice OF Freedom IN Zimbabwe

LAST MOYO

This chapter critically engages with citizen journalism's nascent forms, content, themes, and epistemes as shaped by the Zimbabwean experience. Specifically it explores the Kubatana political bloggers and the Presidential election run-off crisis in 2008 that was characterised by the murder and torture of civilians. The Pan African Observer Mission (PAOM) reported that during the election "the prevailing political environment throughout the country was tense, hostile and volatile" (PAOM, 2008:54). It stated that there were "high levels of intimidation, hate speech, violence, war rhetoric, displacement of people, abductions, and loss of life and many abuses of other rights and freedoms" (Ibid.:34). These claims were corroborated by, among other African regional groupings, the SADC Parliamentary Forum Observer Mission. In the context of a very closed political environment, Kubatana became one of the few virtual spaces for the practices of freedom and political engagement for citizens (others included Zvakwana/Sokwanele which is more of a radical underground online and offline social movement).

Kubatana is a civic organisation involved in cyber activism to highlight democracy and human rights issues in Zimbabwe. Using the Kubatana bloggers as a case study, this chapter explores the following questions: To what extent is citizen journalism as seen through the lens of Kubatana giving ordinary people a voice? What challenges exist for citizen journalism in Kubatana to be a space for subaltern agency so that the oppressed can speak back to power in Zimbabwe? In analysing citizen journalism through these questions, the chapter contends that the practice calls for a radical rethinking of the concept of citizenship. Citizen journlism practices imply a paradigm shift from citizenship as just political membership based on a codification of rights and responsibilities to citizenship as a discursive, deliberative, dialogic, and transformative space and identity. To that degree, citizen journalism recasts citizenship as a practice of freedom and a medium of communication that allows ordinary people to tell their stories free from publishing monopolies. Online spaces like Kubatana therefore epitomise a resurgence of subaltern agency and a return to locality as a key space of democratic interaction and conversation. They bring about local agency in politics which amounts to a development from a monitorial citizen who only keeps an eye on the accountability and transperacy of the state to a participatory citizen who actively engages with the state through self-generated media and social action.

A THEORY OF CITIZEN JOURNALISM AS AN ALTERNATE- SUBALTERN SPACE AND PRACTICE

In this section, I will discuss and develop a theory of alternate-subaltern spaces and practices that will serve as a critique of Kubatana's forms of citizen journalism. Citizen journalism as a discursive space and practice for ordinary people ideally must not be part of the mainstream media. At its very best, it must stand ideologically counter posed to elite media that often advance the interests of the state or capital. However, as stated earlier, citizen journalism is complex and multi-faceted because it takes many forms depending on time and space. In most liberal societies, its development is not mutually exclusive from the mainstream because of its fluidity that gives it the flexibility to conflate, intersect, or even be appropriated by mainstream journalism. However, in authoritarian contexts like Zimbabwe that provide the context of the development of this conceptual framework, citizen journalism is often likely to develop independently as a voice of the subaltern. It enjoys some degree of free access and participation by everyday people, provides a space for divergence and freedom, and most profoundly is autonomous from any form of power that might curtail free speech.

Citizen journalism must be a source of subaltern agency. Subalterns are the oppressed people who lack a voice in the mainstream public spheres and "the basic right of participation in the making of local history and culture" (Louai, 2011:5). They often represent the marginalised in terms of class, race, gender, or even sexuality. Their marginalisation in making their history and culture means that they lack control of the tools of representation of the self. Consequently, they are often constructed as the incomprehensible "other" in the mainstream media spaces. The alternate and subaltern spaces essentially mean moving the centre from the spaces of domination of their collective identities as prescribed by the powerful media to narratives that liberate and empower them as the oppressed and marginalised classes. Hence, alternative-subaltern spaces and practices can be conceptualised in terms of their counter-hegemonic and social change agenda in society that is often expressed in news values that embrace political or cultural radicalism (Atton, 2002; Bailey et al., 2008). They are normally spaces and practices of freedom for everyday people and usually represent ideologies of the underdog that rarely form part of the mainstream discourse in the elite media.

In the struggle to attribute meaning to events and social experience, alternate-subaltern spaces are supposed to be a platform for citizenship that is active, discursive, critical, and transformative. They represent forms of citizenship that are both monitorial and participative as they monitor excesses by power while also clearly engaging with it to demand transparency and accountability (Papacharissi, 2010). In other words, citizenship implies a commitment toward the public good in the political sphere. It represents a "set of civic attitudes, an emblem of civic participation, an arena where right-bearers unfold their personalities [and responsibilities]" (Alejandro, 1993:9). The alternate-subaltern spaces therefore provide modalities that are a sinew for democratic citizenship, political engagement, and political action. They are not simply media spaces that deepen the illusion of inclusive debate and choice from more of the same media forms and content, but serve to articulate a different worldview and orders of reality as reflected in their radical ideological orientation.

Alternate-subaltern spaces must also be organically structured to serve grassroots communities that share common conditions of domination and subordination. Online media as one such space, for example, has brought about virtual communities that transcend the limits of time and space thus making them potentially democratically inclusive. Although such communities are disparate, translocal, deterritorialised, and mediated, they are bound by "sharing a common condition or problem [from which they seek freedom]" (Popple, 1995:4). Subalternism therefore implies resistance and fighting power from below through forms of communication that seek to mobilise, network, and advocate ideologies of emancipation and freedom for the oppressed. In Africa such struggles have been articulated through nationalism, human rights, women's rights, sexual minority rights, among other discourses. In all these, technology has been central in both the advancement of subaltern interests, and the creation of spaces that allow the subaltern to speak.

One of the significant factors underpinning community service in alternate and subaltern spaces is participation, especially one that is organised from below (Atton, 2002; Bailey et al. 2008; Waltz, 2005). Alternate and subaltern media spaces are organised primarily "to enable wider social participation in the creation, production, and dissemination of content than is possible in the mass media" (Atton, 2002:25). This implies ordinary people's direct and autonomous involvement in producing stories and mediating their social experiences through their own media. Bailey et al. (2008:11) argue "participation in the media and through the media sees the communicative process not as a series of practices that are often restrictively controlled by media professionals, but as a human right that cuts across societies." Participation is not only a human right, but also a means to an end or social action by the subaltern. It is highly interwoven with freedom of expression that is fundamental to the practice of various civic responsibilities that are expressed through the broader right to communicate (see Dakroury, 2006; Hamelink & Hoffmann, 2008).

Hence, citizen journalism as a form of alternate-subaltern space and practice has been referred to as "participatory journalism," "citizen-generated media," "we media," "grassroots media," "self-service media" so as to emphasize the notions of inclusion and participatory communication that are often embedded in alternative media spaces (see Atton, 2002; Gillmor, 2006; Kalodzy, 2006). Yet participation in subaltern spaces and practices must not be seen as structureless, as power relations still exist and continue to influence who says what, where, when, and why. To that degree, it is important to understand citizen journalism as a spatial practice that is shaped in many ways by technological and social contexts of actors. However, one must take cognisance of the fact that subaltern and alternate spaces must, remain normatively premised on a democratic participant philosophy that discourages a top-down model of public communication in favour of a bottom-up or lateral model that prioritises effective participation and involvement of citizens in generating media content.

Alternate-subaltern spaces must be autonomous from the state and any form of authority or power. Citizen journalism as one such space, for example, must enjoy institutional autonomy and editorial independence from both the state and market forces so as to allow "subordinated groups [...] to produce non-conformist and sometimes counter hegemonic representations of the views of those marginalised, misrepresented, and underrepresented in the public sphere" (Bailey et al., 2008:17). Autonomy from the state or market forces implies that citizen journalism practices should be more amenable to civil society especially as "a sphere of public life beyond control of the state" (Colas, 2002:25). If construed as a "bulwark against the state" (Keane, 2002:67) and "a buffer zone strong enough to keep both state and market [...] from being too powerful and dominating" (Giddens, 2001:15), then civil society is clearly "the infrastructure that is needed for the spread of democracy" (Anheier et al., 2003:3). In principle, civil society's vanguard role in democratisation as a process that is born from participatory communication, locates it within the same counter-hegemonic framework as subaltern spaces and practices of citizen journalism. However, in Africa the autonomy of civil society must not be taken at face value especially within the broader context of neoliberal globalisation. Most civic organisations depend on Western donor organisations that tend to influence an advocacy agenda that promotes very restrictive free market–based conceptions of democracy as freedom for capital and not necessarily citizens.

To that extent, citizen journalism is theoretically a more meaningful practice of freedom if embedded on social movements. As opposed to civil society, social movements are far less hierachical and can be described as informal, grassroot, or community-based networks that rovide ideal spaces for subaltern agency (see McAdam, McCarthy & Zald, 1996). Social movements and citizen journalism are about locality. Locality as a subaltern space has the potential to create democratic bottom-up transformative agency. To achieve this, the twin processes of the de-institutionalisation and deprofessionalisation of journalism are important to create a form of journalism that is free from newsroom monopolies and professional filtering.

KUBATANA BLOGGING AS CITIZEN JOURNALISM AND A PRACTICE OF SUBALTERN RESISTANCE AND FREEDOMS

In the previous section, I argued that theoretically alternate-subaltern spaces are characterised by the practice of freedom. They are an embodiment and articulation of uncurtailed speech that holds power to account. An examination of the content of Kubatana's bloggers demonstrates that citizen journalism played an important role not only in mediating Zimbabwe's 2008 electoral crisis, but also in exposing human suffering during the run-off election. In a very restrictive media context, the Kubatana citizen journalists were central in the advancement of free speech whose value was the exposure of the violations of human rights by the state. News stories ranged from allegations of vote rigging, violence, rape, abductions, torture, murder, to the general hunger and starvation that were characteristic of the crisis as it deepened. Bloggers wrote of "old men in burnt clothes who were escaping violence from the village," "battered villagers recovering in hospitals," "farm invasions" and farmers "with broken collar bones and fingers," "ruling party officials forcing people to rallies and forced confessions," "acute water and electricity shortages," and "empty supermarket shelves around the country."[1]

The concerns of most of these stories locate Kubatana's citizen journalism within a counter-hegemonic framework that presupposes the development of a nascent subaltern space for the development of discourses that are critical of the state. While Gramsci (1971) regards hegemony as particularly strong because it involves the willing and active consent of the dominated, he also believes that domination has to be constantly negotiated and renegotiated because ordinary people can still engage in a struggle over ideas and even mobilise themselves to resist. According to Fiske (1994: 291), "consent must be constantly won and rewon," because "people's material social experience constantly reminds them of the disadvantages of subordination and thus poses a threat to the dominant class." The rise of Kubatana citizen journalism and the concomitant free flow of information on sensitive issues about real or perceived state atrocities in Zimbabwe have exposed the temporal and unstable nature of hegemony. Citizens, no matter how tight the shackles of oppression, always develop other means of contesting the construction of "commonsense" and "the natural" through which the hegemonic knot of domination and subordination is tied. In Zimbabwe's authoritarian environment, citizen journalism has developed to represent spaces of individualised and collective freedoms that do not necessarily defer to authority especially in the form of state sanctions of civil and political liberties. The Kubatana bloggers, as demonstrated in the 3 cases below for example, disseminated news that potentially shook the foundations of Zimbabwe's hegemonic project by foregrounding the anti-establishment discourses that threatened the survival of the political elite:

Case 1: Last night at about 9:30pm, I heard a lot of noise coming from Tshovani Township near Chiredzi. The next morning I asked several people what it was about and they all said that the residents were shouting for Mugabe to go … A similar action has taken place in Zaka constituencies and I believe Masvingo also. There is a lot of despondence and also a lot of anger, people want to fight now (SMS to Kubatana blogged by Clarke, B. 28 April 2008).

Case 2: I've just been to the CABS (bank) queue. The limit on cash withdrawals is Z$1 billion a day. At the supermarkets and the wholesalers, shelves are empty except for a few packets of chips and rotting vegetables. Maize meal, sugar, milk and soap are non-existent! (Clarke, B. 28 April 2008).

Case 3: Hon. Mahlangu, the MDC MP for Nkulumane constituency in Bulawayo … is battling for his life in intensive care unit after armed Zanu PF militia attacked him yesterday as he and other MDC youth were on their way to the Glamis Arena for a star rally. Over a thousand Zanu PF thugs were bussed to the venue (Atwood, A.B. 23 June 2008).

What is clear from some of these cases is not only concern over human rights and freedom abuses, but also the mundanely written and stylised news narratives that fall within the idiom of everyday speech. The "deprofessionalised" nature of the news is one of the primary characteristics of alternate-subaltern media discourses. In most cases, Kubatana bloggers use no sources, no by-lines, and have no pretentions towards objectivity and balance. Although they perform an important role of informing and educating the public about the crisis, they however do not perceive themselves as objective and disinterested mediators because they tell stories about everyday life of which they are part. For example, extrapolations from the above cases show that just like ordinary people they queue in banks, they are affected by food shortages, and as the quote below shows, they mediate the crisis in which they are also active agents who fight against the violation of citizen rights:

> This morning four of us (bloggers) piled into a car and went to observe a Women of Zimbabwe Arise (WOZA) gathering down town. They wanted to deliver a petition to the Zambian embassy requesting SADC to get more involved in helping to solve the Zimbabwean crisis… Public actions like WOZA's give me hope. But their actions need to be multiplied and replicated all over Harare and other parts of Zimbabwe to create pressure on the illegitimate Mugabe regime (Clarke, B. 7 May 2008).

Unlike the mainstream journalists, citizen journalists do not operate on an illusion of freedom from society, but their storytelling is based on a realistic acceptance of the fact that they are situated interpreters of reality in terms of class, gender, race, and ethnicity. In that sense, citizen journalism, as opposed to mainstream journalism, is reflexive and conscious of its ideological baggage. This has resonance with the age-old truth propounded mostly by the Marxian scholarship, that while the so-called professionals falsely claim to be neutral and distanced observers and critics of events, they are "actually locked into a power structure [and] act largely in tandem with the dominant institutions in society" (Curran, Gurevich & Woollacott, 1995:2). As Schudson (2003:154) explains, "political institutions and media are…so thoroughly engaged in a complex dance with each other, that it is not easy to distinguish where one begins and the other leaves off." Media, as he further argues, "do not define politics anymore than political structures dictate news" (Ibid.: 154). The news text itself does not mirror or reflect the truth, but is always a site for ideological contestation in the struggle for attributing meaning and salience to events (see Fiske, 1994).

The language mostly used by Kubatana bloggers reflects the anger of the writers against the status quo, and seems to be subservient to the greater cause of resisting authoritarianism rather than to upholding some journalistic standards as prescribed by the mainstream. From this perspective, alternate-subaltern spaces and citizen journalism can also be activist in orientation. O'Sullivan (1994), for example, posits that activism in the sense of advocating radical social change must be a key attribute of alternate and subaltern media spaces. While Kubatana bloggers do not always advocate specific political action for citizens against state power, one can still detect elements of advocacy journalism in which they castigate electoral violence as "thuggish" and chastise government for the shortage of food and other basics. At this level, Kubatana bloggers appear to have multiple identities. They are essentially active citizens, concerned activists, but also "journalists" as they also produce and distribute news about the happenings in their communities. They represent "unfiltered," "raw," and "unaestheticised" journalism in the sense of negating the institutional dictates and professional myths of objectivity, balance and accuracy. However, this is not to suggest that citizen journalism and mainstream journalism are mutually exclusive. On the contrary, their relationship is a dialectical one where they sometimes complement, compete, or even challenge one another (Lowrey, 2006; Thorsen, 2008). For instance, Thorsen's study of

Wikinews, a website that allows people to write, edit, and publish news in a collaborative manner, noted that this form of citizen journalism extensively appropriated the stylistic devices of mainstream journalism, such as objectivity, accuracy, and neutrality. In fact, Case 3 in this article demonstrates that even though the Kubatana bloggers write about emotive subjects, such as political violence to which they are susceptible, they also borrow the styles and tones of the mainstream.

In contrast to the election in 2008, the recent 2013 national poll was fairly peaceful and uneventful. Consequently, the blogging by Kubatana citizen journalists during this election was relatively low-key thus confirming the view that citizen journalism thrives mostly during crisis moments. Faced with a low-incident poll, Kubatana bloggers recast their reporting from the previous events-oriented format to more analytical, critical, and issue-oriented reporting which, however, maintained the visual convention as citizen journalism's most dominant mode of telling stories. In 2013, Kubatana citizen journalists took pictures of deserted polling stations as a testimony to how most voters had been turned away from voting because they could not find their names in the voters' roll. They reported that the voters roll had been maniplulated to favour Mugabe. More interesting stories were published in the post-election period where bloggers, almost in a sarcastic way, reported how Zimbabweans seemed disillusioned and disefranchised by a Mugabe victory. For example, Bev Reeler, a popular blogger on the blog, wrote a story titled "Zimbabwe's post lection." In her typical poetic style, the story stated that "A deep silence has settled/ No jubilant cheering crowds/ No smiled greetings from vendors at traffic lights/Just a stunned disbelieving quietness/ Just deep, tired lines etched on the kind, caring faces around me" (Bev Reeler, 2 August 2013). In line with the dominant visual code of citizen journalism, the story was accompanied by pictures of an empty, deserted, quiet, and forlon streets in which the journalist tried to capture the sombreness of the post election period. A key observation is how Kubatana in 2013 continued to be a platform of advocacy for citizen's freedoms through citizen journalism. It consistently endeavoured to be a voice of the people.

THE QUESTION OF ACCESS IN THE EMERGING "FIFTH ESTATE"

The Kubatana blog is a practical manifestation of how the internet as an alternative medium and citizen journalism as a counter-hegemonic practice express the people's right to communicate and to civic virtuousness. Unlike the Fourth Estate of corporate and state media, the Fifth Estate is the property of the people as reflected in terms such as "citizen-generated media," "we media," "participatory journalism," "grassroots media," "doing media" and "self-service media" (see Gillmor, 2006; Kolodzy, 2006). No longer are Zimbabwean citizens helplessly bombarded with messages by the mass media: they are actually actively producing news and initiating news flows among themselves, and, between themselves and the outside world. Through the Kubatana blogs, we are witnessing a paradigmatic shift from the Fourth Estate of news and information consumers to the Fifth Estate of news and information "prosumers" defined by interactivity that entails "a more powerful sense of user engagement with media texts [and] a more independent relation to sources of knowledge" (Lister et al., 2003:20). If Zimbabwe's Fourth Estate was characterised by mass audiences and top-down communication, its Fifth Estate is characterised by what Bell (2009) refers to as "doing media"—media that enable users the freedom to participate in formulating, modifying and disseminating its content (Steur, 1992:84). Hence, the Fifth Estate marks the nascent stages of participatory digital politics in Zimbabwe where freedom of information, as opposed to censorship, is the basis of storytelling by the blogging "netizens." The deinstitutionalisation of journalism and alternative media is not only a harbinger of the freedom of the storyteller, but also of the news narrative whose codes and conventions and content are now left to the individual to decide. For example, Kubatana bloggers presented pictures of human rights victims and victim accounts that would easily have been censored in the institutionalised mainstream media in Zimbabwe. Oftentimes, full victim accounts were lumped together with bloggers' news thus circumventing the problem of selection and exclusion that is often endemic to mainstream representation of the news (Schudson, 2003). This free flow of information is corollary to the health and effectiveness of the Fifth Estate as an alternative space for uncensored news.

Yet Zimbabwe's Fifth Estate is undermined by problems of access and participation. Only 1.4 million people are consistent users of the internet (see World internet Statistics, 27 August 2013). Hence, while enjoying significant levels of autonomy, Zimbabwe's Fifth Estate essentially remains a peripheral public sphere in terms both of agenda setting and influencing the national public opinion. This frustration was evident from the bloggers who were interviewed who felt that while their news was produced purely in the public interest, their impact was however undermined by low levels of access. One blogger feels that "the state sees bloggers as too insignificant to cause any trouble" (Msoza, 29 January 2010), while another argues that this is because "the state in Zimbabwe recognises that the internet is generally only reaching a small portion of society, and for the most part, reaching those that are already firmly against its policies" (Pietrzyk, 3 February 2010). However, the central point of the democratic-participant theory (and by implication the Fifth Estate) lies not only with access and participation, but also with citizens' "needs, interests and aspirations of [being] active 'receivers' in a political society" (McQuail, 2003:122). While the internet has enhanced the right to communicate for a significant minority of Zimbabweans, what is important is how citizens have appropriated that right to create a vibrant Fifth Estate whose political worth is reflected through its contribution to substantive democracy and active citizenship.

Conceptually, alternative media can be "considered the 'third voice' between state media and private commercial media" (Bailey et al., 2008:23). Structurally, it occupies the same space as civil society, thus invariably making it part of civil society's democratisation agenda. While the conceptualisation of civil society is by no means simple nor straightforward, my definitions in the preceding sections described it as a space between the state and the market, a "space in which alternative conceptualisations of the political and economic systems can develop and thrive" (Ibid.: 21). The protracted crisis in Zimbabwe created an environment conducive for civil society to ask questions about the social order and be in the vanguard for social change. The crisis required civil society to demonstrate independence of thought and judgement that goes beyond the state's official nationalism narrative and neoliberalism's populist free-market sound bites.

Given the above, it is interesting then to examine the extent to which Kubatana's bloggers or citizen journalists had the capacity to think outside the "ideological boxes" of the state and the market. Nearly all the bloggers who used the Kubatana blogosphere were critical of the state and the political order, though not of the market and its economic order. Elsewhere I discuss other Kubatana discourses that existed outside the blog and I argue that while Kubatana's role in democratisation in Zimbabwe is unquestionable, its focus on political rights and political governance at the expense of economic rights and economic governance is however problematic (Moyo, 2009). As other scholars have observed, Zimbabwe's electoral crisis was merely a symptom of a much broader problem that had its roots in the land dispute and the 1990s neo-liberal economic reforms, which created an economy based on serious racial disparities and further also nurtured a culture of state violence mostly on Zimbabwe's working classes and students (Saunders, 2000). However, none of the bloggers embedded to the Kubatana civic group criticised either the subversion of economic democracy by post-colonial settler capital or the global forces that resisted agrarian reform and administered the failed economic reforms of the nineties (see Sachikonye, 2003). Hence, while citizen journalism in Kubatana appears to be operating in a deinstitutionalised environment, it is in reality operating in institutionalised virtual environments where the general aura of discourses is neoliberal. In other words, Kubatana bloggers as citizen journalists take the thematic cues from the website and their stories are based on institutional modelling whose template goes beyond Kubatana to embrace nearly the entire civic movement in the country. As embedded journalists, they speak the hollow human rights language of Zimbabwe's civil society, which reflects a selective articulation dictated by Western donors who exercise allocative control of the entire counter-hegemonic establishment. Consequently, the bloggers criticise political violence of the state, but not the economic violence of local and global capital against ordinary Zimbabweans. As such, while the problem of selection and exclusion is not immediately apparent in Kubatana's embedded bloggers, it lies at the core of how the crisis is framed and represented in their blogs. Thus, the major problem with embedment of alternate spaces to civil society in Zimbabwe is that it is developing a form of citizen journalism that lacks financial and editorial autonomy. Zimbabwe's civil society itself is in a state of crisis resulting from financial dependence and ideological bankruptcy because of neoliberalism's grip on young and emerging democracies globally (see Hassan, 2004). From this perspective, while civil society and alternative media certainly play a counterhegemonic role against the state, they are also simultaneously a site for domination and market hegemony. Where the state uses violence, Kubatana and its bloggers use the efficacy of consent that disciplines citizens to be loyal subjects of an equally violent and emasculating capitalist order.

CONCLUSION

Further characterisation of the transformations that are taking place in Zimbabwean media systems helps conclude this chapter. First and foremost, the radicalism—in form and content—of citizen journalism as an alternate-subaltern space must be seen as one informed by the political context of state repression and violence. This radicalism is enhanced by the internet, which not only provides spaces for the practice of freedom by the bloggers, but also for criticising power. While I have used the term counter-hegemonies consistently throughout the chapter, which subsequently locates Zimbabwe's alternative-subaltern online spaces and citizen journalism within a Marxian discourse, it has to be explained that Kubatana bloggers are not revolutionaries seeking to uproot Zimbabwe's political and economic status quo. They are only moral reformists who are concerned with the particular leadership style or the grammar of politics, and not the entire edifice of the ruling elite. I risk the term "liberal-counter hegemonies" to describe what Cohen (1985:664) calls the "self-limiting radicalism" that is characteristic of civil society of a unipolar dispensation. In a sense, one could argue about the crisis of the alternate-subaltern spaces that is reflected in the conservatism of the alternative media and its alternative journalisms. One of the reasons for this weakness is that both are saddled on the idealism of the neo-liberal discourse that empowers them as the watchdog of the state, while simultaneously depriving them of the capacity to conceptualise the alternative in truly radical ways. As such, while there are changes in the way the new journalism articulates itself, the journalist essentially remains of a middle-class orientation where elite continuity in governance is not seen as a problem. The right to communicate through the internet (both its form and content) as a practice of freedom has arguably been appropriated and monopolised by the blogger whose legitimacy is based on working with civil society. Yet, while civil society in Zimbabwe justifies itself on grounds of inclusivity, it is in reality exclusive because working classes and the rural population are not bloggers due to lack of access to online technologies.

NOTES

1. See http://kubatana.net/index.html

REFERENCES

Allan, S., and Thorsen, E. eds. (2009). *Citizen journalism: Global perspectives.* New York: Peter Lang.

Alejandro, Roberto (1993). *Hermeneutics, Citizenship, and the Public Sphere*. New York: State University of New York Press.

Anheier, H., Glasius, M., and Kaldor, M. (eds.). (2001). *Global civil society.* Oxford: Oxford University Press.

Atton, C. (2002). *Alternative media.* London: Sage.

Bailey, O.G., Cammaerts, B., & Carpentier, N. (2008). *Understanding alternative media.* London: Open University Press.

Barker, C. (2000). *Cultural studies: Theory and practice.* London: Sage.

Bell, D. (2009). On the Net: Navigating the World Wide Web. In G. Creeber & R. Martin (Eds). *Digital cultures: Understanding new media*. London: Open University Press.

Castells, M. (2000). *The information age: economy, society and culture.* Oxford: Blackwell.

Cohen, J. (1985). Strategy or identity: New theoretical paradigm and contemporary social movements. *Social Research,* 52(4), 663–716.

Colas, A. (2002). *International civil society.* London: Polity Press.

Curran, J., Gurevitch, M. and Woollacott, J. (1995). *Approaches to Media: A Reader*, Chicago.

Curran, James, Michael Gurevitch & Janet Woollacott (1995). 'The study of the media: theoretical approaches', in Boyd-Barrett, O., & Newbold, C. (eds) *Approaches to media: A reader*. London: E. Arnold.

Dakroury, A. (2006).Pluralism and the right to communicate in Canada. *MediaDevelopment: Journal of the World Association for Christian Communication,* 55(1), 36–40.

Falk, K. (2000). *Citizenship*. London: Routledge.

Fiske, J. (1994). *Televisionculture*. London: Routledge.

Giddens, A. (2001). *The global third way debate.* London: Polity.

Gillmor, D. (2006). *We the Media: Grassroots Journalism by the People, for the People*, New York:O'Reilly.

Gramsci, A. (1971). *Selections from prison notebooks of Antonio Gramsci*. New York: International Publishers.

Habermas, J. (1985). *Theory of communicative action*. Boston: Beacon Press

Hamelink, C.J., & Hoffmann, J. (2008). The state of the right to communicate. *Global Media Journal,* 7(13): Retrieved February, 28, 2010, from http://lass.calumet.purdue.edu.

Hassan, R. (2004). *Media, politics and the network society*. London: Open University Press.

Held, D. (1996). *Models of democracy*. London: Routledge

Jenkins, H. (2006). *Convergence culture: Where old and new media collide.* New York: New York University Press.

Kalodzy, J. (2006). *Convergence journalism: Writing and reporting across news media.* Lanham, MD: Rowman & Littlefield.

Kavada, A. (2005). Exploring the role of the Internet in the movement for alternative globalisation: The case of the Paris 2003 European Social Forum. *Westminster Papers in Communication and Culture*, 2(1), 72–95.

Keane, J. (2002). 'Global civil society?', in Anheier H. et al. (eds) *Global Civil Society*. Oxford: Oxford University Press.

Lister, M., Dovey, J., Giddings, S., Kelly, K., & Grant, I. (2003). *New media: A critical introduction.* London: Routledge.

Louai, E.H. (2008). Retracing the concept of the subaltern from Gramsci to Spivak. *African Journal of History and Culture*, 4(1)4–8.

Lowrey, W. (2006). Mapping the journalism-blogging relationship. *Journalism*, 7(4): 477–500

McAdam, D., McCarthy, J.D, and Zald, M.N. (1996). *Comparative perspectives on social movements.* Cambridge: Cambidge University Press.

McQuail, D. (2003). *Media accountability and freedom of publication.* Oxford: Oxford University Press.

Mosco, V. (1996). *The political economy of communication.* London: Sage.

Moyo, D. (2007). Alternative media, diasporas and the mediation of the Zimbabwe crisis. *Ecquid Novi*, 28(1–2), 81–105.

Moyo, D. (2008). Citizen journalism and the parallel market of information in Zimbabwe's 2008 election. *Journalism Studies,* 10(4), 551–567.

Moyo, L. (2003). Status of the media in Zimbabwe. *Encyclopaedia of International Media and Communication.* Wartham, MA: Academic Press.

Moyo, L. (2009). Repression, propaganda, and digital resistance: New media and democracy in Zimbabwe. In O.F. Mudhai, W.J. Tettey & F. Banda. *African media and the digital public sphere.* New York: Palgrave.

Moyo, L. (2010). The dearth of public debate: Policy, polarities and positional reporting in Zimbabwe's news media. In D. Moyo & W. Chuma. (2010). *Media policy in a changing Southern Africa: critical reflections on media reforms in a global age.* Pretoria: Unisa.

O' Sullivan, R. (1994). *Communication methods to Promote grassroots participation*, Paris: Unesco.

PAOM (2008). Report of the Pan African Parliament Election Observer Mission: Presidentail run-off election and house of assembly by-elections, Republic of Zimbabwe, June 27, 2008.

Papacharissi, Z.A. (2010). *A Private Sphere: Democracy in the Digital Age*, Cambidge: Polity Press.

Phimister, I., & Raftopoulos, B. (2004). Mugabe, Mbeki & the politics of anti-imperialism. *Review of African Political Economy*, 101, 385–400.

Popple, K. (1995). *Analysing community work.* Buckingham: Open University Press.

Ranger, T. (2003). Histriography, patriotic history and the history of the nation: The struggle over the past in Zimbabwe. *Journal of Southern African Studies*, 30(2), 215–234.

Sachikonye, L.M. (2003). From growth with equity to fast track reform: Zimbabwe's land question. *Review of African Political Economy*, 96, 227–240.

Saunders, R. (2000). *Never the same again: Zimbabwe's growth towards democracy*, Harare: HPP.

Schudson, M. (2003). *The sociology of news.* New York: W.W. Norton and Company.

Steur, J. (1992). Defining virtual reality: Dimensions determining telepresence. *Journal of Communication* 6(42): 73–93.

Tettey, W.J. (2009). Transnationalism, the African diaspora, and the deterritorialised politics of the internet. In O.F. Mudhai, W.J. Tettey, & F. Banda. *African media and the digital public sphere.* New York: Palgrave.

Thorsen, E. (2008). Journalistic objectivity redefined? Wikinews and the neutral point of view. *New Media and Society*, 10(6): 935–954.

Waltz, M. (2005). *Alternative and activist media.* Edinburgh: Edinburgh University Press.

SECTION FOUR

New Crises, Alternative Agendas

CHAPTER TWENTY

"Blade AND Keyboard IN Hand": Wikileaks AND/AS Citizen Journalism

LISA LYNCH

Positioned at the edge of the media field, and dedicated to the dissemination of anonymously submitted leaked materials revealing government or corporate corruption, Wikileaks has been described as incorporating citizen journalism's disruptive tendencies (Beckett, 2012) and its emphasis on civic witnessing (Allan, 2013) and mobilization (Bruns, 2011). However, the shifting structure and mission of the organization has meant that, though Wikileaks has at times relied extensively on citizen participation, their relationship to the idea of the citizen journalist has been ambivalent and inconsistent. In this chapter, I will discuss Wikileaks' conflicted relationship with citizen journalism, tracing the site's inception as a project reliant on citizen input, to its current incarnation as a publisher working in collaboration with a range of professional media outlets. I will also look at the efforts of citizens and other non-journalists to catalog and interpret the organization's leaked material, concluding by considering the influence of Wikileaks on the project of citizen media.

WIKILEAKS' WIKI ROOTS

In late 2006, when a collective of online activists and journalists discussed the shape and scope of what would eventually become Wikileaks, they settled on developing a platform that would facilitate citizen participation. Since the project depended on the public submission of leaked material, they planned to create a secure, anonymous system that would allow for both online and offline contributions. And in order to publish the leaks they selected for distribution, they would build their website using MediaWiki, an open-source software originally used to develop Wikipedia. This meant that, like Wikipedia, each web page authored by the group would be linked to a discussion page. Email exchanges between Wikileaks' founding members describe the expected conversations on these discussion pages as vital to Wikileaks' mission, as those who were most affected by the contents of a given leak were expected to come forward to provide authentication, context and interpretation. In reference to the organization's first leaked document, a report revealing corruption in Somalia, Wikileaks founder Julian Assange predicted that Somalis would vet the report and explain its political significance:

> When WL is deployed, feedback will be, like Wikipedia, an act of creation and correction; the Alweys document and those like it will eventually face one hundred thousand incensed Somali refugees, blade and keyboard in hand, cutting, cutting, cutting apart its pages until all is dancing confetti and the truth (Assange, 2006).

When the site launched in early 2007, the group added another means of participation for those who wanted to write longer, structured commentary on leaked documents, authoring a "Writers Kit" for potential authors. This kit provided a detailed framework for evaluating leaks and guidelines for submission, as well as the offer of payment for "professionally written articles." Despite the gesture towards professionalism, Wikileaks was clear that the organization welcomed submissions from non-journalists who were interested in working for the group. Asking the rhetorical question, "Who is a Wikileaks Writer?" the document advised:

> A Wikileaks writer is something like a journalist. Something like an intelligence analyst. Something like an academic. Something like a fact-checker. Something like a research assistant. Something like a human rights lawyer. Something like a political activist working towards a better world. But we push no agenda except that of truth and exposing corrupt power; and we do so on the basis of revealed fact. Call yourself what you like. Call yourself a journalist, citizen journalist, citizen intelligence analyst, citizen academic, scholar, activist, seeker after truth...however you're inclined (Wikileaks, 2007).

As this excerpt suggests, while the role of public participation was central to Wikileaks' original design, what was being asked of the public was both more and less than an act of journalism. The description of potential contributors muddied the terms of engagement, suggesting that Wikileaks was looking for a hybrid genre of author who could balance journalism, scholarship, and intelligence work. While likely intended to be inclusive, the description was daunting; one erstwhile contributor noted on the "Talk" page connected to the "Writer's Kit" that Wikileaks had not thought through how to provide protection for those who might perform intelligence work with the same rigor they had thought through the protection of whistleblowers (Wikileaks, 2007).

Aside from the challenges posed by the "Writer's Kit," those who came to the Wikileaks website may have been frustrated by the group's modifications to the MediaWiki platform, which prevented them from adding original topics or modifying leak pages. Though the modifications were understandable given Wikileaks' desire to eliminate identifying marks from leaks and vet them before publication, they added to the perception that the group was too secretive and closed in comparison with the more open, collaborative nature of citizen-media projects. As Wikileaks built up a follower base, they also drew criticism for their desire to centralize control. On Reddit, for example, some Wikileaks supporters noted that since Wikileaks seemed to have a hard time keeping afloat (its servers were constantly crashing due to outdated equipment and high bandwidth demands) the organizations' material might be better served by a more decentralized, peer-to-peer distribution model than the hierarchical, centralized model promoted by the site design (Reddit, 2009).

Whatever the causes—the site's relative obscurity; the intimidating requirements of the editorial process detailed on the talk page; fear of reprisal; or simply lack of engagement—citizen participation on the Wikileaks website was more limited than the organization had anticipated. Only a handful of leaks attracted significant commentary on the "Talk" pages, and articles that accompanied leaks, if they appeared at all, were largely bylined "Staff," "Wikileaks" or "Assange." Aware that their efforts at citizen engagement had met with only middling success, Wikileaks began to discuss abandoning the Wiki format as early as the summer of 2008 (Singel, 2008). In the end, they retained the platform until 2010, but placed less and less emphasis on the "Wiki" nature of Wikileaks, shifting their focus to the broad distribution of leaked materials and thus to engagement with the professional press.

COURTING THE PROFESSIONAL PRESS: THE MEGALEAKS

In retrospect, Wikileaks' decision to abandon their initial approach seems somewhat abrupt; after only eighteen months of operation, the site could have explored further means of encouraging participation or simply waited until they had a larger follower base. But the choice stemmed from both Wikileaks' keen sense of disillusionment with the work produced by non-staff contributors and a changing sense of what the organization was intended to do. In an April 2010 panel discussion, Assange referred to his early hopes for citizen collaboration as "bullshit," as those who came to the site were unwilling to engage with source documents before someone else had done preliminary analysis (Assange, 2010). Shortly afterwards, in an interview with *Time* magazine, he suggested that journalism was best produced by those who worked for—and were funded by—news organizations:

> The analytical effort which we thought would be supplied by internet citizens around the world was not…You can't expect to get news-style articles out of people that are not funded after a career structure in the same way that news organizations are (Stengle, 2010).

Assange's remark suggested Wikileaks was as displeased by the quality of submissions as they were discouraged by the quantity. This disenchantment made the organization increasingly set its sights on coordination with the professional press in order to find wider distribution for their leaks. Redefining themselves on their site's "About" page as a "not for profit media organization" and revising their site to incorporate references to news and journalism (Savage, 2010), they increasingly reached out to news organizations for ways to more efficiently circulate leaked material.

At the same time, Wikileaks did not completely abandon the idea of citizen participation. Rather, they moved from seeing their follower base as a source to analyze leaks to seeing them primarily as possible sources for leaked materials. In 2009, they published a list of "Most Wanted Leaks," encouraging the group's followers to seek out and submit documents ranging from Bilderburg group minutes to records of applications for Bahraini citizenship (Wikileaks, 2009). The list proved successful beyond Wikileaks' hopes, as it prompted one Wikileaks follower to deliver the organization its most significant coup. Concerned about the direction of US foreign policy in the Middle East, and hoping to "spark a debate," US Army Sergeant Bradley Manning downloaded data off of SIPRNET, a military intelligence network, and provided Wikileaks with a set of US military and diplomatic documents that included a video showing US soldiers firing on allegedly unarmed Iraqi civilians, thousands of field reports from Iraq and Afghanistan, and over a quarter million US diplomatic cables (Savage, 2013).

The potential significance of these disclosures—collectively labeled the 'megaleaks' by Wikileaks—convinced the organization that new methods of delivery might be needed in order to bring the leaks to the widest possible audience. After de-encrypting the US military video and sending their own reporting team to Iraq to investigate the shooting incident, Wikileaks invited US and international media to a screening of the video at the National Press Club in Washington DC. Following the well-attended press conference, the video was published online by Wikileaks on a proprietary website with the dramatic URL *collateralmurder.com*. Then, beginning in July 2010, Wikileaks began to coordinate with a series of major media outlets including *The Guardian*, *The New York Times*, *Der Spiegel*, and *El Pais* on a gradual release of the military field reports and diplomatic cables.

Depending on one's definition of the term, the publication of the "Collateral Murder" video by the Wikileaks arguably represented the closest Wikileaks ever came to acting as a citizen media venue in its own right. The self-published video, accompanied by an investigative report that sympathetically explored the fate of the victim's families, was an attempt to bypass legacy media to deliver not only the leak, but Wikileaks' own journalistic narrative of events, directly to the public; in turn, Wikileaks was recognized by journalists as a possibly significant emerging force in the media ecosystem (Cohen and Stelter, 2010). But the tone of that narrative produced dissent within the organization itself, as it deviated from the restrained tone Wikileaks itself had suggested to potential contributors in its "Writers Kit" and harshly condemned the US military. Responding to accusations that the group had created a biased account of events, Assange asserted that since the original source document was available to viewers, a subjective interpretation of the video was permissible since viewers were able to draw their own conclusions (Chen, 2011). Other members of the group felt differently, however; in interviews after he left the group, former Wikileaks spokesperson Daniel Domschiet-Berg noted that the aggressive packaging of the video as "murder," as well as the framing of the investigative report, alienated members of the group who felt a more neutral tone was appropriate (Domscheit-Berg, 2011).

Though it seemed to represent a new direction for Wikileaks as an organization, the Collateral Murder publication instead turned out to be something of an anomaly, the only major effort made by the group to produce its own reporting around a leak. Subsequent leaks were published through collaboration (or as Wikileaks would have it, partnership) with media outlets that were granted advance access to material through embargo agreements. Justifying their decision to withhold material from the public before it had been analyzed and filtered by the press—a sharp break from prior procedure—Wikileaks argued they had acted strategically to ensure the broadest possible audience for their leaks. Media observers, in turn, emphasized the importance of professional journalists' ability to sift through the data and provide "added value" such as redaction, contextualization, and perspective (Jarvis, 2010; Usher, 2010). Though in practice the collaborations proved brittle and ultimately untenable, they also facilitated instances of simultaneous publication that were unprecedented in the history of the professional press and spurred global media interest in both Wikileaks and the substance of its megaleaks.

Whatever advantages this collaboration provided, one effect was that the public was largely excluded from the journalistic process. Since the leaks themselves were pre-packaged and delivered via legacy media, the public was positioned as an audience for the documents instead of as interpreters, verifiers, or co-creators of meaning. However, the megaleaks period was not entirely devoid of citizen participation, as Wikileaks turned to crowdsourcing for the last chunk of its US disclosures, hoping that "citizens and journalists" might draw attention to material their media partners had neglected.

"THE WORLD COMMUNITY WILL CROWDSOURCE": THE PUBLICATION OF THE CABLE ARCHIVE

In 2009, Wikileaks made its first effort to prompt a public discussion of one of its leaks on a platform other than its own website. Over the course of 24 hours, they staged a real-time release of over a half-million pager messages from the New York City area that had originally been recorded during the time of the 9/11 attacks. To publicize this event, they urged those who were reading the pager messages to discuss significant finds by using the #911txts hashtag on Twitter or visiting a dedicated Reddit page. The resulting online conversation involved over 1,000 participants who retweeted, analyzed and responded to the messages; as well, the pager messages attracted the attention of programmers and data visualization specialists, who quickly developed tools to search and visualize the large number of messages (Lynch, 2011).

Though Wikileaks had been experimenting with Twitter for over a year prior to the pager leak, the resulting publicity suggested to the organization the possibilities of using social networks to draw attention to leaked material and organize their follower base. In August of 2011, nine months after the release of the diplomatic cables, Wikileaks drew on the lessons they had learned and made another effort at crowd-sourcing. This time, they had a sizable Twitter base to draw on; over 1 million followers, up from around 5,000 in 2009. They also had far more sensitive, and more consequential material to crowdsource; a zip file containing the entire, unredacted set of diplomatic cables.

Between November 2010 and August 2011, Wikileaks had kept their promises to media outlets, coordinating cable publication on the Wikileaks website with an initial set of five outlets and then a broader, more regionally diverse group of partners (Lynch, 2013a). However, a combination of accident and intent caused them to shift their approach yet again. The most immediate cause of the shift was that a file containing the complete set of cables was discovered online by an outside party and decrypted. In a convoluted set of circumstances best described by *Der Spiegel* journalist Christian Stöcker (2011), the encrypted cable set was accidentally sent out via Bit Torrent by well-intentioned volunteers; it then circulated unnoticed on the web until Domscheit-Berg, in an effort to disparage the Wikileaks' security precautions, mentioned to a journalist that the file was discoverable and could easily be decrypted with a password that was "out in the open." At that point, what Stöcker describes as "hobby investigators" deduced that the password Domscheit-Berg referred to was the same password that Assange had given to

Guardian journalist David Leigh when the cables were first handed over to that publication—and which Leigh, perhaps incautiously, had noted in his book about Cablegate. The end result of this fiasco was that the complete, unencrypted set of cables was in the hands of those who made the connection between Leigh's book and the BitTorrent file, rendering Wikileaks' embargo agreements moot.

In their public statement concerning the release, Wikileaks asserted that any blame for the publication of the full cable set lay not with them, but with the *Guardian* and Domscheit Berg. At the same time, other language in the statement suggested that they were not entirely displeased with the release. Expressing frustration with how few cables had been published, they suggested that working with a network of smaller outlets had increased the global circulation of the cables, but logistical difficulties and censorship meant that significant cables had still passed through the filter of the professional press:

> Readers are discovering that even the media organisations with the most resources, WikiLeaks' original partners, do not have the capacity to sift through all the cables nor report on all the big stories. It is a shared responsibility, then, for citizens, journalists, and researchers to comb through the material and find its local and global significance. Those stories that established media organisations are unable or unwilling to report on due to fear of being sued, or conflict of interest, or both, should nevertheless be in the public domain and available for everyone to access (Wikileaks, 2011a).

This line of argument suggests that Wikileaks had, in a sense, come full circle. While their initial failures at citizen engagement had left them disillusioned with citizen journalism, their experience with Cablegate had left them equally disillusioned with the mainstream press, and convinced that the best way to distribute the remaining cables was through direct public engagement. On Twitter, the group announced, "The world community will crowdsource all 251k cables using the method developed this month" (Wikileaks, 2011b). The "method" was described in a subsequent tweet: "Tweet important cable discoveries with #wlfind. The entire world press does not have enough resources and there are substantial biases (Wikileaks, 2011c)."

Wikileaks' decision to release the entire cache of cables proved highly controversial. Their publication was roundly condemned by many of those in support of the initial Cablegate releases, including Wikileaks' initial media partners, who had long emphasized the importance of their efforts at redacting and filtering (Ball, 2011). But others turned eagerly to the cables, hoping to find material that had been suppressed or ignored. In the weeks following the release, the #wlfind hashtag drew attention to cables discussing the US view of Venezuela, US speculations about the health of Zimbabwean president Robert Mugabe, and India's concerns about Monsanto's interference into Indian food production. Building on the work of interested citizens who blogged or tweeted about the newly released cables, journalists wrote yet more cable stories, often for media outlets who had not had access to the cables beforehand. For the most part, reporting on this final wave of cable stories was regional in nature, but a few international stories emerged as well. For example, CNN reported on a cable about a 2006 raid by American troops in Iraq, saying its emergence in the unpublished cable cache had strained negotiations between that country and the United States and may have prompted early troop withdrawal (CNN, 2011).

While that cable—given its content and impact—was one that likely would have been published had it not escaped the attention of Wikileaks' media partners, other cable discoveries suggested that the media outlets chosen by Wikileaks may have had limited interest in certain topics. Though a December 2010 story in *The Guardian* had reported on Pfizer's dubious activities in Africa (Boseley, 2010), there were few other reports on cables connected to the pharmaceutical industry. Convinced that there was more to be found on the subject, Knowledge Ecology International, an advocacy group concerned with misuse of intellectual property rights, searched the full database and found cables discussing the fixing of drug prices and the loosening of testing regulations. On KEI's blog, director James Love provided a gloss on these cables and argued in defense of the mass release of the cable cache, noting that the pharmaceutical stories were newsworthy, even if they were not of interest to news organizations:

> Many people could care less—including reporters and editors of newspapers. How much of this ends up in the *Washington Post*, *The New York Times* or the *Guardian* these days? But others who do care now have more access to information, and more credibility in their criticisms of government policy, because of the disclosures of the cables (Love, 2011).

While not technically a "citizen journalist," Love, an economist and nonprofit director, made a strong case that removing the cables from the sole custody of media organizations was a matter of public interest, a point also raised by others concerned about how the professional media might limit the scope and impact of the leak (Sifry, 2011). Thus, while Wikileaks' release of the entire cable set may have permanently quashed any chance that it could continue to work with its initial set of media partners, it also brought into clearer focus the limitation of such partnerships, and the tension between the journalism's professional and market imperatives and its public service function.

RIGHTING THE WRONGS OF A "POISONOUS" PRESS: CITIZEN JOURNALISM CATALYZED BY WIKILEAKS

If it can be argued that Wikileaks had, ultimately, an ambivalent relationship with the citizen press, it can also be said that the citizen press embraced Wikileaks, writing extensively about the organization in 2010 and 2011 and continuing to report on its activities and their impact after Wikileaks had faded from the headlines of the mainstream press. During the initial weeks of Cablegate, a sizable percentage of blog and Twitter activity was taken up with discussion of the substance of the leak and the ethics behind it (PEJ, 2010). In the right-wing US blogosphere, sharp condemnation of Wikileaks dominated the discussion, with some bloggers calling for Assange's assassination (Hawkins, 2010). Among US left-leaning bloggers, the reception was far more positive and sustained, as well-read figures such as Glen Greenwald and popular left news portals such as *Firedoglake* devoted substantial space to the consideration of the leaks and to chronicling the resulting legal struggles of Julian Assange and Bradley Manning. In Britain and in European countries where Wikileaks had established partnerships with major media outlets, there was also a brisk discussion of Wikileaks in citizen media, including the pan-European site *Blottr*, the German discussion board and citizen reporting site *Gulli*; and the French citizen journalism platform *Rue 89*.

Outside of the US, Europe and Britain, Wikileaks was also a focus of discussion among bloggers and on citizen sites, though the reactions to Wikileaks and their disclosures were tempered by local realities. At times, leaks which attracted attention in the United States were given little coverage by bloggers in the country they pertained too, either because of fear of government reprisals (Lynch, 2013a) or a sense that the contents were not as novel as US readers imagined (Petrossian, 2010). For the former group—especially in African countries, where media restrictions often prevented the professional press from partnering with Wikileaks—diaspora media blogs played an important role in reporting on cables that described government corruption (Mawando, 2011; Lynch, 2013b).

Beyond this occasional coverage, the megaleaks phenomenon also inspired two citizen media collectives dedicated to reporting solely on Wikileaks; *WL Central* and *WikiLeaks Press*. Both are still producing stories on the continued fallout from the cable disclosures and other leaks, as well as the broader issues raised by Wikileaks, including surveillance, censorship and freedom of information. These sites have asserted their independence not only from Wikileaks, but also from the institutional media, which is frequently described as ideologically driven and biased against Wikileaks. Contributor's attitudes toward institutional media are suggested by the editor's statement posted by a member of the *WLCentral* collective, describing herself as "in a semi-permanent state of embitterment by contemporary journalism's pathological reliance on shameless entertainment, and poisonous PR manipulation, which is the key agent in creating and maintaining this aforementioned disconnect" (JLo, 2011).

This frustration with the end product of the professional press has not prevented some of the citizen journalists on the Wikileaks beat from embracing the rigor and meticulousness that characterizes good reporting. One notable "alumnus" of WL central is Alexa O'Brien, known to Twitter followers as @carwinb. O'Brien worked at the site between January 2011 and June 2012, covering the megaleaks and their fallout, then left to independently cover the Bradley Manning trial. Due to the unusual nature of the Manning case, there has been no public docket made available, and O'Brien has taken on the task of transcribing proceeding and assembling a docket—a task which has led her to be described as the trial's "one-woman court records system" (Sledge, 2013).

O'Brien's professionalism places her in an emerging category of "citizen" journalist who nonetheless asserts her place within the journalistic field. In July of 2013, a *The New York Times* article on Manning described O'Brien as an "activist," a label that prompted a vigorous online discussion about journalistic boundary-work. O'Brien herself asked for, and received, a correction in *The New York Times.* Displaying its own discomfort with the possibility of a politically engaged journalist, the *Times* acknowledged that "while Ms. O'Brien has participated in activist causes like Occupy Wall Street and US Day of Rage, she also works as an independent journalist; she is not solely an activist" (*The New York Times,* 2013). In the online article itself, the *Times* labeled O'Brien an "activist *and* independent journalist," (emphasis mine) tipping its hat to those who might still consider the two categories to be mutually exclusive.

CONCLUSION

With its founder trying to avoid legal proceedings and its coffers largely drained, Wikileaks has nonetheless persisted in the post-megaleaks period in attempts to coordinate with the press in the circulation of leaked documents. Now, however, its media partners are predominantly non-Western publications or advocacy groups, instead of the mainstream press; from trying to reach the broadest possible audience, Wikileaks has come to understand that its disclosures might best be placed where they will reach the most sympathetic readership. As well, the leaks themselves are no longer sourced from the broader public. After internal dissent within the group led to the destruction of online Wikileaks' submission system, Wikileaks began coordinating with first a group of privacy advocates and then the cyber-activist collective Anonymous to obtain materials, including a series of documents about private surveillance contractors and email correspondence between the Syrian government and Western governments and businesses. Most recently, it has taken on the task of archiving and organizing diplomatic cables from the Kissinger era already publicly available in the US National Security Archives, combining these documents with the Cablegate cables to create what they label the "Public Library of US Diplomacy," or "PlusD."

In its current incarnation, Wikileaks has thus established itself as a library for those interested how US soft power has been used and abused over the past decades: in other words, as a public-minded but read-only archive. Though the most recent version of its website encourages journalists, academics, and bloggers to write about Wikileaks' material, the closed structure of the site indicates that Wikileaks itself is not a platform for such writing, but rather hopes to serve as a catalyst for writing that appears elsewhere. Given the existence of *WL Central* and *WL Press*—and given the increased US government scrutiny of those directly affiliated with Wikileaks—it is unlikely that the organization will once again embrace citizen contributions.

But even if Wikileaks has lost its initial faith in citizens with "blade and keyboard in hand," the organization has still made a considerable contribution to the project of citizen media. As conflicted, irascible and flawed as Wikileaks has been, it has drawn attention to the gap between what the twenty-first century press is willing or able to write about and what the public demands to know. In doing so, they have inspired other groups and institutions—some existing in opposition to the professional press, and some hoping to ensure its survival—to step forward and try to bridge that gap. With some persistence, daring, and good luck, the end result of such efforts might be both an energized citizen press and a more responsible and accountable media.

REFERENCES

Allan, Stuart (2013). Citizen Witnessing: Revisioning Journalism in Times of Crisis. New York: Polity.

Anonymous (2010). "Operation Leakspin." YouTube Video posted on December 11, 2010.

Anonymous (2013). "Your Anon News." Indiegogo, April 16, 2013. http://www.indiegogo.com/projects/your-anon-news.

Assange, Julian (2006). Email to Wikileaks List. December 25, 2006. Online at http://cryptome.org/wikileaks/wikileaks-leak.htm.

Assange, Julian (2010). Remarks at the Logan Symposium on Investigative Journalism, The University of California at Berkeley, April 18, 2010.

Ball, James (2011). "Wikileaks Publishes Full Cache of Unredacted Cables." *The Guardian*, September 2. http://www.guardian.co.uk/commentisfree/2011/sep/02/leader-wikileaks-unredacted-release#ixzz2UhtqkS2l

Beckett, Charlie (2012). *WikiLeaks: News in the Networked Era*. 1st ed. New York: Polity, 2012.

Boseley, Sarah (2010). 'WikiLeaks cables: Pfizer "used dirty tricks to avoid clinical trial payout"', *The Guardian*, http://www.theguardian.com/business/2010/dec/09/wikileaks-cables-pfizer-nigeria, 9 December.

Bruns, Axel (2011). "Towards Distributed Citizen Participation: Lessons from WikiLeaks and the Queensland Floods. In Parycek, Peter, Kripp, Manuel J., & Edelmann, Noella (Eds.) *CeDEM11: Proceedings of the International Conference for E-Democracy and Open Government*, Edition Donau-Universität Krems, Danube-University Krems, Austria, pp. 35–52.

Cohen, Noam, and Brian Stelter (2010). "Airstrike Video Brings Attention to WikiLeaks Site." *The New York Times*, April 6, 2010, sec. World. http://www.nytimes.com/2010/04/07/world/07wikileaks.html

Chen, Nadeemy (2011). 'Wikileaks and its Spinoffs: new models of journalism or the new media gatekeepers?', *Journal of Digital Research and Publishing*, Digital Publishing: what we gain and what we lose, pp. 157–167.

CNN Wire Staff (2011). "Obama; Iraq War Will Be Over by Year's End; Troops Coming Home." *CNN*, October 22, 2011. http://edition.cnn.com/2011/10/21/world/meast/iraq-us-troops/

Domscheit-Berg, Daniel (2011). *Inside WikiLeaks: My Time with Julian Assange at the World's Most Dangerous Website*. New York: First Edition Thus. Crown.

Hawkins, John (2010). "Reactions to the Wikileaks Document Dump from Around the Right Side of the Blogosphere." RightWing News, November 29, 2010. http://www.rightwingnews.com/war-on-terrorism/reactions-to-the-wikileaks-document-dump-from-around-the-right-side-of-the-blogosphere/

Irien (2013). "Cables: How the US State Department Promotes the Seed Industry's Global Agenda." Wikileaks Press, May 14, 2013. http://wikileaks-press.org/cables-how-the-u-s-state-department-promotes-the-seed-industrys-global-agenda/

Jarvis, Jeff (2010). "Value Added Journalism." *Buzz Machine*, July 27, 2010. http://buzzmachine.com/2010/07/27/value-added-journalism/

JLo (2011). "Overcoming Mendacity." *WL Central*, November 15, 2011. http://wlcentral.org/node/166

Love, James (2011). "Notes from the Wikileaks Cables." Knowledge Ecology International, August 27, 2011. http://keionline.org/wikileaks

Lynch, Lisa (2010). "We're Going to Crack the World Open," *Journalism Practice* 4:3, 309–18, 2010.

Lynch, Lisa (2011). "'Pls Call, Love, Your Wife': The Online Response to Wikileaks' 9/11 Pager Messages," *Global Media Journal: Australian Edition* 5: 1, 2011.

Lynch, Lisa (2012). "That's Not Leaking, That's Pure Editorial: Wikileaks, Scientific Journalism, and Journalism Expertise." *The Canadian Journal of Media Studies*, Fall: 40–68.

Lynch, Lisa (2013a) "The Leak Heard Round the World? Cablegate in the Evolving Global Mediascape," in B.Brevini, A.Hintz & P.McCurdy, *Beyond Wikileaks*, London: Palgrave.

Lynch, Lisa (2013b) "Cablegate in the Congo," *Radical History Review*, TK

Mawando, Mandalitso (2011). "Zibabwe: Deluge of Online Reactions to Latest Wikileaks." GlobalVoices, September 8, 2011. http://globalvoicesonline.org/2011/09/08/zimbabwe-a-deluge-of-online-reactions-to-latest-wikileaks/

The New York Times (2013). Editorial Staff, "Corrections." June 26, 2013.

Petrossian, Fred (2010). "Afghan Bloggers on the Wikileaks War Logs. *Global Voices*, July 29, 2010. http://globalvoicesonline.org/2010/07/29/afghanistan-quiet-in-blogs-since-wikileaks-war-logs/

PEJ New Media Index (2010). "Leaked Docouments Drive The Online Conversation." Journalism.org, November 29, 2010. http://www.journalism.org/index_report/leaked_documents_drive_online_conversation

Reddit Discussion Forum (2009). "Since Wikileaks is obviously in bandwith trouble…" Reddit, March 21, 2009. http://www.reddit.com/r/reddit.com/comments/86bvy/since_wikileaks_is_obviously_in_bandwidth_trouble

Savage, Charlie (2010). "US Tries to Build Case for Conspiracy by Wikileaks" *The New York Times*, December 15. http://www.nytimes.com/2010/12/16/world/16wiki.html?_r=3&

Savage, Charlie (2013). "Soldier Admits Providing Files to Wikileaks." *The New York Times*, February 28, 2013. http://www.nytimes.com/2013/03/01/us/bradley-manning-admits-giving-trove-of-military-data-to-wikileaks.html?pagewanted=all

Sifry, Micah (2011). "Diplomat Carne Ross Asks: Are the Cables Too Important to Leave to WikiLeaks, the NYTimes, and The Guardian to Sift?" *TechPresident*, Friday, February 24, 2011. http://techpresident.com/blog-entry/diplomat-carne-ross-asks-are-cables-too-important-leave-wikileaks-nytimes-and-guardian-si

Singel, Ryan (2008). "Immune to Critics, Secret-Spilling Wikileaks Plans to Save Journalism—And the World." *Wired*, July 3, 2008.

Sledge, Matt (2013). "Alexa O'Brien Is Bradley Manning Trial's One Woman Court Records System." *The Huffington Post*, April 16, 2013. http://www.huffingtonpost.com/2013/04/16/alexa-obrien-bradley-manning_n_3086628.html

Stengel, Richard (2010). TIME's Julian Assange Interview. *Time*, Dec 1, 2010. http://www.time.com/time/world/article/0,8599,2034040,00.html

Stöcker, Christian (2011). "A Dispatch Disaster in Six Acts." *Spiegel Online*, January 9, 2011, http://www.spiegel.de/international/world/0,1518,783778,00.html.

Usher, Nikki (2010). 'Why WikiLeaks' latest document dump makes everyone in journalism—and the public—a winner', *Nieman Journalism Lab*, http://www.niemanlab.org/2010/12/why-wikileaks-latest-document-dump-makes-everyone-in-journalism-and-the-public-a-winner/, 12 December.

Wikileaks (2007). Discussion Page, "Writer's Kit." 2007. https://www.wikileaks.org/wiki/WikiLeaks_talk:Writer%27s_Kit

Wikileaks (2009). 'The Most Wanted Leaks of 2009', WikiLeaks, https://wikileaks.org/wiki/Draft:The_Most_Wanted_Leaks_of_2009, 21 December.

Wikileaks (2011a) "Global – Wikileaks Statement on the 9 Month Anniversary of Wikileaks." Wikileaks.org, August 29, 2011a. http://wikileaks.org/Wikileaks-Statement-on-the-9-Month.html

Wikileaks (2011b) Twitter Message, September 2, 2011b. "The world community will crowdsource all 251k cables using the method developed this month"

Wikileaks (2011c) Twitter Message, September 2, 2011c. "Tweet important cable discoveries with #wlfind. The entire world press does not have enough resources and there are substantial biases."

CHAPTER TWENTY-ONE

Beyond Journalism: The New Public Information Space

NIK GOWING

What is so freely referred to as the proliferation of "social media" is creating acute new vulnerabilities for business, governments and systems in what I label the new "Public Information Space" (Gowing, 2009). The new issues and challenges go well beyond those associated with the pressures created on executive power and governance by the traditional framing of "media" and "journalism."

The near ubiquity of digital empowerment by ad hoc communities of mutual concern coupled with rapidly available video of unfolding events is forcing new levels of instant accountability. This is done through vivid, ever-changing matrices of immediate and usually hyper-transparent data. Such ubiquity exposes slow and questionable judgements by highest-level executives, ministers and public servants. It also exposes, helps track and identifies the nature of criminal activity—like that of the Boston Marathon bombers, or the two men seen on mobile phone after they murdered the British army soldier Lee Rigby in Woolwich, London (see Allan's chapter, this volume). And it lays bare the slowness of the law and legal systems to adapt to these new, cutting-edge digital realities created in the largely unregulated new information space

But few at the highest level are willing to concede the enormity of change, along with the urgent and profound implications for them to modify the core principles of political and corporate governance. Such slowness, reluctance and frequent institutional resistance carry a very high price, especially when the alerts and growing evidence are so unambiguous (Gowing, 2009).

Everything in this new information space is so obvious. But the most obvious is also the most elusive. The oft-used phrase "citizen journalism" is proving far too narrow a description for the profound impact on power, governance and what tends to be called "the media space", with its new and ubiquitous digital information dynamic especially in moments of tension and crisis. This goes far beyond the usual assumptions about the traditional definition of journalism, and the often convenient but superficial use of the catch-all phrase "citizen journalism." A far more profound new relationship is being defined between instant digital revelations that can be labelled partly journalistic in nature, and the credibility of leadership and power.

In this chapter I outline the transformation of the new digital Public Information Space, then crucially how time and again leaders get caught out because they fail or refuse to adapt to it. The chapter highlights two notable but rare examples. Institutions of both corporate and political power at the highest levels were willing to admit publicly how they were forced to confront vulnerabilities they had never foreseen. They then confirmed the price they paid for their failure to adapt in a smart, timely fashion.

This chapter also highlights an emblematic selection of my fast-growing accumulation of case studies which confirm the gap between old and new realities. All of them reinforce the relevance of alerts raised four years ago (Gowing, 2009). But they also confirm the reluctance or inability of political and corporate leaderships to embrace the implications of the new scale of digital disruption, then to modify their understanding and structures of governance.

NEW THREATS TO TRADITIONAL ASSUMPTIONS OF POWER

The new digital Public Information Space has turned the traditional Media Space on its head. Six billion mobile phones have created a new bottom-up empowerment almost anywhere, usually defying the most authoritarian of leadership structures. This challenges and disrupts head-on traditional assumptions of power at every level from the global to the most local. Vivid public perceptions are created by the instant capacity of this vast number of "information doers", "digital natives" or "born digitals" to firstly bear witness from almost anywhere, then to upload both images and evidence almost instantaneously. Webcams and sufficient bandwidth increasingly mean this can now effectively be done live and in real time without any mediation. In a fast-increasing majority of cases traditional media and journalism platforms are bypassed to the point where they often have a marginal role or none at all.

Critically this is all often before those in the corporate C-suite or government systems even know something has taken place, let alone been mobilised to respond. This gulf in capacity to both know and respond in a timely manner prompts searching questions about the resilience, relevance and credibility of both leaders and institutions. It also raises questions about their will to adapt when faced with this increasingly predictable digital bullet train of disruption. This chasm of credibility frequently reaches a point where there is a public perception of a deficit of legitimacy both for the leaders and the organisation they lead. This deficit is because the proliferation of instantaneous real-time information being consumed on smartphones or tablets often provides a dramatic early flow of often incomplete impressions. But they instantly fix perceptions, and usually well before political or corporate leaders are aware, let alone fully informed of fast moving developments.

Despite sharp alerts and warnings—including my own—about the inevitability of what they now face (Gowing, 2009:1–3), few at the highest levels of power have been willing to realise or concede that this new capacity to disrupt threatens their institution's reputation and brand, along with the careers of executives and leaders at the highest levels. Instead of gripping decisively these new realities and adjusting systems accordingly, they largely resort by default to status quo thinking. Most leaders are more comfortable finding refuge in actively cursing "the damn meejah!" or blaming allegedly "subversive forces" instead of understanding and accepting the profound new recalibration of power demanded in the new information space, and with it the new vulnerabilities exposed (Naim, 2013 reaches a similar conclusion with a different analysis). Responding by creating a Facebook or Twitter account is no substitute for a deep understanding of the change in power dynamics now taking place at considerable speed, along with the implications for executive behaviour.

Despite a few lone voices urging smart and timely engagement,[1] even more remarkably, there remains denial and a reluctance to learn from the destabilising "oh shit!" experiences of others during moments of acute crisis in multiple fields and locations.[2] For example, in 2011 Tunisia's President failed to learn from the new digital empowerment revealed in Iran's Green Revolution two years earlier. In turn, Egypt's President Mubarak failed to draw the implications for his regime's survival from Tunisia and Iran. Then Colonel Ghaddafi in neighbouring Libya was equally blind or arrogant to the near inevitable consequences of the new digital realities unfolding. It has been the same for President Bashar Al-Assad in Syria since early 2011. In June 2013 Turkey's Prime Minister Erdogan fundamentally failed to realise the new political power generated by "Facebook flash mobs" over the emotive issue of felling trees in a park in Istanbul. Instead he defaulted to old thinking by blaming inebriated, copulating demonstrators and unnamed "terrorists" for plotting anti-state protests and even the BBC for fomenting them by reporting the events globally. Simultaneously in that same week, Brazil's political class suddenly confronted similar but far larger numbers of instant protesters. As in Turkey, Brazil's national leaders found themselves exposed and wrong footed by millions mobilised digitally and speedily by just a common sense of disgruntlement initially sparked by a rise in bus fares. As one Brazilian poster put it: "the people do not need parties." This confirmed the new challenge to power. It came from an organically created, leaderless digital crowd whose united, spontaneous e-commitment became the superglue that bound together everyone in a single, instant cause.

Together these examples further confirm that the core principle I highlighted in my own Black Swan alert is even sharper and more relevant now than ever (Gowing, 2009). But the leadership class is largely in denial. To re-quote Nassim Nicolas Taleb's definition of Black Swan moments: "We don't learn what we don't learn" (Taleb, 2007). That is certainly largely the case for government ministers and corporate executives alike, with only a few rare exceptions. One of them is China. Up to the very highest levels, Communist Party officials continue to wrestle with the implications for one party rule. They understand fully the potential inverting of power underway, and are alarmed by their conclusions. As a result, during the second half of 2013, under advice from the Central Party School, they took firm action to limit significantly the ability of the new Public Information Space to challenge the Party's authority. (Author's meetings in Beijing, October 2013)

And this current scale of disruption must be viewed as only the start. The overall message for the largely laggardly leadership class around the world is that "You ain't seen anything yet." So far they, and we, have all experienced maybe only 0.001% of the technological change coming down the track.[3] That looming scale of change will have even more profound impact on all we assume about the nature of power, who exercises it and how. That power is being first challenged then steadily sapped by the overwhelming realities of the new Public Information Space. This is how BBC Director General Tony Hall described those realities in an upbeat alert on the future implications of this rapid pace of change for staff and their work. "Expectations are changing. When something is happening, people don't want to wait. The public want news they can trust about what is going on, wherever it is in the world. Now. In the palm of their hand. And then they want to share it, and talk about it" (October 8th 2013).

FAILURES OF LEADERSHIP TO ACCEPT THE NEW DIGITAL REALITIES

Overall the new matrix of real-time information flows and transparency created in particular by the explosion of social media has created a new executive fragility. The explosion of digital connectivity and new IT realities are disruptive game changers. They challenge mercilessly the inadequacy of the structures of power to respond both with effective impact and in a timely way. This has significant policy implications for government ministers, civil servants, defence and security agencies plus corporate executives, their corporate institutions and NGOs. As Sunil

Mittal, Chairman and Global CEO of Bharti Enterprises, including the telecom giant Bharti Airtel, warned one thousand global executives at the Confederation of Indian Industry conference in Delhi on 8th April 2011: "to people who have built their business on information arbitrage, [my message is] those days are over. Wake up!"

Overall the impact and direction of travel are inevitable, inexorable and above all irreversible. But in a majority of institutions of power the inherent conservatism of the "flat earthers" who resist the new realities tends to succeed in holding and dominating the high ground. After a crisis they even return to the *status quo ante* (Lee, Preston and Green, 2012). Mistakenly, they believe that the acute stresses were merely a temporary aberration, and that fundamentally their organisation's systemic architecture is sound. But the new realities mean they delude themselves. The systems are no longer sound. This is as much because of a determined institutional resistance to first concede then embrace the new realities. Despite being recognised as having the traditional executive skills and qualifications for high office, the evidence shows that many are either unwilling or unable to confront the enormity of digital change, along with its new impact on the nature of their responsibilities and implications for policy responses when it should most matter. And that is in a moment of crisis when the public often believes that because of its instant digital access to the new real-time information flow it knows far more than those in charge. At least, that is the brutal impression.

This is evidenced by that fast-multiplying proliferation of examples: from the pro-democracy uprisings in Burma then Iran, through Tunisia, Egypt, Libya and now Syria, to BP after the Gulf of Mexico oil well failure in 2010, police during street protests and violence in London in 2011, the Tokyo Electric Power Company (TEPCO) after the massive March 2011 earthquake in Japan, and even for the governing political class during elections in a highly developed nation like Singapore. Other examples include accountability forced on NATO by air strike errors in Afghanistan, the instant anti-corruption street mobilisation by Anna Hazare in India, the internet campaign in the US that swiftly halted the Stop Online Piracy Act (SOPA) in just a week, the way travellers marooned at Heathrow airport during two snow crises could vent their anger publicly through social media, or Britain's Prince Harry revealed naked in Las Vegas in a picture taken by a fellow frolicker. There are plenty more in a fast-expanding arsenal of examples. At the time of writing the list continues to grow with ever-increasing momentum, as the ongoing audit by this author confirms.

Each example is marked by the core failing of leaderships to pre-empt the impact of a crisis moment in the new Public Information Space by both comprehending then transmitting in advance the implications of an increasingly disruptive undermining of their power, and how profound the scale of change really is. In particular—and remarkably—there is denial of the digital direction of travel and its merciless, immediate implications for both leaders and the systems they represent. Most are unwilling to accept it, let alone review how best to either prepare for or handle the typical enormity of its inevitable and irreversible impact. Yet to repeat: that direction and those implications are not just inevitable, and inexorable. Most crucially, they are irreversible.

This institutional unwillingness is sharply illustrated by comparing the response of Myanmar's ruling generals to the leaked mobile phone video of mass protests by monks in September and October 2007, to that of the Syrian authorities during 2012 to emerging smartphone video of atrocities committed on civilians. The Myanmar military junta dismissed the video of monks on the streets as a "Skyful of Lies" (see also Gowing 2009). The dismissive language of this official put down suggested that the monks were not protesting, and the images of large gatherings had been fabricated or manipulated. At the UN General Assembly on 8th October, Myanmar's Foreign Minister tried to dismiss what had been seen worldwide as "small protests hijacked by political opportunists." The powerful images and video uploaded at great risk for global circulation told a very different story, and many of those who risked recording them ended up being convicted and facing long prison sentences.[4]

Yet five years later the Syrian government took a similarly preposterous line of denial when confronted by the overwhelming daily flow of social media video evidence of violence and deaths that emerged from its country starting in March 2011. The sharpest illustration of this was in late May 2012. Verified video emerged of 107 bodies lying in new white shrouds in a burial pit in Houla. This followed reports of a massacre which the emergence of the video then confirmed. But on 26th May, Syria's ambassador to the UN in New York dismissed the reporting and video as just part of a "Tsunami of lies." Just like the Myanmar junta five years earlier, he suggested that the video had been faked and the massacre never took place. So the official instinct of denial first confirmed by the more modest "Skyful" of lies from Myanmar's government was now repeated as a much more intense "Tsunami" of lies!

This highlights once again that as the vulnerabilities are increasingly exposed by the new Public Information Space, remarkably the institutional mindsets and systemic behaviour learn nothing from the clear trend of what other leaderships have experienced in crises. Indeed they resist it, which means that they lag woefully way behind the new realities. A core reason for this is that the career risks of challenging the top-down imperative to conform to traditional, embedded assumptions of who controls power are too great. The old, embedded systems of power largely dislike change. This is often to the point of resenting and even despising it because of its threat to the status quo. This is especially when it challenges long-standing assumptions about the power systems that promoted most leaders to the top. "How many get it?" I have asked some of the small number of more visionary leaders who do. "Twenty per cent if we are lucky," conceded one leading adopter at a highest leadership level who spoke with a grimace of despair. This assessment reflects those shared with me by others.

FRANK ADMISSIONS OF INSTITUTIONAL VULNERABILITIES AND THE PRICE PAID

Some leaders humiliated by a failure to grasp the implications of the new Public Information Space are willing to share details off the record of how they failed and why. But overall it is hard to get most to speak on the record because of the way such failings reflect on them and their reputation. This section highlights the blunt words of two who did speak publicly after they lost their high-profile public jobs. First is a high-level corporate leader. Second is a former government minister. Both reveal this institutional inability to grasp the enormity of change in the Public Information Space and its implications for both power and leadership. They know this because of the price they and their institutions of corporate or political power paid professionally and reputationally.

Tony Hayward was the Chief Executive of BP during the undersea blowout of the Macondo deep-sea oil well being drilled in the Gulf of Mexico in April 2010. BP has, of course, long been in the high-risk business of exploiting energy reserves worldwide. But Macondo showed how even the crisis and disaster-management response systems of this global corporate giant were inadequate and underprepared for the torrent of anti-BP social media traffic that swiftly overwhelmed the Public Information Space.

Before oil first leaked then blighted the US Gulf coast, BP's most senior executives failed to even conceive the scale, let alone the imperative to have the appropriate mindsets and capacities in place to handle it. But within the first few days of the blowout, a staggering 22% of all social media traffic in the US had become consumed by only one issue—BP's drilling rig blowout (sourced to a private conversation with a senior oil executive). That in turn created the overwhelming public pressure on President Obama to act decisively against BP. Even though risk had long been a core constant of their business, neither BP corporately, nor Tony Hayward personally, had the necessary level of institutional or behavioural preparedness to cope. This was conceded by Mr Hayward during an interview on BBC2 (9th November 2010), after he had been forced to leave BP:

> We tried to be open and transparent. We gave access to the operation. But the reality is [that] we were completely overrun and just not prepared to deal with the intensity of the media scrutiny.

The second example is from Singapore. George Yeo was a long-serving minister in the PAP-led government, which for decades had assumed its almost unchallenged natural inheritance to power. In the elections of May 2011 the party received a shock from an electorate increasingly attracted by alternative forms of making politics and challenging the PAP via social media. Mr Yeo lost his seat and his cabinet job as Foreign Minister. This was in large part because of what he described as the new digital information phenomenon "sweeping the world," and which—as he later conceded publicly—Singapore's political class was not prepared for. Speaking at the World Economic Forum on 16th September 2011 in Dalian, PR China, he talked publicly about the scale of disruption to Singapore's political traditions of a near one-party state because of the new Public Information Space.

> With the iPhone and the Blackberry in the pocket, the disintermediation of hierarchies is a relentless, inexorable process. Institutions are all coming under attack because the myths, the hypocrisies which protected them in the past are all being punctured by the access to information. So whether you are the Catholic Church or a big government or a big corporation, there is growing distrust. And Singapore is part of the same process that is going on around the world. From hierarchies to networks; from leadership appropriate to hierarchies, to leadership styles more appropriate to networks, there is a shift that we are seeing. And those who are able to make that shift will succeed, while those who are less able to will find themselves cast aside.

By any standards of candid revelations from a leader, these were extraordinary admissions. Yet George Yeo's rare but blunt public assessment confirms explicitly the theme of this chapter and my ongoing analysis.

Despite the intellectual or career qualifications that mark leaders out for appointments to the highest levels, those mindsets remain largely mired in a behavioural time warp of complacency and denial. Central to this is the misplaced assumption that existing systems are adequate or can be modified incrementally at the margins in order to handle and embrace the new digital realities. They are not and cannot. The old approaches or assumptions are not just inadequate. They are wrong.

A fundamental re-assessment must therefore be made of institutional vulnerabilities in times of crisis and tension to what might now be recorded digitally, then transmitted in the ways most people now seem to believe they have the right to do. Much can be learned from a tweet from Rupert Murdoch after the online exposure of Prince Harry's naked torso in a photo taken surreptitiously by a fun-seeking friend during a strip billiards game in Las Vegas (23rd August 2012). "Beware playmates with cameras," Murdoch tweeted after websites—not traditional media platforms—who published the image reportedly received 150 million hits in the first 24 hours.

Corporate executives and ministers or public servants at all levels are most unlikely to be caught naked at a game of strip billiards. But they should note the nature of Mr Murdoch's slightly tongue-in-cheek warning, when their instinct is to assume that they or their organisations will never have to face such revelations, and such opportunistic revelations will never affect them. They will and they do, and usually in a brutally revealing and therefore inevitably destructive way.

So the question which must be confronted by leaders in this new, fast-changing Public Information Space of dramatic digital and social media developments is: how prepared are you? How well do you understand its relentless new impact on your power?

The implications are vital leadership issues. Governments, public servants or security officials plus commanders and corporate leaders need to look through their digital telescope from the other end. What instinctively they see as a threat must be viewed as a new leadership opportunity that will serve them well in a crisis that few ever believe will happen to them and their organisation. The growing list of examples confirms there is a fast-growing likelihood that it will happen. To assume otherwise is irresponsible and a dereliction of the new requirements of executive leadership.

To create a new state of preparedness they must open their eyes. They must shed that traditional C-suite prism which currently defines their assumptions of power with levers that they will always control. As an executive duty, they must do considerably more to understand how to embrace effectively the new realities of today's media and Public Information Space. They are far broader and more multi-dimensional than the vast majority are prepared to realise, let alone concede. They challenge all the conventional assumptions about the nature of power.

The Hayward and Yeo case studies already reported and discussed here are not isolated examples of the lack of leadership awareness and preparedness. In the following section I detail a small selection of the fast-growing number of recent examples that build on previously cited cases (Gowing, 2009). They should convince the "flat earthers" and skeptics, and demonstrate beyond doubt:

- the *acute vulnerability for all power, politics and systems* imposed by the new public information space.
- the new levels of *instant accountability* being imposed,
- the *challenges to institutional legitimacy* if responses are belated, untimely and out of sync with the proliferation of information circulating instantly on social media.

THE FAST GROWING EVIDENCE

Despite the frequently heard belief that "it will not happen to us," often it does. And those whose mindset is denial—which the evidence suggests remains the vast majority—are those most exposed, most vulnerable and therefore increasingly humbled and probably humiliated, with their brands and reputations damaged, and careers destroyed. No leader can take comfort from the examples outlined here. But each should take away lessons.

Public disorder: embarrassments created for London's Metropolitan Police Gold Command.

There is a growing proliferation of examples of how social media video routinely embarrasses the commanders of police forces that are meant to have a superior real-time knowledge of events unfolding on their streets. On 1 April 2009 at the start of the G20 heads of state and government meeting in London, newspaper seller Ian Tomlinson died after being inadvertently drawn into a confrontation between police and protesters in The City (See also Gowing 2009). The Metropolitan Police said Mr Tomlinson died from a heart attack. But the incident tragically highlighted how police attitudes later described by the Chief Inspector of Constabulary Sir Dennis O'Connor as needing to return to the principles of "approachability, impartiality, accountability and minimum force" were out of sync with the new social media reality that now has the potential to hold to account every and any police action. As Duncan Campbell wrote in *The Guardian* "wherever you turned someone seemed to be pointing a camera." The police's initial "heart attack" version would have endured unchallenged had *The Guardian* newspaper not tracked down a 38-year-old American investment banker. He happened to be passing the incident. Crucially he was videoing as a police officer pushed Mr Tomlinson to the ground (see also Greer and McLaughlin, this volume).

That 41 seconds of video was broadcast six days later. Its impact was profound. Over the following four years of tortuous legal process the credibility of evidence submitted by police officers, plus the postmortem and inquest system, Home Office pathologists, the Crown Prosecution Service, witness statements and police conduct in general were all found wanting. After an inquest ruled that Tomlinson died from "internal bleeding" and he was "unlawfully killed," in 2012 the Home Office pathologist Freddy Patel was struck off for manipulating postmortem evidence and processes. Following months of legal manouvering, police officer Simon Harwood—whose identity and role in pushing over Tomlinson was only revealed by the video—was acquitted of manslaughter. Subsequently he was dismissed from the police for gross misconduct. However, without any admission of police failures, the Tomlinson family continued to reject the process as a "whitewash."

Eventually the indisputable revelations from the investment banker's mobile phone video secured justice. On 5 August 2013, a Deputy Assistant Commissioner apologized publicly and "unreservedly" for the "unlawful and excessive force" used by PC Harwood. The police paid undisclosed damages to the Tomlinson family. There is a high probability that such a police admission of guilt would never have happened had it not been for the accountability forced by those vital 41 seconds of video shot by chance.

11 March 2011: ineffectual public information handling after Japan's devastating earthquake

The devastation caused by Japan's 3/11 earthquake and tsunami was made even worse by the dysfunctional way both the government and especially the Tokyo Electric Power Company (TEPCO) coped with the acute public expectation that it would at least provide timely, accurate information to a deeply anxious public. They did not. Despite the well-known core risks of being the operator of nuclear power stations in a nation vulnerable to massive earthquakes at any moment, TEPCO was grossly ill-prepared.

The vulnerability of both was exposed by social media and real-time imagery emerging from around the Fukushima power plant and devastated towns like Sendai. It trashed the reputations and credibility of government officials, especially the most senior TEPCO executives right up to the company's president himself. This systemic failure to bring together and give out the real-time information which Japanese citizens expected aggravated the scale of the tragedy and intensified public anger.

At the height of the crisis when the public expected reassurance or at least clarity on the scale of danger, the TEPCO Managing Director Akio Komori was seen publicly in tears. The government then took control of informing the public in a statement on 25th April 2011. But still the Cabinet Secretary could not get from TEPCO the information which he and his government both expected and needed. Here is a taste of language in the reporting from Japan and resulting headlines:

- "Public patience wears thin as TEPCO's bungling far exceeds 'inconvenience,'" David Pilling. *Financial Times*, 16th March 2011.
- "'Behind the scenes there were signs of the government's plummeting faith in the plant's operator TEPCO. Kan [the Prime Minister] was overheard reading the riot act to executives for failing to inform him of the blast, media said….The TV reported an explosion, but nothing was said to the prime minister's office for more than an hour'. The Kyodo agency quoted Kan as saying. 'What the Hell is going on?'" *The Guardian*, 16th March 2011.
- "TEPCO, particularly in its communications, has looked more like the Keystone Cops than is desirable for an organisation struggling to prevent a nuclear meltdown. TEPCO's attempt to impart information has left the public mostly confused and incredulous. At press conferences, anxious looking junior executives hang their heads like naughty schoolboys and apologise for 'causing inconvenience,' a stock Japanese phrase. In matters of substance they appear to know little." David Pilling. *Financial Times*, 16th March 2011
- "Everything is a secret. Japan universally condemned for its response to the unclear crisis," former Japanese nuclear power engineer, cited in *Evening Standard*, 17th March 2011.

- "At present the people simply do not trust the information they are given. Their doubts and anger are justified—as is the growing storm of international condemnation of Japan's failure of communication during the crisis." Editorial, *Evening Standard*, 17 March 2011.
- "Chronic lack of leadership costs nation dear in its hour of need," Leo Lewis Commentary. *The Times,* 19th March 2011.

The state of dysfunctional relations on even getting basic information to the public was highlighted by the open row after TEPCO stated on 27 March that radiation levels were "ten million times higher than normal." The next day TEPCO had to admit in public that it had been wrong. Its spokesman Takashi Kuritau said: "The number is not credible. We are very sorry." This admission from the company led to an immediate and devastating public put down from the government to TEPCO that same day. Chief Cabinet Secretary Yukio Edano, who was by now the public face of the government, described TEPCO's error as "absolutely unforgiveable!"

Why had TEPCO and the government been so vulnerable in the Public Information Space? A remarkable insight came two weeks into the crisis from Ambassador Masa Ishii, Director for Policy Planning and International Security Policy at the Japanese Foreign Ministry. His explanation is reminiscent of the confession Tony Hayward made about BP's failings in the Gulf of Mexico in 2010. It underlines the Black Swan problem that "we do not realise what we do not realise." It confirms how institutions learn little from the acute failings of others and the inevitable price they pay:

> We are supplying information. But it did not reach…partly our fault…partly it is what they did. And yes we had a problem. And we will learn lessons from the way we did [it]. We are too good to cover the area we are supposed to cover. Our information is sometimes too deep to be understood by the public. Each ministry does that. Pile of paper. But what it means for the public is quite unclear. So it is up to the political leadership to pick up the right to information, sort it out and send the right sober message to the public. It was not done quite well at the beginning, I must admit. (Public remarks by Ambassador Ishii to the Brussels Forum of the German Marshall Fund, 26th March 2011)

The public handling, especially on information, cost TEPCO's president Masataka Shimizu his job. He confessed publicly that "I am resigning for having shattered public trust about nuclear power, and for having caused so many fears and problems for the people." In its conclusions, the official enquiry commission into the nuclear disaster said: "Across the board, the Commission found ignorance and arrogance unforgivable for anyone or any organization that deals with nuclear power. We found a disregard for global trends and a disregard for public safety."

But how many other major corporations facing similar inherent risks in the new Public Information Space will assume they can pre-emptively defy the inevitable and embedded Black Swan tendencies? How many of the most senior executives in the C-suites are willing to dive deeply into the systemic shortcomings of their own institutions which leave them ill prepared both practically and behaviourally for the new information realities? The research and interviewing of senior executives by this author confirms that the grim answer is "regrettably very few."

April 2011: e-mobilisation of 10 million Indians in a spontaneous campaign to end corruption

A 74-year-old social activist, Kisan Baburao "Anna" Hazare, decide to take on India's political establishment to force decisive measures to end corruption, especially at the highest levels of public life. The man some chose to call a "21st-century Ghandi" began a public fast in Delhi in front of several hundred followers on 5th April 2011. Within hours the protest went viral on the internet. In three days it accumulated up to ten million e-supporters. This was unprecedented and an historic first in public life for India. It revealed a new vulnerability of India's political elite, which was caught off guard and at a loss over how to respond. Just 96 hours after Anna Hazare's fast began, the country's Congress-led political leadership in the person of its Mr Fix-it, the Finance Minister Pranab Mukherjee, gave in to the unprecedented pressure generated by both Hazare and the e-crowd. Mr Mukherjee "agreed to all his demands" to open a dialogue. The country's political class had been wrong footed. It reeled from the instant success of this e-challenge to the traditional power they assumed was always theirs to control, and for which there had always largely been unquestioning respect.

Whether at federal or state level, Indian politics would never be the same. Yet there was a return to type by political leaders. This meant that when the promised progress on anti-corruption measures faltered a few weeks later, Hazare felt forced to start a second fast. Prime Minister Manmohan Singh symbolised how out of touch the political elite were in the fast-changing realities of the new public information space. On 17th August he described the protest as "totally misconceived." The government then ordered that Hazare be imprisoned, which merely further inflamed anti-corruption campaigners. Millions again took to the streets and keyboards. This time their e-anger forced Hazare's release as a public hero.

Hazare's e-campaign made a big initial impact. But it lost public traction because of infighting among the odd rainbow of senior public figures who signed up to it. However, that that is not the point. The critical issue was those two moments of e-power in April and August 2011. Firstly, large numbers mobilised. Then they challenged the leadership assumptions of the political class in a nation whose understanding of the implications of social media was barely even immature. As the BBC's India correspondent Soutik Biswas wrote: "Mr Hazare's movement has humbled India's entire political class by bringing the issue of corruption under a dazzling spotlight" (BBC News, 28th August 2011).

18 January 2012: e-mobilisation for the Stop the Online Piracy Act campaign

The instant power of e-protest to expose a new vulnerability for the political class was not confined to a developing country like India. In the US Congress in January 2012, Congressman Lamar Smith introduced a bill to expand the ability of U.S. law enforcement to combat online copyright infringement and online trafficking in counterfeit goods. Significant opposition through traditional lobbying activities was reinforced by yet another extraordinary e-mobilisation online. Opponents were determined to kill the proposed legislation because it threatened free speech and innovation. It would also allow law enforcement agencies to block access to entire internet domains.

Such was the intensity of pressure that internet blackouts and tens of millions mobilised online forced Congress to shelve the bill just two days later. The campaign's own website described the "largest internet protest in history." It went on: "January 18th was unreal. Tech companies and users teamed up. Geeks took to the streets. Tens of millions of people who make the internet what it is joined together to defend their freedoms. The network defended itself. Whatever you call it, we changed the politics of interfering with the internet forever—there's no going back" (http://sopastrike.com/).

12 January 2012 transmission: US Marines urinating on dead Taliban in Afghanistan

A "trophy" video taken in Afghanistan in late 2011 of US Marines urinating on dead Taliban emerged on the internet. It swiftly went viral on YouTube, TMZ and other websites. The incident was not unique. Urinating on those they have killed has been known as a US Marine tradition, although speaking to this author subsequently, two former US Marine Corps Commandants denied this. They expressed outrage at what the video showed. How could military personnel at any level be in denial after the well-documented, seismic impact on the US military of the obscene pictures of prisoners being abused at Abu Ghraib prison in Iraq in late 2003?

But the video created vulnerability for the image and reputation of NATO's ISAF operation in Afghanistan. On 12th January 2012, Deputy Force Commander Lt-Gen Adrian Bradshaw—himself British—was forced to stand uncomfortably in front of a TV camera to issue an official condemnation of what the US marines had done. He did not question the video's authenticity. He described the urinating "as utterly unacceptable" and not representing "the standards we expect from coalition forces." Eventually one staff sergeant was sentenced to a reduction in rank and forfeited $500 in pay. Other cases are outstanding. Two more soldiers were detained in February 2013.

But the greatest damage was to the reputation of ISAF and US operations, and the US Marines in particular. Overall, it must be accepted that for whatever reason, the likelihood of a video recording of inappropriate and probably unlawful behaviour becoming public is now very high.

13 January 2012: Costa Concordia disaster off Italy

The cruise liner Costa Concordia hit rocks and half capsized after the captain, Francesco Schettino, set course to bypass the island of Giglio. He wanted to show off his ship and helmsmanship to both his crew and a retired former skipper living on the island. The captain had invited onto the bridge a lady passenger friend. Sitting at the rear, she recorded on her mobile phone events unfolding on the bridge. As she did so she could not have known that she was recording the drama as the ship hit rocks and captain Schettino struggled to assess how to handle the dreadful events unfolding because of a devastating navigational misjudgement. The video ends with her running along the listing deck as the ship starts to capsize.

The video taken by the 24-year-old Moldovan dancer became important evidence in the subsequent trial in late 2013 which investigated the accountability of the captain, some crew members and the ship owners. Because of what the video recorded, there was less doubt than there might have been about who said what to whom at critical moments on the bridge.

The West Bank in the Middle East : activities of all sides under constant video scrutiny

In 2007 the Israeli Human Rights group B'Tselem decided to issue a large number of video cameras to Palestinians of all ages on the West Bank. The aim was to monitor 24/7 the activities of the Israeli Defence Force (IDF) against Palestinians and make them more accountable, especially legally. Day after day B'Tselem's "Shooting Back" programme has produced often damning footage about the behaviour of IDF soldiers. Most graphic was the shooting in the legs of a Palestinian activist at an Israeli roadblock (posted online 20th July 2008). The video of the shooting led to a military investigation of the commander's actions.

But remarkably, while the imagery often highlighted questionable Israeli military actions against Palestinians, for a long time it seemed to make no difference to IDF behaviour. More recently, however, it has exposed a new command vulnerability that has led to a more robust official response to any apparently improper behaviour. For example, the video published on 28th August 2012, filmed from inside a window and showing a 9-year-old Palestinian boy being kicked by an IDF solider led to a disciplinary hearing by the Department for the Investigation of the Police in the Ministry of Justice.

Now on the West Bank, there is the extraordinary sight of all sides involved in many incidents deploying video cameras to record the behaviour of each other, whether Palestinian or Israeli. Almost nothing goes unrecorded by social media. However uncomfortable some soldiers or activists find this, it is creating a new level of both reputational vulnerability and accountability for all sides, whether Israeli or Palestinian.

Late 2012 : worker unrest in Foxconn factories in China which assemble I-Products for Apple

Apple was embarrassed in September 2012 when mobile phone video emerged of worker protests at the Foxconn factory in Taiyuan, PRC. Two thousand mainly male workers had reportedly been involved in fights rooted in complaints about their dormitory living conditions. Usually such claims about protests in a Chinese workplace emerge belatedly by word of mouth, with often questionable hearsay evidence. The process of verification and interviewing of witnesses would always take time.

But this video of yelling, fighting and large numbers of men moving around the plant was both vivid, dated and contemporaneous. Having been verified, it confirmed the workers' anger about the hitherto unreported conditions in which they have to assemble Apple's high value iPhone and iPads. What we did not know about before was thereby confirmed instantly by way of the uploaded phone video. Instead of being a journalistic footnote based on hearsay, rapidly there was enough evidence to justify leading news programmes with the graphic images from Foxconn (see for example *The Hub* on BBC World News, 24th September 2012). This was an embarrassment both to Foxconn and the global reputation of Apple.

CONCLUSION

This chapter is an abridged summary of this author's extensive, ongoing *wiki* process of logging, research and analysis of cases as they take place. I could cite many other examples. Like those detailed in this chapter they all confirm the inexorable direction of travel with its new impact on a leadership cohort that remains in denial. This is inspite of the fast-growing number of incidents, many of which could have been mitigated if leaders had accepted the scale of disruption and challenge that would one day threaten them from the new Public Information Space.

The UK's Chief of the Defence Staff General Sir David Richards was one senior figure who embraced the issue. But within Whitehall he faced considerable institutional resistance. He went public to confirm how out of sync most modern military thinking had become with the new Public Information Space realities identified here. In late 2011 he challenged conventional thinking and urged it be turned on its head. He told those who served him that instead of the military strategy setting the information narrative as was traditionally the case, the preferred information narrative and campaign should set the military strategy.

General Richards' suggestion is a fundamental reversal. It was completely at odds with traditional institutional thinking and training. Many senior colleagues—both civilian and uniformed—were both sceptical and scornful. They defaulted to largely retaining their traditional prism of thinking. But the radical position taken by General Richards highlights why looking through the other end of the telescope is vital to cope with the uncharted, often overwhelming pressures from the new social media empowerment in the Public Information Space. To achieve irreversible traction will require a similar fundamental reversal of thinking. Yet the evidence confirms that this remains uncomfortable for most leaders-whether in public service or the corporate executive C-suite.

Both the potential and the realisation of the need for a mindset change are confirmed by one important section in the latest version of British military doctrine known officially as JDN-1/12. Using often radical language that was contested through much of the drafting process the document sharply defines the tensions and possible solutions for all leaders, and not just senior military officers. The words emerged from a tortuous internal process of review and argument which saw significant resistance. That is why the outcome and words deserve close reading for all leaders seeking to grapple with the new tensions between their long-assumed power and the new Public Information Space.

Under a sub-heading "Empowerment" for paragraph 309, JDN 1/12 states:

> Speed of response is often critical; therefore, commanders should *be prepared to accept risks* in accuracy in order to enter the information space early. Effective strategic communication recognises that commanders, and their people at all levels, must be confident to use all means of communication to exploit opportunities. Indeed, the best communicators are often commanders and soldiers at the tactical level. To exploit this we must provide them with appropriate training and guidance, and *relax overly restrictive control mechanisms; we must be prepared to lose control to gain control*. (The UK MoD Development, Concepts and Doctrine Centre, 2012, author's emphasis)

Preparing to lose control to gain control runs counter to every instinct of leadership. Yet the spirit and detail of this new principle drill to the heart of the kind of new mindsets needed to stand a chance of embracing successfully the realities of the new Public Information Space. It almost goes without saying that the "modern technology" that produces the new social media empowerment cannot be disinvented. Neither can the profound, inexorable impact on the assumptions of power be reversed. Instead, amongst those with highest-level leadership responsibilities there needs to be a new determination to embrace it then ecalibrate the institutional architecture accordingly.

Such are the acute new vulnerabilities of power, politics and systems in the new Public Information Space. The future price of denial is likely to be even greater than endured so far.

NOTES

1. See the two citations of *Skyful of Lies and Black Swans* by General Sir David Richards, Chief of the General Staff in his presentation to the IISS: *Future Conflict and Its Prevention: People and the Information Age on* 18 January 2010.

2. Having concluded a 3-year Cabinet Office study for the UK government's official guidance on crisis management PAS 200, security consultant Peter Power concluded: "Being trained to promote your company is quite different to being trained to defend it." See "Drama School" by Peter Power, in *Continuity, Insurance and Risk cirmagazine.com* September 2012 pp. 18–19.
3. Vivek Kundra, Chief Technology Officer at the White House in Washington DC until mid-2011 speaking at the 25th anniversary celebration conference for the Joan Shorenstein Barone Center for Press and Public Policy, Harvard University on 14 October 2011.
4. See the Oscar-nominated movie *Burma VJ* (2008 dir. Anders Ostergaard) which vividly recorded the great risks taken by underground video journalists.

REFERENCES

Gowing, Nik (2009). *Skyful of Lies and Black Swans*. Oxford: Reuters Institute for the Study of Journalism.

Lee, B., Preston, F., and Green, G. (2012). *Preparing for High-Impact, Low-Probability Events. Lessons from Eyjafjallajokull*, London: Chatham House.

Naím, Moisés (2013). *The End of Power: From Boardrooms to Battlefields and Churches to States, Why Being in Charge Isn't What It Used to Be*, New York: Basic Books.

Taleb, Nassim Nicolas (2007). *The Black Swan: The Impact of the Highly Improbable,* London: Allen Lane 2007.

The UK MoD Development, Concepts and Doctrine Centre (2012). "Strategic Communications: The Defence Contribution," *Joint Doctrine Note 1/12*, Shrivenham: The UK MoD Development, Concepts and Doctrine Centre.

CHAPTER TWENTY-TWO

The Evolution of Citizen Journalism in Crises: From Reporting to Crisis Management

HAYLEY WATSON AND KUSH WADHWA

No longer in its infancy, our understanding of the role of citizen journalism in the news production process is well established, if a continuing subject of debate nonetheless. This chapter demonstrates that citizen journalists are not only contributing to the construction of news, however, they are also increasingly playing a central role in crisis communication. This shift, we will argue, has important consequences, ones that require careful consideration for enhancing crisis management.

Researchers have provided numerous examples of the presence of members of the public participating in the construction of news in relation to the reporting of a crisis. In recent years, there has been a noticeable growth in the use of citizen journalism material to inform others, as well as the increasing presence of citizen journalists within crisis communication. In 2009, during the Winnenden massacre in Germany when a mentally unstable offender killed 16 people and wounded 11 more, one of the first alerts to the general public was issued via Twitter, with a tweet advising people not to enter the town because the offender had not been apprehended (Tretbar 2009). Further reports claim that residents used Twitter to report "minute-by-minute sightings of the gunman" (Rayner and Hall 2009). Elsewhere, Taylor et al. (2012) surveyed members of the public following Cyclone Yasi in Australia and New Zealand in 2011; results showed that individuals were turning to social media for updates regarding real-time information relating to the crises. In addition to a host of others, these instances provide evidence of not only the reliance of the news media on the public for information, but additionally, the activities of members of the public acting as crisis reporters, informing others of unfolding events in a crisis situation and contributing to the construction of news.

Citizen journalists are not limited to acting as eye witnesses and reporters at the scene of a crisis; in addition, their firsthand contributions from the scene are being harvested to complement crisis management efforts. Accordingly, as our understanding of citizen journalism within the reporting of the news develops, it is necessary to consider the multiplicity of ways in which this material is being used to advantage. This chapter draws upon our current research in an EU-funded project, COSMIC, relating to the contribution of social media to crisis management.[1] We aim to provide a wider understanding of the factors shaping these processes, paying particular attention to the challenges those involved in responding to crises should consider in relation to citizen crisis coverage.

CRISIS REPORTING TO CRISIS MANAGEMENT

Advances in technology and the proliferation of social media are helping to transform the news media, enabling members of the public to participate in the production and distribution of news to an increasing extent. This trend came to the fore during the London bombings on 7 July 2005, which saw the city's transport network attacked by four suicide bombers during peak morning commuter time. During the attacks, several different types of citizen content were submitted to the BBC, including eyewitness statements and accounts, survivors' diaries, and personal videos and photographs (see Watson, 2012). Two years later, on 30 June 2007, the attempted attack at Glasgow airport—where a suspect drove a Jeep Cherokee into the airport terminal, loaded with propane canisters—generated similar responses. Hermida's (2007) analysis pointed out that once again the BBC was inundated with footage of the attacks, with over 70 images and videos sent in by bystanders, it was "the public, rather than professional journalists" who were "increasingly recording the first draft of history." Helen Boaden (2008), BBC Director of News at the time, maintained that "at the BBC, we knew then that we had to change. We would need to review our ability to ingest this kind of material and our editorial policies to take account of these new forms of output" (Boaden 2008; see also Allan and Thorsen, 2009). In addition to complementing existing news organisations, citizen journalists were praised for providing up-to-date information regarding how best to respond to these crises as they unfolded. The value of such information has become ever more significant as social media have grown in popularity. For example, following the Mumbai attacks in November 2008, citizen journalists were commended for their contributions; the UK's *Daily Telegraph* declared that throughout the Mumbai attacks the "social web came of age" (Beaumont 2008).

Citizen journalism has not always been met with praise, however. In the wake of the 2005 London attacks, for example, Glaser (2005) criticised members of the public taking images at the scene for acting like "citizen paparazzi." In the aftermath of the attempted Glasgow attacks, citizen journalists were criticised for placing themselves in danger by trying to record evidence, and potentially ignoring their civic duty to offer assistance to others (Welsh 2007). During the 2008 Mumbai attacks, citizens were criticised for reporting sensitive information. At one point, when Twitter was relaying information about the military and police responses to the siege, it was rumoured that the Indian authorities feared that such details might be "useful" to the terrorists (Beaumont, 2008). At the same time, critics insisted that the BBC relied too heavily on citizen journalism (via Twitter) to provide information, raising questions about trust and reliability in the absence of independent verification. In responding to criticisms, one of the BBC's editors, Steve Hermann (2008), argued that the Corporation was forced to make "quick judgements" in "selecting" what they thought was relevant and informative. In his view, Tweets provided "a strong sense of what people connected in some way with the story were thinking and seeing." Since the Mumbai attacks, the BBC and other news organisations have been making a conscious effort to identify approaches to verifying information they receive in order to uphold their journalistic standards (Murray 2011), as well as developing training for staff in the use of social media in the news gathering process (Hughes 2013; Walton 2012). Further efforts include initiatives such as the recently launched "GuardianWitness," intended to enable readers of the *Guardian* to have their videos, photos and stories featured on its pages (see also Geary 2013).

Deserving of greater attention than it has typically received to date, however, is the role citizen journalists can play for efforts to manage the very crises in question. Crisis management involves a myriad of actors, not least response organisations and authorities dealing with potential threats prior to their occurrence, handling the threat once it has occurred, and responding accordingly following the crisis to reduce its impact as far as possible (see also Coombs, 2007). Vitally important in this regard is crisis communication, which involves relaying pertinent warnings to people in advance, preparing them to act by disseminating timely and efficient information, and co-ordinating response and recovery efforts (ITU, 2005).

Liu et al. (2009) describe this field relating to the analysis of public involvement in crisis communications as "crisis informatics," where citizens are perceived to be central to planning, warning and responding to a crisis. Recent events such as natural disasters, including for instance Hurricane Sandy in the US in 2012, suggest that the tools used by citizen journalists are playing an increasingly crucial role in preparedness for, and emergency response to, associated threats, risks and hazards. Technologies such as crowdsourcing, remote sensing and data mining offer new opportunities that support officials and first responders in gathering information so as to optimise their response efforts. At the same time, social media can potentially enable citizens to share information with each other, as well as with emergency officials, more efficiently.

Examples of how emergency officials (including government agencies and aid organisations) incorporate news and information provided by citizen journalists to improve emergency response efforts continue to grow in number. Such activities enable crisis mangers to bridge the gap between "official" and "civilian" efforts in emergency preparedness and response. The following figure illustrates the scope of publicised uses of ICT by members of the public and crisis managers to assist with crisis management.[2]

Fig 1. Citizen journalism & crisis management

The use of a wide variety of information communication tools by citizen journalists can assist in performing a series of roles. The following table illustrates the various capabilities of different technologies:

Table 1: ICT capabilities in crisis communication

	Communicate	Inform	Assist	Relay	Request	Recover
Blogger	✓	✓	✓	✓	✓	✓
Facebook	✓	✓	✓	✓	✓	✓
Flickr	✓	✓				
FourSquare	✓	✓	✓	✓	✓	✓
Google Form			✓			✓
Google Maps		✓	✓		✓	✓

Table 1: continued

	Communicate	Inform	Assist	Relay	Request	Recover
Google Person Finder		✓	✓			
Google Spread sheet		✓	✓		✓	✓
SMS	✓	✓	✓	✓	✓	
Tumblr	✓	✓	✓	✓	✓	✓
Twitter	✓	✓	✓	✓	✓	✓
Ushahidi	✓	✓	✓	✓	✓	✓
Wordpress	✓	✓	✓	✓	✓	✓
YouTube	✓	✓				✓

The tools for participating in citizen journalism are not restricted to social media; instead, there are multiple tools that enable a citizen to capture and transmit news and information. The following figure provides further information regarding the process, including aspects that must come together for citizen journalists to contribute to crisis management effectively:

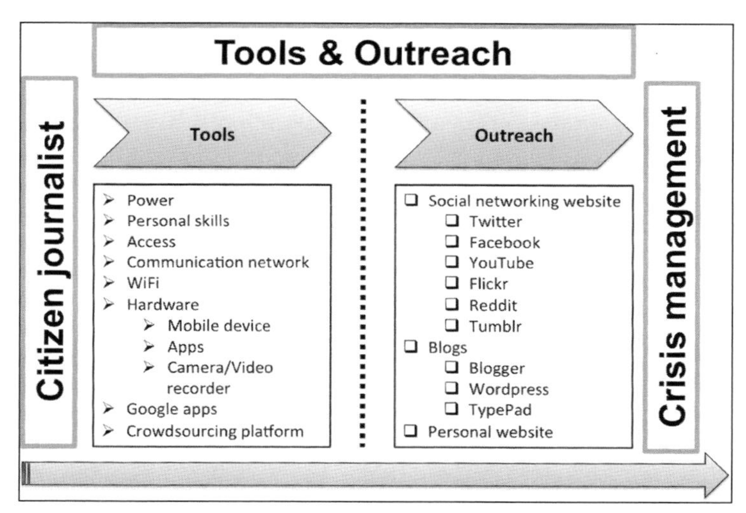

Fig 2. Tools & outreach for citizen journalism material to be of use to crisis management

To illustrate further, in the aftermath of the catastrophic earthquake in Haiti in January 2010 which affected an estimated 3 million people, initially traditional disaster-response systems were employed by response organisations to share information between each other. However, the system was not able to take into consideration localised information coming from within the Haitian community. The subsequent use of crowdsourcing, via open-sourced crisis-mapping software, Ushahidi, enabled response organisations to communicate with Haitians to "capture, organize and share critical information coming directly from Haitians." This sharing of information greatly benefitted response efforts and enabled citizens to be actively involved in the response stage of the disaster (Heinzelman and Waters 2010). Similarly, the use of crowd mapping through the Ushahidi platform (among others) to aid disaster recovery was also seen following the February 2011 earthquake in Christchurch, New Zealand, which killed 185 people (Moses 2011). Crowd mapping is also being currently used to optimise resources to respond to the ongoing humanitarian crisis in Syria, where civil war has broken out between those that support the Syrian Ba'ath Party government, and those opposing it (Syria tracker, 2013). Mexico City is planning to use mobile phone networks and social media to create a seismic alert system to provide warnings about earthquakes to their citizens, who can then use this information, via acts of citizen journalism, to continue to spread information to their wider social networks (Smith 2012).

Whereas the conventional approach to emergency response often favours a top-down co-ordination of search and rescue operations, these examples suggest that the successful use of new information technologies typically require officials to assume other roles, such as facilitator of decentralised networks, or require them to use crowd-sourced information for coordination purposes.

Following the attacks at the Boston marathon in April 2013, where two bombs exploded in backpacks left at the finishing line, some members of the public joined the authorities and news organisations rushing to the web to provide and share information alerting others to what was happening. This information ranged from help seeking, to visual imagery of the incidents, to further details shown in the figure below:

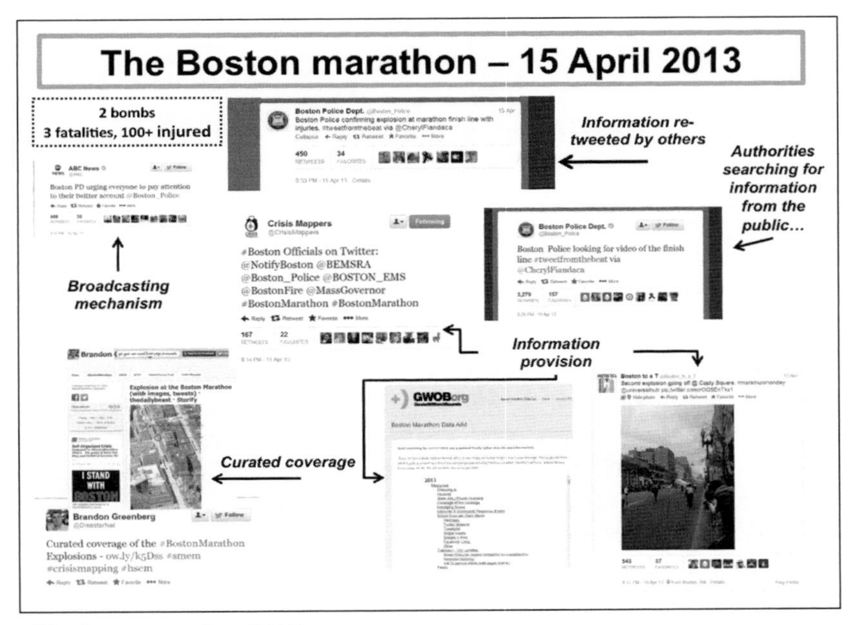

Fig 3. The Boston marathon (2013)

Importantly, information being reported and shared by those witnessing these events proved to be significant for those responsible for managing and responding to the emergency while, at the same time, revealing the potential dangers associated with unregulated citizen involvement. For instance, following the Boston attacks there were numerous incidents of citizens, such as some Reddit users, searching through citizen photography to identify and track those who they believed were responsible for the attacks. This activity—much, but not all of it, well-intentioned—resulted in innocent individuals being wrongfully identified, targeted and fearful of their safety (Lee 2013; see also Meikle, this volume).

Other studies have investigated the utility of information generated by the public for emergency responders. Like the studies presented above, these studies have employed a one-way communication paradigm focusing on dissemination of information. For example, during the October 2007 California wildfires, citizen use of social media was crucial to the supply of local information. The wildfires led to nine people being killed and many others injured, while more than 1,500 homes were destroyed and over 500,000 acres of land burnt. Novak and Vidoloff (2011) studied members of the public and their efforts in using social media applications to generate a citizen journalism platform for the dissemination of localised information news. These activities during the wildfires provided up-to-date information to community members and emergency responders. While the study also recognised that some responders were displeased with the posting of raw information, they nonetheless conceded that it performed a "helpful role in meeting the voluminous requests for on-going information" (Novak and Vidoloff, 2011: 191).

Several studies have highlighted further reasons why decision makers might usefully integrate content from citizens into their crisis management activities. Participation in the sharing of information during a disaster, for example, may fill a psychological role in building and promoting community-centred resilience. For instance, evidence from the use of the internet and social media after Hurricane Katrina in August 2005 in the US indicates that online media were instrumental not only in terms of coordination but also in terms of reducing uncertainty, connecting geographically dispersed communities and building social capital (Procopio and Procopio 2007). Elsewhere, Vieweg et al.'s (2008) study of the Virginia tech shooting led them to observe that social media helped individuals form an altruistic community in an online setting. Additionally, Taylor et al.'s (2012: 24) study of Cyclone Yasi in February 2011, which caused severe damage to affected areas in Queensland, Australia, argues that the use of social media helped promote community, centred resilience by building a public sense of security. More specifically, those using social media experienced a "sense of connectedness and usefulness, felt supported by others and felt encouraged by the help and support being given to people." These and other studies underline the role that social media may play in enhancing social capital, citizen well-being and citizen involvement in a crisis.

These and related examples suggest that the evolution of communication technologies provides ample opportunities for citizen involvement, including the activities of citizen journalists in contributing to the management of crises. It is worth noting that some individuals may not identify their activities as acts of citizen journalism, however; rather they may see what they are doing as simply supplying information to others (and thereby not deserving of a particular label, let alone a journalistic one). The growth of mobile, social media technologies have increased the public's ability to organise in concerted terms, share more information with each other, and even challenge the monopoly over facts official sources have typically wielded. However, recent research indicates a number of challenges ahead.

THE CHALLENGE AHEAD

In order to optimally use material created by citizens during a crisis, there is a range of complex considerations stakeholders must first identify and then overcome. How best to employ new information communication tools and software is a continuous process of negotiation, demanding constant learning on the part of stakeholders themselves, as well as an awareness of the various uses of these systems by citizens. In addition to more practical issues such as whether individuals have access to mobile networks, or even electric power for that matter, it is also important to consider the social implications associated with pertinent developments. Such considerations include matters relating to ethical issues regarding citizen journalism and the potential misuse of information, as noted above, as well as inequalities in access to informational sources and means of communication (see also Lindsay, 2009; Watson and Finn 2013).

Further obstacles complicating stakeholders' engagement with citizen journalists include the difficulties news organisations face when seeking to verify information. This process of verification is vital, Coyle and Meier (2009) claim, because the material gleaned can be used in all stages of the crisis management cycle: preparation, response and recovery. However, as they point out, stakeholders must be aware of the problems associated with inaccurate information, such as the possibility of placing harmful rumours into wider circulation, leading to uncertainty and, potentially, panicked responses hindering response efforts. A study by Briones et al. (2011) examining the use of social media by the American Red Cross (ARC) revealed that although social media enables dialogue between response organisations and citizens to occur, a series of internal barriers may arise in times of crisis: for example, the availability of social media resources, together with the expertise to use them effectively, is unevenly distributed, both within relief organisations as well as with respect to members of the public. Regarding the latter, there is a danger that older individuals, for example, may feel alienated or discriminated against. Additional issues include data protection, privacy infringements, and impacts related to profiling—even placing pressure on members of the public to assist with crisis response might lead to the exploitation of their labour.[3] Crisis managers' implementation of such strategies, therefore, raises concerns about effective operationalization, suggesting further attention needs to be focused on how best to reap the associated benefits while avoiding the danger of reinforcing divisions within the community they are assisting.

Furthermore, it is essential that the authorities, crisis managers and response organisations ensure that they are aware of the public dialogue surrounding a crisis event, so as to restrict the sharing of valuable, tactical information—particularly where citizen journalism and social media are concerned—within appropriate parameters. Whilst in many natural disaster situations, for example, citizens can be engaged to help promote public resilience, in those instances where their personal security is threatened sufficient care needs to be exercised so as help protect them from becoming too caught-up in events. Accordingly, proper training and preparation in relation to the use of citizen media in a crisis is essential. Existing training includes programs offered by FEMA (2012), "Social Media in Emergency Management," and papers relating to good practice, such as the International Federation of Red Cross and Red Crescent Societies "Beneficiary Communication" (IFRC 2013). It is necessary for those involved in crisis management to prepare for the challenges posed by citizens' engagement so as to optimise the use of this added resource.

In conclusion, much work remains to be done to better understand how and why so many ordinary citizens are prepared to adopt the role of crisis reporters, and how their efforts to gather and share news and information can effectively support crisis management and response initiatives in emergency situations. As illustrated in this chapter, these functions include: 1) their abilities to capture, transmit and communicate up-to-date, localised information from the vicinity of the crisis; 2) informing others of evolving events in real time, particularly via social media; 3) co-operation with those responding to a crisis by helping them to better manage the allocation of resources, in part by bringing to bear localised expertise and knowledge; 4) relaying news and information updates to distant others via their social networks; 5) requesting help and assistance from those with more adequate resources, including relevant, practicable information; and 6) assisting the wider community to cope psychologically with their ordeal by promoting greater resilience and self-sufficiency. Further research will enhance our awareness of the social, ethical, privacy and operational issues associated with these functions, particularly where they relate to the verification of news or information and the maintenance of security, but there can be little doubt that there is a pressing need to optimise the contributions of citizen journalists occupying a unique—and increasingly important—position in the human chain of crisis communication.

NOTES

1. The Contribution of Social Media in Crisis management (COSMIC) under grant agreement no. 312737.
2. This figure was created with the assistance of two tools that are available online—iMapbuilder and Woordle.
3. Data protection issues need to be taken into consideration, for example, when using material created by citizens reporting at the scene of a crisis. As noted above, such material needs to be handled in a manner that is transparent, with proper consent secured regarding its intended use (see also Watson and Finn 2013). For instance, following the earthquake and tsunami in Japan in 2011, eight firms (e.g., Google, Twitter, Honda, Rescuenow Inc. JCC Corp) are working together to understand what they can learn from the data they collected during the disaster, to "explore ways to support disaster survivors more effectively" (Kyodo 2012). Whilst this is an important venture, it also highlights some of the implications for citizens not necessarily aware of how their contributions would be used in future as public information. The privacy and data protection issues engendered by ICTs in crisis situations have also been identified by the United Nations Office for the Coordination of Humanitarian Affairs (OCHA 2013, p. 57), which recommends that humanitarian organisations should "develop robust ethical guidelines around the use of information"

REFERENCES

AFP (2012). "US Disaster Agency Tries to Dispel Social Media Rumors," *AFP Press*, 3 November.

Allan, S., and Thorsen, E. (eds) (2009). *Citizen Journalism: Global Perspectives*. New York: Peter Lang.

American Red Cross (2010). "Web Users Increasingly Rely on Social Media to Seek Help in a Disaster," *American Red Cross Press Release*, Washington DC, 9 August.

Barton, D. (2011). "People and Technologies as Resources in Times of Uncertainty," *Mobilities*, 6(1), 57–65.

Beaumont, C. (2008). "Mumbai Attacks: Twitter and Flickr Used to Break New," *The Telegraph*, 27 November.

Boaden, H. (2008). "The Role of Citizen Journalism in Modern Democracy," Keynote speech at the e-Democracy conference, RIBA, London, 13 November.

Boston to a T (2013). *Twitter feed.* https://twitter.com/Boston_to_a_T

Briones, R.L., Kuch, B., Liu, B.F., and Jin, Y. (2011). "Keeping Up with the Digital Age: How the American Red Cross Uses Social Media to Build Relationships," *Public Relations Review*, 37(1), 37–43.

Burgess, J., Vis, F., and Bruns, A. (2012). "How Many Fake Sandy Pictures Were Really Shared on Social Media?' *The Guardian Data Blog*, 6 November.

Coombs, W.T. (2014). *Crisis Management and Communications*. http://www.facoltaspes.unimi.it/files/_ITA_/COM/Crisis_Management_and_Communications.pdf

Coyle, D., and Meier, P. (2009). "New Technologies in Emergencies and Conflicts: The Role of Information and Social Networks," *United Nations Foundation & Vodaphone Foundation*, London.

FEMA. (2012). "IS-42: Social Media in Emergency Management," *FEMA Emergency Management Institute*.

Geary, Joanna. (2013). "Introducing GuardianWitness, Our New Platform for Content You've Created," *The Guardian*, April 16, 2013.

Glaser, M. (2005). "Did London Bombings Turn Citizen Journalists' into Citizen Paparazzi?" *Online Journalism Review*. 12 July.

Guiver, J. and Jain, J. (2011). "Grounded: Impacts of and Insights from the Volcanic Ash Cloud Disruption," *Mobilities*, 6(1), 41–55.

Gutner, T. (2012). "Hurricane Sandy Grows to Largest Atlantic Tropical Storm Ever," *CBS Local, Boston*, 28 October 2012. http://boston.cbslocal.com/2012/10/28/hurricane-sandy-grows-to-largest-atlantic-tropical-storm-ever/

Heinzelman, J., and Waters, C. (2010). "Crowdsourcing Crisis Information in Disaster-Affected Haiti," *United States Institute of Peace*, Special Report 252, October.

Hermann, S. (2008). "Mumbai, Twitter and Live Updates," *BBC: The editors*, 4 December.

Hermida, A. (2007). "Citizen Journalists, the Media and the UK Terror Attacks," *Reportr.net*, 2 July.

Hughes, S. (2013). 'Social media newsgathering', *BBC College of Journalism*.

IFRC. (2013). "Beneficiary Communications", IFRC.

ITU (2005). "Handbook on Emergency Telecommunications," *International Telecommunications Union*, Geneva, Switzerland.

Kyodo (2012). "Google, Twitter to Mine internet Disaster Data," *Japan Times*, 13 September.

Lee, D. (2013) Boston bombing: How internet detectives got it very wrong, *BBC News—Technology*. http://www.bbc.co.uk/news/technology-22214511

Lindsay, B R. (2011). "Social Media and Disasters: Current Uses, Future Options, and Policy Considerations," *CRS Report for Congress*, Congressional Research Service.

Liu, S.B., Palen, L., Sutton, J., Hughes, A.L., and Vieweg, S. (2009). 'Citizen Photojournalism during Crisis Events'. In Allan, S., and Thorsen, E. (eds) *Citizen Journalism: Global Perspectives*, pp. 43–63. New York: Peter Lang Publishing, Inc.

Moses, Asher. (2011). "Web Acts as Virtual Crisis Centre for Christchurch Quake Victims." *The Sydney Morning Herald*, February 23, 2011.

Murray, A. (2011). "BBC Processes for Verifying Social Media Content", *BBC College of Journalism*, 18 May.

Novak, J. M. and Vidoloff, K. G. (2011). "New Frames on Crisis: Citizen Journalism Changing the Dynamics of Crisis Communication," *International Journal of Mass Emergencies and Disasters*, 29(3), 181–202.

OCHA. (2013). Humanitarianism in the Network Age: Including World Humanitarian Data and Trends 2012. Office for the Coordination of Humanitarian Affairs (OCHA).

Palen, L. (2008). "Online Social Media in Crisis Events," *EDUCAUSE Quarterly*, 31(3), 76–78.

Procopio, C. H., and Procopio, S.T. (2007). "Do You Know What It Means to Miss New Orleans? internet Communication, Geographic Community, and Social Capital in Crisis," *Journal of Applied Communication Research*, 35(1), 67–87.

Rayner, Gordon, and Alan Hall. (2009). "Germany Shootings: Gunman Shot Dead by Police After He Kills 16," *Telegraph.co.uk*, March 11, 2009.

Smith, D. (2012). "Mexico Introduces New Seismic Alert System for BlackBerry Phones," *AWARE*, 6 April.

"Syria Tracker", 2013, *Crowdmap*, Accessed April 17, 2013. https://syriatracker.crowdmap.com/.

Taylor, M., Wells, G., Howell, G., and Raphael, B. (2012). "The Role of Social Media as Psychological First Aid as a Support to Community Resilience Building," *The Australian Journal of Emergency Management*, 27(1), 20–26.

Tretbar, C. (2009). "Winnenden—Das erste Twitter-Ereignis in Deutschland" [Winnenden—the first twitter event in Germany], *Der Tagesspiegel*, 11 March.

Vieweg, S., Palen, S., Liu, S. B., Hughes, A. L., and Sutton, J. (2008). "Collective Intelligence in Disaster: Examination of the Phenomenon in the Aftermath of the 2007 Virginia Tech Shooting," *Proceedings of the 5th International ISCRAM Conference*, University of Colorado, Washington DC.

Vis, F. (2009). "Wikinews Reporting of Hurricane Katrina." In Allan, S., and Thorsen, E. (eds) *Citizen Journalism: Global Perspectives*, pp. 65–74. New York: Peter Lang Publishing, Inc.

Walton, C. (2012). "Social Media Training is Getting Results for the BBC," *BBC College of Journalism*, 25 January.

Watson, H. (2012). "Dependent Citizen Journalism and the Publicity of Terror." *Terrorism and Political Violence*, 24(3), 465–482.

Watson, H. and Finn, R.L. (2013). "Privacy and Ethical Implications of the Use of Social Media During a Volcanic Eruption: Some Initial Thoughts', *Proceedings of the 10th International ISCRAM Conference*, Baden-Baden, Germany, May.

Welsh, P. (2007). "Disaster Movies," *The Guardian*, 2 July.

Yardley, J. (2012). "Panic Seizes India as a Region's Strife Radiates," *The New York Times*, 17 August.

CHAPTER TWENTY-THREE

Citizen Journalism IN THE Age OF Weibo: The Shifang Environmental Protest

LEI GUO

Rapid advancement of global communication technologies has helped transform forms and practices of citizen journalism across the world. Chinese citizens have for some time been seeking to bypass state censorship by sharing their stories and thoughts through SMS (Short Message Services), blogs, BBSs (Bulletin Board Systems), and online discussion forums. In recent years, however, these have been surpassed by China's Twitter-like micro-blogging service, Weibo (微博), which has become one of the country's most popular platforms for disseminating citizen journalism.

In late 2009, Sina Corporation launched the first micro-blogging site in China, Sina Weibo. After that, several other companies such as Tencent and NetEase began to provide similar services. In just a few years, the micro-blogging medium has witnessed a number of significant examples of citizen journalism. For example, in 2011 a young woman named Guo Meimei, who claimed to be a "manager of the Red Cross Commerce," published a few Weibo posts to show off her wealth. The Weibo community soon responded with numerous posts questioning the relation between Guo and the Red Cross Society of China—the country's biggest charity organization, which helped lead to a yearlong investigation of the charity's alleged corruption and misuse of public donations. In addition to exposing scandals, Weibo has also provided a platform for ordinary citizens to share information and monitor government performances during accidents and natural disasters. In July 2012, for example, two high-speed trains collided in the eastern Chinese province of Zhejiang, killing at least 40 passengers. The story first broke on Weibo and millions of posts were published exchanging information on rescue operations, spreading sympathy for the victims, and urging the government to rigorously investigate the cause of the train crash.

From the so-called Red Cross China scandal, to the fatal high-speed train collision, and to the Shifang environmental protest to be analyzed in this study, Weibo serves as an alternative public sphere for ordinary Chinese citizens to report and discuss issues—particularly social conflicts that would otherwise be silenced. The public opinion on Weibo is so influential that even China's newly named Prime Minister Li Keqiang remarked on the social medium at one of his first work meetings:

> There are hundreds of millions of Weibo users in China. If the government does not communicate information in a timely manner, the public will become easily suspicious of the government's work, in turn creating a negative impact on the society (China News Service, 2013).

While some scholars and journalists have suggested that the revolution in China might be micro-blogged, others remain reserved about Weibo's potential to engender regime change (Hassid, 2012; Sullivan, 2012). With respect to the government's speech control, Weibo probably fares no better than other forms of online media. Indeed the popularity of Weibo in China is premised on the country's ban on Twitter and other foreign social media services. It is not surprising therefore to see localized micro-blogging sites in China closely adhering to government regulations. This begs the question of how Weibo-mediated citizen journalism differs from that based on other online media platforms in China? Moreover, what do people talk about on Weibo and to what extent can the platform facilitate free speech?

This chapter reviews citizen journalism in contemporary China and, in particular, Weibo-mediated citizen journalism based on a case study of the 2012 Shifang incident. In this case, thousands of residents in Shifang (什邡), a southwestern city in China, went to the streets to protest against a government-approved copper plant for fear of its potential environmental damage to the area. Not only did the citizen journalism on Weibo raise public awareness of the protest nationwide, but it also successfully pressured the local government to suspend the project. This chapter quantitatively and qualitatively analyzes a sample of Weibo posts about the Shifang protest and examines what and how people talked about the issue on Weibo.

CITIZEN JOURNALISM IN CHINA

China features a unique Party-State media system, in which the government maintains control of the fundamental press structure, and ordinary citizens are not allowed to launch their own alternative media in the traditional mediascape (Huang, 2003; Pan, 2000; Winfield & Peng, 2005; Zhao, 1998). In recent years, the emergence of diverse digital communication tools has helped partially loosen the controlled speech environment. A growing number of Chinese citizens began using SMS, blogs, BBSs, and more recently Weibo to break news, expose government wrongdoings, and share their thoughts on a variety of public issues. Moreover, the collective of China's citizen journalists serves as an independent media watchdog, effectively pressuring the mainstream news media to investigate and report on social controversies that would otherwise be censored or under-reported (Guo, 2012; Nip, 2009; Reese & Dai, 2009; Xin, 2010).

Still, the government's control over speech remains powerful in the realm of the internet. Laws and administrative regulations, together with a number of advanced technological measures, work efficiently to filter content distributed online. As a result, many online media platforms self-censor information such as contentious expressions and sensitive facts (MacKinnon, 2008; Tsui, 2003), posing a challenge to the development of citizen journalism in China. To circumvent internet censorship based on keyword detection, people have adopted a large number of coded words and metaphors (Xiao & Link, 2013). Moreover, a variety of new styles and genres of citizen-generated social critiques such as *E Gao* (i.e., video spoofs) have emerged in China that contest the dominant perspectives in an indirect and creative manner (Meng, 2011).

In addition to government control, citizen journalism in China is also limited in terms of the quality of the content. Based on her analysis of citizen journalism during the 2008 Sichuan earthquake, Nip (2009) suggests that most Chinese citizen journalists worked in an individual manner without affiliating to any group or organization, thus making it hard for them to provide coherent and credible information or investigative reporting of the disaster. In addition, most internet users were found to just share their sympathy for the victims, as opposed to reporting new information. Xin's (2010) overview of several citizen journalism practices also concludes that a good number of individuals express sentiments and sometimes even delivered hate speeches in the online environment. Here it should be noted that most studies on citizen journalism in China were conducted in the pre-Weibo era, and as such do not take account of the opportunities and challenges afforded by this new platform.

Weibo has risen rapidly to become one of the most popular online communication platforms in contemporary China. The latest CNNIC[1] statistical report shows that Chinese Weibo users exceeded 300 million by the end of 2012, accounting for more than half of the internet users in the country. The report also described Weibo as "the center for online public opinion" (CNNIC, 2013:36).

With respect to its features, Weibo functions as a combination of Twitter and Facebook. Like Twitter, Weibo allows its users to type up to 140 characters in a post. But 140 Chinese characters—compared with 140 English letters—enables Weibo users to write much more nuanced messages, thus making it easier to participate in online conversations (Sullivan, 2012). Weibo users can also attach images and videos in their posts, much like Facebook users. Furthermore, in this social medium, Weibo users can interact with one another by "following" other users, including organizations and celebrities, and by sharing and commenting on others' messages.

As noted above, all Weibo services are subject to government regulations and censorship requirements. However, information spreads more rapidly on Weibo than previous communication tools because of the large number of users on Weibo and the relative ease of reposting messages. Distribution of information on Weibo can therefore sometimes occur faster than censors can keep up (Sullivan, 2013). According to the chief editor at Sina Weibo, it became a "big headache" for the company to keep monitoring the content on their site (Ansfield, 2010).

This chapter will now turn to examine citizen journalism on Weibo during the 2012 Shifang incident. It will detail how Weibo users communicated with one another, and to what extent Weibo facilitated the dissemination of online speech.

THE 2012 SHIFANG INCIDENT

The southwest Chinese city of Shifang was among the most severely impacted areas during the 2008 Sichuan earthquake. On June 29, 2012, the municipal government of Shifang announced the construction of a copper smelting plant—one of the largest in the world—and claimed that the project would revitalize the city's economy after the quake. Fearing that the plant would cause environmental damage to the area, and also charging that the Shifang government failed to consult local residents on the project, thousands of people participated in the anti-pollution protest on July 1–3.

Participants and many other local residents used Weibo to report the protest via text, photos, and videos—showing scenes of police officers firing teargas and protesters being injured. As the conflict escalated, people nationwide helped distribute information about the protest and posted their comments. In response to the protest as well as to the numerous messages on Weibo, the Shifang government promised to suspend the project on July 2 and confirmed this on July 3.

The Shifang incident provides a particularly interesting case study of citizen journalism in China for at least two reasons. First, "mass incidents" such as the street protest in this Shifang case is a typical topic that internet censors would target. However, to many people's surprise, millions of messages about the protest were disseminated on Weibo without being removed. Second, the Shifang government halted the project just two days after the protest broke out, making them the fastest in the country to respond to citizen outcries.

My case study focuses on Sina Weibo, the first micro-blogging site and one of the most influential Weibo service providers to date. Figure 1 shows the number of Sina Weibo posts that mentioned "Shifang" between June 30 and August 15, 2012. On an average day before the protest, dozens or at most hundreds of Weibo posts can be found about Shifang—a very small and under-developed city. During the protest, however, millions of posts about Shifang were published on Weibo, and the number peaked around 2.3 million on July 3. Just a few days after the protest, the number of Weibo posts plunged to its normal level. Interestingly, tens of thousands of people posted about Shifang again between July 27 and July 30, when a similar environmental protest broke out in Qidong, another Chinese city located at the mouth of the Yangtze River. In fact, the Shifang incident became a milestone for Weibo-mediated citizen journalism with many subsequent Weibo discussions, especially those on social conflicts, referencing the case.

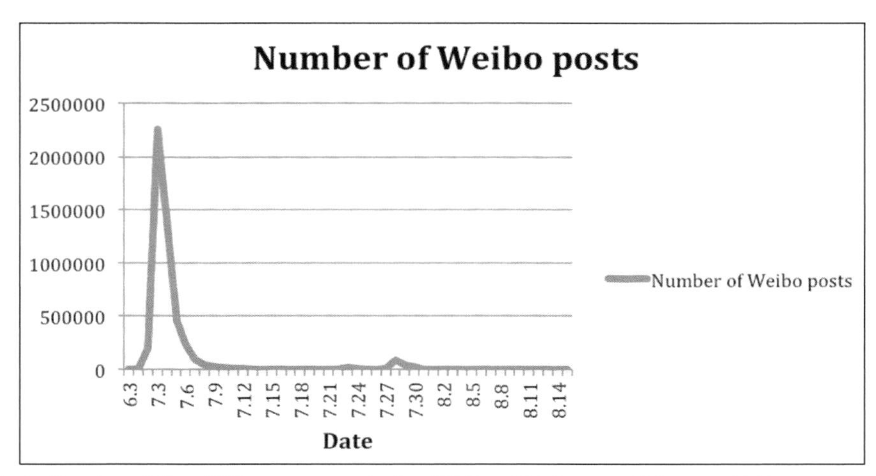

Fig 1. Number of Sina Weibo posts mentioning "Shifang,"30 June to 15 August, 2012

To produce a manageable sample and focus on the Weibo posts that specifically discussed the Shifang incident, the hashtag-keyword combination "#Shifang incident#" (i.e., "#什邡事件#") was used to search Sina Weibo's archive during the three protest days: July 1–3, 2012. Like on Twitter, people use hashtags to label their Weibo posts so that other users can easily identify relevant posts on the site. In other words, Weibo users who include hashtags in their posts intend to distribute information beyond their circle of friends and followers, and thus are more likely to perform citizen journalism.

The search retrieved a total of 293 posts containing "#Shifang incident#" published during July 1–3, 2012. Notably, the search result excluded "posts with similar content" and those that were not shown "according to the relevant laws and regulations."[2] As the first step of analysis, all of the 293 Weibo posts were coded and categorized into a list of reoccurring themes.[3] The results show that the majority of the identified Weibo posts were published via computer (72.7%), and the rest were published via smartphone (25.6%) or tablet (1.7%). Most were original posts that talked about the Shifang incident (76.8%). About one-fourth of the posts (24.2%) were reposts, half of which included user comments. An average post contained 55 characters. In other words, most Weibo users provided some information or commentary on the incident rather than merely sharing messages from others.

In addition to text, over one-fourth of the Weibo posts (25.2%) contained images or included images as a part of the repost. The majority of the images were scenes from the protest. For example, a group of images that went "viral" on Weibo showed a police officer chasing several protestors with a baton. Many other images focused on ordinary citizens who were injured in the protest. Compared with the number of posts with images, a much smaller number of Weibo posts—only three of them—included videos in my sample.

Twenty Weibo posts (6.8%) included a "long Weibo" (长微博)—a type of text attachment in the form of an image, which provides a way to post longer messages beyond Weibo's 140-character limit. Notably, nine out of the 20 posts were reposted from Han Han (韩寒), a well-known writer and commentator, who published two essays by attaching long Weibo in the posts during the protest. In these essays, Han discussed the Shifang incident in a satirical manner, relating the incident to the ongoing conflict between economic development and environmental protection in China. Because of Han's large influence—more than 16 million followers on Weibo, his posts played a significant role in spreading the news about Shifang. One of his essays "The Liberation of Shifang" was shared at least 298,173 times within one day (Qian, 2012). In addition to Han, several other well-known social critics published posts with long Weibo to participate in the online discussion of the Shifang incident. Unfortunately, many of these celebrities' posts, including Han's two essays, were removed from Weibo right after the posts became popular.

While the Weibo posts created by Han and other influential people led to a great number of reposts and comments, those posted by ordinary citizens were shared less and attracted fewer comments. The results show that in the sample a post was on average reposted twice, and generating only one comment.

With respect to government censorship, it turned out that Sina Weibo did not censor speech regarding the Shifang incident as extensively as they did other sensitive and contentious issues. Still, in my sample, 21 posts—including the aforementioned ones that reposted Han's essays—contained some content that was deleted, leaving a line that says "Sorry, this Weibo post is not appropriate for public viewing."

HOW WEIBO USERS FRAMED THE SHIFANG INCIDENT

Weibo users discussed the Shifang incident mainly from three perspectives—the government (32.1%), ordinary citizens (18.8%), and media and celebrities (17.7%) (see Table 1). In addition, quite a number of people (12.6%) provided comments beyond the single incident, and argued that the conflict reflected the country's social problems in general. Ironically, while the protest originated from the environmental concerns about a copper plant, only a total of nine posts (3.1%) perceived the issue as an environmental problem. Surprisingly few people on Weibo paid attention to China's environmental challenges and the problems of environmental policies as the incident reflected. In other words, most Weibo users were more concerned about how the government, citizens, and the media performed in this incident rather than the incident itself. The following section details how Weibo users framed the incident.

Table 1.

	Themes	N of posts	Percentage
Government (94, 32.1%)	1. Charged that the government failed to consult Shifang residents on the project beforehand	10	3.4%
	2. Criticized the government's use of violence to disband the crowd	68	24.2%
	3. Challenged the official *wei wen* (i.e., maintain stability) discourse	9	3.1%
	4. Contended that the government should be held accountable even after the protest	13	4.4%
Citizens (55, 18.8%)	5. Celebrated the power of collective actions	23	7.8%
	6. Published on Weibo to raise awareness	10	3.4%
	7. Expressed sympathies for the injured protestors	20	6.8%
	8. Suggested that the protest should be more rational	6	2.0%
Media (52, 17.7%)	9. Discussed how traditional news media covered the incident	14	4.8%
	10. Discussed how Weibo treated the incident	34	11.6%
	11. Discussed how celebrities treated the incident	7	2.4%
System (37, 12.6%)	12. The incident reflected the political and societal problem in general	37	12.6%
Environment (9, 3.1%)	13. The plant would cause environmental damages to Shifang	9	3.1%

Note:

1. For each Weibo post, we coded whether each of the 13 themes was present. Therefore, one Weibo post might include more than one theme.
2. Some Weibo posts included content that cannot be categorized into these 13 themes.

The Government

About one-third of the Weibo posts (94 posts, 32.1%) discussed the Shifang incident from the perspective of the government. Among them, ten posts charged that the Shifang government failed to inform the local residents about the construction of the copper plant. Some even suspected that the project was an "under the table deal," in which government officials possibly benefited financially from the developer of the plant. One Weibo user questioned the government:

> Was it really that hard to communicate with us before you decided to approve such a big project? Do you [the government] indeed serve us citizens or do you work for the companies? (piaobaoyi, 2012–7–3 02:16)

More attention was concentrated on how the local government dealt with the protest. A total of 68 posts—almost one-fourth of the sample—criticized the government for its use of violence against ordinary citizens who participated in the protest. Many argued that it was ironic to see the government spending taxpayers' money on teargas and grenades and then using the weapons against taxpayers. Others commented on the contrast between the government's "soft" foreign policies and its tough position against its own citizens in Shifang. For example, one Weibo user noted:

> After a series of incidents, such as the United States continuing to sell weapons to Taiwan and Japan's attempt to occupy our Diaoyu Island… our government finally realized that oral dispute couldn't solve any problems. So they finally resorted to the use of weapons—however this time the targets are their own citizens. They wanted to show off their power by using violence. (yangguangshibeishang, 2012–7–3 02:08)

In this post, the user supported his/her argument by attaching a photo that showed hundreds of police officers lined up with guns and shields. Critiques about the government's use of violence were prevalent on Weibo; this was also found to be the most salient theme in my sample.

Nine posts were found to question the government's discourse of *wei wen* ("维稳," i.e., maintaining social stability), an emphasized and repeated theme in government reports and addresses in recent years. Similar to how other governments on different levels responded to social conflicts, the Shifang government issued several public notices right after the protest broke out, advising citizens that their priority was to maintain social stability, and reminding the protestors that "a handful of people with ulterior motives might take advantage of the conflict to hurt the community." In fact, the saying of "a handful of people with ulterior motives" was not unfamiliar to many Weibo users. One challenged the discourse in a satirical way:

> …Whenever mass incidents occur, some mysterious people with "ulterior motives" always show up. Are there really that many people with "ulterior motives," or are there just too few government officials who have sincere motives to serve people? (Feiyiji, 2012–7–3 09:28)

In addition, in official discourse, "the handful of people with ulterior motives" often refers to foreign anti-China forces. Thus, another Weibo post mentioned the U.S. Embassy in China,[4] and asked:

> Would today's CCAV evening news refer you as "the handful of people" involved in the Shifang incident? (wubingsanguo, 2012–7–3 04:44)

As many other experienced internet users, this one used the coded expression "CCAV" to mock "CCTV",[5] the official state television broadcaster, which is widely charged for producing propaganda-like programs. In this post, the Weibo user also sought to be sarcastic by assuming that CCTV would frame the Shifang incident using a *wei wen* discourse in a similar fashion to that for other social conflicts.

In response to the protest as well as numerous Weibo posts, the Shifang government, using its official Sina Weibo account, posted its decision to suspend the plan for the copper plant on July 2 and July 3 consecutively. However, few Weibo users praised the government for its quick response. Instead, 13 posts in my sample argued that the government announcement should not be considered the ending point of the incident, and called on government officials to be accountable for the case. For example, one Weibo user raised several questions he/she felt the government should answer:

> … 1. The project was ended now, but is it temporary or permanent? 2. Who approved the project? What kind of responsibility should he/she take? 3. Who decided to use violence against protestors? (yipingkele, 2012–7–3 15:48)

However, the 13 posts only accounted for a small percentage of the posts. In fact, after the government announced the suspension of the project, Shifang-related discussion on Weibo and elsewhere faded away quickly and few people concerned themselves with the government's follow-up actions. As it turned out, no government official has been publicly held accountable since then.[6]

Ordinary Citizens

Another group of Weibo posts (55 posts, 18.8%) focused on the role ordinary citizens played in the Shifang incident. Twenty posts expressed sympathy for the protestors who were injured during the protest. One user said:

> After seeing images of those barehanded people who got injured, I really want to cry! We just don't want our homeland to be polluted, nothing more than that. Do we really ask too much?! (sibadasini, 2012–7–2 22:42)

In contrast to these emotional posts, others were more rational. Six posts called for non-violent resistance, suggesting that the citizens should protest in a peaceful and mature manner. Moreover, 23 posts celebrated the power of civic actions, similar to this Weibo user:

> …I would like to thank those who took the risk to safeguard the homeland. It is they who stood up. It is they who raised the civic awareness in China. (Stephanie_Chow_Is_Here, 2012–7–3 16:46)

Similarly, many people believed that the greatest achievement of the Shifang incident was the awakening of civic awareness among Chinese citizens particularly among the younger generation.

Media and Celebrities

An important part of the Shifang-related discussion (52 posts, 17.7%) focused on how the media and celebrities treated the incident. Among those, 14 posts blamed the traditional news media for being silent on the incident, or assumed that traditional media would "sing and dance to extol the good times" as always without even a mention of what happened in Shifang.

Not only did many Weibo users remain suspicious about the country's mainstream media, they were also concerned about the extent to which Weibo could help to spread the news about Shifang. In fact, the theme—the role of Weibo in this incident—was found to be the second most frequently discussed theme in my sample (34 posts, 11.6%). Five out of the 34 posts acknowledged that Weibo played an important role in informing the public. One post compared the traditional news media and Weibo, thus comparing the two generations:

> I just asked my dad—who was watching the news on TV—whether he has heard about the protest in Shifang, and it turned out he knew nothing about it. Because of Weibo, our new generation knows the truth ahead of others (shuangyuzhazhabujiaoqingQAQ, 2012-7-3 12:25).

On the other hand, many others criticized or assumed Weibo's censorship on this issue. For example, quite a few Weibo users questioned why Sina Weibo did not make the Shifang incident one of the hot topics on its front page even though millions of people were talking about the issue on the site. Others presumed that Weibo would remove their comments about Shifang sooner or later. One user criticized Weibo for being increasingly entertainment-oriented, and overlooking important social issues like the Shifang incident:

> Sina [Weibo], I am not sure if you received any pressure from the government. Why don't you include the Shifang incident as one of the hot topics? Isn't this topic hot enough? Isn't the topic about ordinary citizens a real topic? You always highlight news about celebrities being involved in love affairs or them breaking up. Today, however, you are silent about the incident that makes every Chinese citizen so angry. You will remove this Weibo, won't you? Then remove it. But people have their own sense of judgments. (zhaoyazhuo, 2012–7–3 05:27)

In light of ubiquitous online censorship practices, Chinese internet users are clear that the host media self-censors and deletes radical and contentious speech on their site. But many Weibo users still posted about the Shifang incident, hoping to disseminate news before it was removed. More interestingly, when it turned out that a good number of posts about Shifang remained on Weibo, many people became confused:

> I am really curious, very curious. Common sense tells us that Sina [Weibo] would remove posts extensively and direct people to the state media's official report. But how did this happen??? …Is this all part of a bigger scheme? (misitedingdingdang, 2012–7–3 05:00)

Some Weibo users were "shocked" to see so many posts about a massive protest, and even suspected that the government might be attempting to cover up a more serious problem.

In addition, some Weibo users also blamed celebrities for not being active in the online discussion of the Shifang incident. Due to the increasing popularity of Weibo among Chinese internet users, a growing number of celebrities including scholars, movie stars, pop singers, and successful entrepreneurs registered for Sina Weibo, and attracted a huge number of followers because of their real-life fan base. Many of them even became opinion leaders on Weibo because their posts have the potential to reach millions of people instantaneously. When it came to the Shifang incident, as mentioned earlier, Han Han and several other influential social critics contributed to the publicity of the case. However, many other celebrities, who are usually active on Weibo, chose not to share news about Shifang on their Weibo pages. Therefore, seven posts in my sample urged influential people to break the silence and help "get the word out."

System

A total of 37 Weibo posts provided comments beyond the single case of the Shifang incident. They analyzed different stakeholders in the incident—the government, companies and ordinary citizens, and argued that the protest reflected China's social conflicts in general. By way of example, one Weibo user analyzed the case from the lens of game theory:

> In today's "game," companies usually lead the game, and determine the progress of the projects. Governments can benefit from those projects because investment leads to the increase of GDP, tax, and employment. Ordinary citizens, however, are the group whose opinions are always unheard. How could such an unbalanced game bring about a harmonious society? (104dashixiong, 2012–7–3 14:49)

Like this post, many people contended that one major reason for the growing number of social conflicts in the country is the political system, and urged the central government to prioritize the citizens' needs over economic development.

In addition, while many Weibo users criticized the government or the mainstream media for not communicating with citizens in a timely and candid manner, others questioned what the truth is in current Chinese society:

> What is truth? Are what you've seen and heard really the truth? Are you really concerned about these people and issues? Or are you just gathering together to create a disturbance? In this era, the government is not credible, nor is anyone. What do we really need to reconsider? (AtrousNico, 2012–7–3 02:25)

As the Weibo discussion illustrates, a good number of people were concerned about the credibility of the government, the media system, as well as Chinese society as a whole.

CONCLUSION

Based on the analysis of the 2012 Shifang incident in which thousands of Shifang residents went to the street to protest against a government-approved copper plant for fear of its potential environmental damages to the area, the present study demonstrates how ordinary Chinese citizens reported and discussed the civic protest on Weibo. While the majority of mainstream news media in China remained silent on the incident, millions of Weibo posts drew nationwide attention to Shifang and successfully pressured the local government to suspend the copper plant project.

Indeed, Weibo as a new type of online media has brought China's citizen journalism to a new stage. First, compared with earlier communication technologies, this social medium reaches far more ordinary citizens, which account for about half of the internet users in China (CNNIC, 2013). Just one day after the protest broke out, millions of posts were published on Weibo, making it a heated issue nationwide almost immediately even though few traditional news media outlets covered it. In this process, celebrity users on Weibo such as Han Han served as important opinion leaders and helped draw public attention to the issue.

Second, Weibo saw more rational commentaries than pure sentiments. As the results show, just a small portion of people primarily expressed sympathy for the victims, with many more Weibo users sharing news about the protest and discussing the issue from various perspectives via text, photos, and long Weibo. The rationality might be explained by the fact that most Weibo users—like Twitter and Facebook users—also utilize the platform to network with their friends and colleagues in real life. Thus, unlike earlier online forums such as BBSs where many anonymous users would deliver hate speeches or even utter curses, most Weibo users in my sample kept a critical stance, but still conversed in a socially accepted discourse.

Last but not least, as this case illustrates, online censors appeared to relax their grip on public speech about social conflicts. This is a positive sign that Chinese leadership is becoming more tolerant of, and more responsive to the public's criticism—especially those on Weibo—provided that the criticism does not hurt the central government's legitimacy. Without a doubt, control over free speech remains powerful in China, but the unprecedented influence of Weibo is pushing the boundaries of what is acceptable to talk about in the public sphere.

Equally important, what was lacking in this study also provides critical insight into Weibo-mediated citizen journalism. In light of an environmental protest, few Weibo posts were found to discuss the issue from the angle of China's environmental problems. It was the street protest, and bloody conflict between the police and protestors that attracted most people's attention. This result is in line with the "protest paradigm" (McLeod & Hertog, 1999) that Western news media usually use to cover collective actions. Similar to those Western reporters who tend to focus more on spectacles or dramatic actions than the rationales behind the protests (Watkins, 2001), many citizen journalists on Weibo also neglected the underlying environmental problem that caused the protest. Therefore, though the project was suspended, the small victory might not be able to create a significant impact on China's environmental policy.

Further, in addition to the environmental problem, many other questions remained unanswered—e.g., whether the copper plant project would be resumed later, who approved the questionable project, who decided to use violence, and which government officials should be held accountable. However, the Weibo discussion on the Shifang incident was so short-lived that few Weibo users continued posting about Shifang soon after the protest. Again, it is true that the citizen journalism on Weibo helped stop a government-approved project, but it was not able to provide further information and stories behind the project or the protest, nor was it powerful enough to push the traditional news media to conduct investigative reports on the case. Indeed while I agree with one aforementioned Weibo user that the new generation in China has better access to truth in the age of Weibo, the citizen journalism on this social medium only uncovered part of the truth—perhaps only the tip of the iceberg.

NOTES

1. CNNIC, or the China Internet Network Information Center, is a government-authorized internet research organization.
2. The search was conducted on January 21, 2013. Sina Weibo filtered the search result, and made some sensitive posts hidden.

3. A Chinese-speaking graduate student and the chapter author independently coded 44 posts, 15% of the entire sample, in order to calculate intercoder reliability. The average intercoder reliability value was 0.94 (Cohen's kappa), with individual variables ranging from 0.70 to 1.00. We discussed all the discrepancies in the initial coding results and refined the codebook accordingly. The chapter author coded the remaining posts and closely read all the posts again for both quantitative and qualitative analysis.
4. Like in Twitter, Weibo user can mention other users by including an @ sign followed by the user's Weibo username.
5. CCTV is on the list of keywords that the government uses to censor internet traffic.
6. On July 5, 2012, in response to the massive protest, the Shifang government released a post on its Weibo page and announced the appointment of a new Party chief for the city of Shifang, whereas the city's current Party chief Li Chengjin was assigned to "assist" the new Party official. On October 30, the Shifang government announced that Li was removed from office (Xu & Zhao, 2012). However, in less than a month, Li appeared in a news story as the Deputy Secretary-general of Deyang, a prefecture-level city that administers Shifang. According to the news, Li's reassignment was a lateral transfer within the province's administrative structure. Li's responsibility in the Shifang incident was never publicly discussed.

REFERENCES

Ansfield, J. (2010, July 16). China tests new controls on Twitter-style services. *The New York Times*. Retrieved from http://www.nytimes.com/2010/07/17/world/asia/17beijing.html

China News Service. (2013, March 26). Li Keqiang: There are hundreds of millions of Weibo users. The government should communication information with the public timely. *Sina News*. Retrieved from http://news.sina.com.cn/c/2013–03–26/193226648402.shtml

CNNIC. (2013). *The 31st statistical report on the Internet development in China*. Retrieved from http://www.cnnic.cn/hlwfzyj/hlwxzbg/hlwtjbg/201301/P020130122600399530412.pdf

Guo, L. (2012). Collaborative efforts: An exploratory study on citizen media in China. *Global Media and Communication*, *8*(2), 135–155.

Hassid, J. (2012). The politics of China's emerging micro-blogs: Something new or more of the same? Presented at the APSA 2012 Annual Meeting.

Huang, C. (2003). Transitional media vs. normative theories: Schramm, Altschull, and China. *Journal of Communication*, *53*, 444–459.

MacKinnon, R. (2008). Flatter world and thicker walls? Blogs, censorship and civic discourse in China. *Public Choice*, *134*(1–2), 34–46.

McLeod, D. M., & Hertog, J. K. (1999). Social control, social change and the mass media's role in the regulation of protest groups. In D. Demers & K. Viswanath (Eds.), *Mass media, social control and social change: A macrosocial perspective* (pp. 305–330). Ames: Iowa State University Press.

Meng, B. (2011). From steamed bun to grass mud horse: E Gao as alternative political discourse on the Chinese Internet. *Global Media and Communication*, *7*, 33–51.

Nip, J. Y. M. (2009). Citizen journalism in China: The case of the Wenchuan earthquake. In S. Allan & E. Thorsen (Eds.), *Citizen journalism, global perspective* (pp. 95–106). New York: Peter Lang.

Pan, Z. (2000). Improvising reform activities: The changing reality of journalistic practice in China. In C. C. Lee (Ed.), *Power, money and media* (pp. 68–111). Evanston, Il: Northwestern University Press.

Qian, G. (2012, July 11). China's malformed media sphere. *China Media Project*. Retrieved from http://cmp.hku.hk/2012/07/11/25293/

Reese, S., & Dai, J. (2009). Citizen journalism in the global news arena: China's new media critics. In S. Allan & E. Thorsen (Eds.), *Citizen journalism, global perspective* (pp. 221–231). New York: Peter Lang.

Sullivan, J. (2012). A tale of two microblogs in China. *Media, Culture & Society*, *34*(6), 773–783.

Sullivan, Jonathan. (2013). China's Weibo: Is faster different? *New Media & Society*, 1–14.

Tong, Z., & Xiao, H. (2012, November 21). Former Shifang CPC Party Secretary Li Chengjin was reassigned as the Deputy Secretary-General of Deyang municipal government. *People's Daily Online*. Retrieved from http://politics.people.com.cn/n/2012/1121/c1001-19651422.html

Tsui, L. (2003). The panopticon as the antithesis of a space of freedom. Control and regulation of the Internet in China. *China Information*, *17*(2), 65–82.

Watkins, S.C. (2001). Framing protest: News media frames of the Million Man March. *Critical Studies in Media Communication, 18*(1), 83–101.

Winfield, B. H., & Peng, Z. (2005). Market or party controls?: Chinese media in transition. *Gazette*, *67*, 255–270.

Xiao, Q., & Link, P. (2013, January 4). In China's cyberspace, dissent speaks code. *The Wall Street Journal*. Retrieved from http://online.wsj.com/article/SB10001424127887323874204578219832868014140.html

Xin, X. (2010). The impact of "citizen journalism" on Chinese media and society. *Journalism Practice*, *4*, 333–344.

Xu, L., & Zhao, J. (2012, October 30). The Shifang CPC Party Secretary Li Chengjin was removed from office. *The Beijing News*. Retrieved from http://www.bjnews.com.cn/news/2012/10/30/230503.html

Zhao, Y. (1998). *Media, market and democracy in China: Between the party line and the bottom Line*. Urbana and Chicago: University of Illinois Press.

CHAPTER TWENTY-FOUR

Little Brother Is Watching: Citizen Video Journalists AND Witness Narratives

MARY ANGELA BOCK

The measure of a man is what he does with power.

—PLATO

On New Year's Eve, 2012, a U.S. war veteran stopped at a convenience store in time to witness police arresting a woman. The veteran, who'll be referred to as "BC" in this chapter, thought the behavior of the police to be unnecessarily rough and bullying, and so he started to videotape the arrest with his smartphone. Police officers demanded that he stop, the altercation escalated, and BC was himself arrested. What neither BC, nor police, knew at the time was that across the street, another citizen was filming the scene. That video turned the incident into a significant local news story. Months later, the most serious charges against BC were dropped.

Reaching for one's smartphone during an emergency is quickly becoming the norm for those living in the digital age. Cellphone videos often provide the only documentation of sudden disasters and crimes, as with the bombings of the London subway in 2007 or the mass shooting on the campus of Virginia Tech in 2007. As tools in the hands of citizen journalists and activists (and individuals who claim both hats) smartphones videos are challenging traditional borders between journalism, source routines, and the public. One phenomenon in particular, the so-called cop-watch movement, offers a site of study that demonstrates how such video both complicates and simplifies the discourses of authority. Cop watchers are often drawn into their activities by happenstance (as was BC) when they become unintentionally involved with police and try to document the incident with video, only to find that their camera exacerbates the situation. Others are activists who've added police-recording to their repertoire as a form of protest defense. Still others say they've joined in order to monitor what they believe is an unfair, oppressive system.

This chapter blends several years of research on video journalism in a variety of contexts with data collected in 2013 from activists involved in cop-watching, a loose term that describes the concerted and deliberate videotaping of police activity in public places. One organization in particular (which here shall be called the Safe Streets Project, or SSP) serves as a case study for understanding how cop-watching combines the work of journalism, activism, and filmmaking. The project also draws material from similar organizations, such as *copblock.org* and *photographyisnotacrime.com,* as well as a general multi-week training class for citizen video journalists called Our City Our Voices (OCOV). With this collection of observations, interviews and videos, it becomes possible to better understand how citizen video fits into the changing journalistic landscape.

VIDEO AND CITIZEN JOURNALISM

Video technologies have been small enough for one person to shoot, write and edit video since before the turn of the recent century. Now with a smart phone, most people can do everything, even broadcast a live shot, with a one piece of equipment they already own. True, editing video in a phone is not as sophisticated as with a full-fledged system, but simple story structures are the mainstay of breaking news. When coupled with the internet, wireless capability and social networks, smart phones allow for unprecedented participation in visual news discourse.

Proponents of citizen journalism point to its victories as monitor of the mainstream press, citing examples of the *Little Green Footballs* blog to dismantle a Dan Rather story on George W. Bush's military experience; or coverage by *TPM* (Talking Points Memo) of then-U.S. Senate Majority leader Trent Lott singing the praises of segregationist Strom Thurmond in 2002. Critics charge that word-based citizen journalism, which often appears on blogs or aggregate websites, often lacks in original reporting.

Citizen journalism *video*, however, is evolving via different trajectories. One root is in activism, through organizations such as the IMC (Independent Media Centers), which started after the "Battle of Seattle" riots in 1999 with the slogan "Don't hate the media, be the media." Another path can be traced back to what television stations once called "viewer video," the precursor to J.D. Lasica's "random acts of journalism" (2003). Many news organizations online now encourage non-professionals to submit video that they obtain when they happen to be in the vicinity of remarkable events, extreme weather, or tragedy. The difference between these two basic paths of citizen journalism is a matter of routine. While "random acts" citizen video journalism operate outside of the so-called "news factory" (Bantz, McCorkle, & Baade, 1980), IMC-style organizations may offer some support, usually in terms of social capitol, for institutional routinization (Ananny & Stohecker, 2002).

Cop-watching exemplifies this sort of collective. This emerging branch of citizen video production asserts itself as a check on police authority. Participants post videos of traffic stops, public arrests, or other publicly visible police activity to websites that contextualize them as police accountability activism. Some participants become involved as part of activist efforts. Others become "accidental" cop watchers by using their smartphone during an emergency, or, in many cases, during their own encounters with police. The latter often adopt activist routines.

In spite of First Amendment protection from the U.S. Constitution and a Justice Department memo (Smith, 2012) that explicitly reminds police departments that citizens have the right to tape police activity in public places, such work is often contested and resisted by law enforcement. One cop-watching blog is devoted almost entirely to videos of police officers demanding that a photographer stop filming, or grabbing at cameras. Still, no matter what is captured on tape, the very act of photographing officers at work is considered by activists to be an effective preventative measure, protecting the rights of people during a police encounter. It purports to provide a counterpoint to traditional journalism's coverage of police activity, which, as many scholars have reported, is often institutionally aligned with the interests of the powerful (Cook, 2005; Sigal, 1973; Tuchman, 1978).

Professional journalists are rarely willing or truly able to contend with the notion that their source networks constitute a form of bias. Journalism scholars have repeatedly noted that a defining characteristic of what could be seen as a professional "ideology" is an ethic of objectivity (Hanitzsh et al., 2010; Mindich, 1998; Schudson, 2001). This blind spot can interfere with a second key aspect of journalistic ideology the ethic of public service. The conflict between the two can be credited, at least in part, as fuel for the citizen journalism movement. A third dimension of journalistic work was identified by Tuchman (1973, 1978) as the routinization of the unexpected. While questions regarding the expressed subjectivity of citizen journalism have attracted considerable attention, the "routinization" dimension has not, perhaps because on its face it seems to be the less politically ideological characteristic. Yet as numerous newsroom ethnographies have shown, the results of routinization are themselves ideological (Boczkowski, 2004; Gans, 1979; Gitlin, 1980; Tuchman, 1973).

Community video websites do not necessarily use balanced language or make claims to journalistic objectivity. They do make claims to public service and truth, however, and purport to give voice to new and discernible points of view (Coffman, 2009). Without the institutional ideology of professional journalism, citizen activists must forge their own path for asserting truth-telling authority.

AUTHORITY IN VIDEO NARRATIVE

Authority, the power to claim, name or present something as "true," is largely a discursive activity, and journalism is a specialized form. Max Weber's (1947) three sources of authority; rational (rooted in law or social sanction), traditional (rooted in established social beliefs), and charismatic (derived from exemplary behavior or heroic acts on the part of an individual) can each be used to describe traditional journalism practices (Carlson, 2006). Citizen VJs usually work outside of conventional institutions and cannot claim rational authority. In fact, their activities often challenge these rational sources of authority. Because they often present themselves as a counterweight to mainstream media, they are also unable to draw upon Weber's notion of "traditional" authority. Charismatic authority, resting in the individual, is all that remains. Coupled with an ancient faith in the eyewitness and a camera's technological wizardry, Weber's notion of charisma imbues citizen video journalism with its own, unique, discourse of truth-telling power.

Routine decisions for storytelling, including fact selection, the use of quotes and sound-bites, the use of a recorded vocal track and the form of literary address are all discursive strategies for representing the authority of a creator (Zelizer, 1990, Allan, 1998, Chatman, 1978, Hall, 1973, Knobloch et al., 2004, Montgomery, 2006, Raymond, 2000, van Dijk, 1985). Video stories blend image, sound and script, and unfold in real time, and therefore require some sort of narrative to be understood (Bird & Dardenne, 1987). Various genres of non-fiction film have clustered around particular narrative styles. Conventional TV journalism has tended to use a correspondent "voice of God" recorded track with a declarative literary address (epitomized by Cronkite's iconic "That's the way it is") (Gitlin, 1980, Molotch and Lester, 1974, Glasgow University Media Group, 1976, Cook, 2005). Documentary filmmakers and still-photographers-turned VJ are often partial to a more observational style reminiscent of cinema verité, blending sounds and scenes without an added commentator (Bock, 2012; Nichols, 1991).

Digital media provide a nearly infinite set of possibilities for structuring a video narrative. Beyond their presentation in the context of a citizen journalism website, three creative choices seem key to the construction of authority *within* a video: The first is the mode of address, or literary voice. Secondly, a producer must choose to be known or unknown to the audience. Finally, a video creator can choose whether, and how, to incorporate testimonial, that is the discourse of witnessing.

Modes of Address

By blending word and image, filmic storytelling, at its most basic, combines showing and telling; roughly akin to what rhetoricians call diegesis and mimesis. Film scholar Bill Nichols (1991) organized documentary according to a continuum between the two, with "observational" documentary representing the most mimetic, as it operates without a seen or heard narrator, and "expository" documentary, which places authority in a narrator's vocal presence, at the other extreme (Nichols, 1991, 2001) TV news generally uses the more diegetic form, which Peter Dahlgren (1987) describes as:

> ...matter of factual and self-assured, with little or no trace of self-doubt, emotionality or uncertainty about the material it presents. It conveys seriousness, and where appropriate, urgency and even light touches of irony. News talk is confident talk, secure in its professionalism. (p. 42)

In contrast, Nichols' mimetic, "observationalist style" has no voiced-over narrative, no formal interviews with subjects, no direct address to the audience, and it utilizes editing that emphasizes real time and spatial realism. It invites the audience to "work" to interpret the story. Some documentary filmmakers argue that this form is *more truthful* in that it locates authority in the scenes and subjects recorded, to be interpreted by the viewer without mediation by a vocal narrator (Stoller, 1992, Rouch, 2003).[1] Time constraints, which beget routines, demand the efficiency of the diegetic/declarative format, which invites attack from cultural critics who contend that TV news is a poor representation of events, while mimetic, long-form documentary claims a more "serious" status because the stories can "speak for themselves." Citizen video journalism generally and cop-watching video more specifically is likely to take on the latter mode, given that their authority is grounded in Weber's third "charismatic" pillar of the individual eyewitness, not an institutionalized profession.

Author Identity

Writing a story allows for myriad choices for establishing the identity of an author; video production multiplies them. Such choices change the audience's perception of the identity of a narrator and author: sometimes they are one in the same, sometimes they are not; sometimes the author is revealed, sometimes not (Chatman, 1978, 1990). For instance, authorship in conventional TV news was once a team effort, wherein a reporter might voice a script written by a producer, and an anchor might read a script over video shot by a photographer in the field who remains unseen. By hiding most members of the team, particularly photographers, a narrative asserts rational authority.

Because they operate outside news institutions, citizen video journalists must make themselves known in order to incorporate Weberian charisma. While journalists sometimes use the concept of witnessing rather loosely, it is the authoritative anchor for those who would claim the visual truth of a bad arrest, an improper traffic stop, or a sudden disaster. Such testimonials constitute the purest form of witnessing, as conceptualized by Peters (2001), in that recording an image requires corporeal work, that is, bodily presence. One simply has to, quite literally, *be there,* and the author's presence must be clear in order to effectively make a witnessing claim.

Discourses of Witnessing

The role of the eyewitness has been held sacred by society for millennia and it remains a core value for journalism (Peters, 2001; Zelizer, 1990, 2007). Conventional journalism, however, applies the term even to situations that do not necessarily involve direct human observation (Zelizer, 2007). Other scholars of journalistic practice have noted that the norms of objectivity, coupled with the routinization of events, could trump eye-witnessing (Molotch & Lester, 1974; Tuchman, 1978). Contemporary newsgathering often relies on second-order witnessing, with journalists relying on the witnessing of others, viewing speeches on government-controlled video feeds or depending on material provided by sources (Bock, 2012; Cook, 2005; Hess, 1981; Sigal, 1973; Tuchman, 1978). Even directly viewed phenomena might go unreported in favor of what a trusted source declares to be important.

COP WATCHING: ROUTINES AND NARRATIVES

While citizen journalists might reject mainstream claims to objectivity, the groups observed for this project have adopted loosely structured routines akin to a form of newsgathering. SSP maintains its own website and communicates its work through social media. Members go out regularly Saturday nights to monitor police activity in the city's nightclub district. The group starts by walking the district and recording video of every officer on patrol in order to document who is on duty. Then they walk and wait for arrests, running to any location where police converge. Most SSP members carry a smart phone; one volunteer uses a larger, more sophisticated and visible video camera. When arrests occur, members of the group surround the situation to record it from multiple angles. A young man who formerly volunteered with local socialists says he believes that the regular cop watching patrols have changed police behavior in this section of town. In an interview, organizer BC explained that he's come to realize that some of SSP's activities could be considered "journalism" because while the work is ideological, it is intended as a form of truthful witnessing:

> ...when we do videotape cops, we are intending to sit there and capture what's going on. And we will give our own narrative, as we're filming, but we're not trying to influence the scene other than letting the cops know that we're there so that they modify their behavior so that they dampen any aggressive behavior that they may have.

Similarly, the leader of a Virginia-based network of cop watching makes a point of announcing that he is there to hold everyone accountable, making sure to assert his right to film but making it clear he is recognizing the police perimeter. Cop-watchers often make proclamations on location that they are there to "protect everyone." Another cop-watcher, DN notes that cameras become tools for balancing power:

> ...cops have always been able to create their own truths when they make an arrest... And they become very accustomed to it they go into court and they testify in their uniforms, and they get the benefit of the doubt because you're supposed to believe cops. But well the camera adds a dimension, where, so it's what really happened, not just what the cop says happened.

While a smartphone camera is crucial to such activism, another cop-watcher, AV, believes that the power of social networking—for distributing information—is the most essential component of his group's success:

> You needed the 5 o'clock news back in the day to get a story out. Now anybody can be the reporter the journalist and with no swing or bias in any way. It's like, turn the camera on here we are my actions, your actions, and you bring everybody to that scenario.

A careful look at cop-watching videos reveals the narrative strategies employed by activists to establish discursive authority. In contrast to mainstream television news narratives, these strategies include:

1. A more fluid sense of literary address
2. Making the creator known through script and/or image.
3. Explicit witnessing testimonies

A More Fluid Form of Address

Most clips posted by SSP have no additional track and are presented "as is," in the mimetic, documentary tradition. They are only contextualized with words to caption and tag the arrests presented, usually without the sourcing used by conventional journalists (from elites) and often with what would be considered a subjective, anti-police tone. Similarly, a story about gun control from the Philadelphia *IndyMedia* project (OCOV) begins with an unseen narrator saying, "There ought to be a law," without any pretense of objectivity. Workshop participations took turns performing in the story, with one providing an expository voice track, then joined midway by other members of the team in vocally eliciting questions from an interviewee before the entire group appeared on camera chanting, "Keep the guns off the street."

This mix of performative and observational style does not assume the consistent declarative voice of professional journalism. It commands the authority of the camera and its operator, pulling in the audience as a secondary witness. In another example, a video posted to a cop-watching site starts with a woman's voice addressing an unknown audience as she goes to answer the door to police, saying that she's already told the police her name, she's barely dressed, and she doesn't want to do this. She puts the camera down, and police are seen entering her home without a warrant and demanding her identification without explanation. The video ends when, apparently, while the woman is in another room looking for her ID, someone (presumably an officer) finds the phone and moves it to face a blank wall. The video's headline states "Papers Please Incident makes cops look like Gestapo" in the context of a cop-watching website.

While traditional journalism declares, citizen video argues, using text and headline as proposition, video as evidence. This discursive strategy is found throughout the cop-watching and activist journalism sites. Because traditional journalism claims the norm of objectivity, the discourse of news relies on video less as evidence and more as illustration. Of course, television news often does invite viewers to "see for themselves," but such an invitation, literal or implied, is *essential* to citizen video journalism.

Creators Seen and Unseen

Conventional journalism often makes objective truth claims about images by occluding the identity of photographers. Images and video are presented as truthful representations of scenes made possibly by the camera's technical perfection, while their creators remain behind the scenes. In the professional realm, these images are generally well composed, properly lit, and artfully edited, which often causes an audience to forget about camera technique. But for citizen VJs, especially in crisis mode, part of their authority is rooted in their persona—the essence, really, of Weber's notion of charismatic authority. Citizen authors must be known, seen or heard if they are to testify to their truth-telling.

Video from cop watching is often dark and occasionally hard to interpret. Sometimes SSP volunteers will record their own voice in real time to narrate what is happening in front of their cameras. Sometimes, when the group streams live, they will provide a running, vocal commentary as they patrol the nightclub district. There is no separation of "correspondent" and "photographer" roles; everyone is a witness, everyone is a narrator; everyone is heard and known, though not necessarily seen since they are more intent on capturing the police activity.

Explicit Witnessing

Traditional journalistic narrative blends testimonies of witnessing with the language of objectivity and relies heavily on powerful sources. Reporters are expected to observe, not participate; to report, not feel. They are taught in journalism school to attribute observations to others—often relying upon the observations of traditional authority figures such as police, politicians and bureaucrats. Citizen journalism videos are more likely to rely on the testimonies of non-elites.

By its very nature, cop watching video questions traditional elite authority and attempts to use camera technology as a counter-balance.[2] Videos on cop-watching sites are often dark, so dark that it is very difficult to discern what is happening without additional context. The sites usually provide captions and claims about what the video depicts. A half-hour video posted by SSP to YouTube goes so far as to provide a timeline with its version of a drunk-driving traffic stop:

> 1:58 The driver of the vehicle repeatedly asks if he has to submit to the cop's tests, and the cop continues to try to coerce him to conduct the test.
>
> 2:53 The driver of the vehicle requests that we continue to film, while Stalker cop *(the site names the officer)* threatens PSP with arrest for exercising our 1st Amendment rights. The driver continues to flex his rights.
>
> ... *(and so on)*

Unlike the previous two characteristics of citizen VJ narrative (the looser conception of "voice" and the opportunistic revelations of a story's creator), the dimension of subjective witnessing seems to be more deliberate. Citizen VJs, it seems, not only desire to create narratives that contrast with the mainstream, they desire to reveal how their personal experience contrasts with what they see in the mainstream.

CONCLUSION

Cop-watching videos are by nature one-sided, and this chapter has not contended with yet another genre, namely videos posted by police officers, nor with video created by police car cash or body cameras. One public relations officer for a mid-size city pointed out the complexity of today's digital environment:

> ...there are sort of eyes on everyone all the time. We have in-court videos going all the time there are uh, every bank, every mall everywhere you go you are on some type of film, usually.

> Technology has just changed the world we live in and we just sort have to understand and respect that a little bit.

As a purposeful, organized documentary activity, cop-watching groups offer a useful site for studying the way citizen video journalists and activists establish their authority in visual narrative. Without the rational and traditional authority identified by Weber, cop-watchers rely more on charismatic authority rooted in their individual experience, and this was reflected in their narrative strategies. Three characteristics of citizen VJ narrative can be identified: a more fluid conception of narrative voice; opportunistic revelations of a story's creator, and greater reliance on the language of witnessing over claims from elite sources.

This chapter could have been devoted to arguing whether or not citizen video journalism generally or cop-watching specifically should be considered "journalism." But that may no longer be the most interesting question, as opportunities continue to multiply for anyone who wants to contribute to the public sphere. It seems more useful to understand how various enterprises establish their discursive authority. Focusing on these strategies may be more useful for our understanding of today's quickly-evolving media environment.

NOTES

1. Filmmakers in the *Cinema Verité* or *Realist Cinema* tradition tended not to make objective/realist truth claims; however, they only made claims to a form of truthfulness that lets the audience do more interpreting than the filmmaker.
2. Even so, their videos are not always accepted by elite authorities as arbiters of truth. In one northeastern U.S. case, a judge simply refused to watch a video that depicted an officer punching a woman, relying instead on witness testimony to exonerate the officer.

REFERENCES

Aitken, I. (1992). *Film and Reform: John Grierson and the Documentary Film Movement* London, New York Routledge.

Aitken, I. (Ed.) (1998). *The Documentary Film Movement: An Anthology,* Edinburgh Edinburgh University Press.

Allan, S. (1998). News from NowHere: Televisual News and the Construction of Hegemony In Bell, A. & Garrett, P. (Eds.) *Approaches to Media Discourse,* Oxford, Blackwell.

Ananny, M., & Stohecker, C. (2002). Sustained, Open Dialogue with Citizen Photojournalism *Development by Design Conference.* Bangalore, India, online.

Bantz, C. R., McCorkle, S., & Baade, R. (1980). The News Factory. *Communication Research,* 7(1), 45–68.

Barnhurst, K. G., & Nerone, J. (2001). *The Form of News,* New York, The Guilford Press.

Barnouw, E. (1974). *Documentary: A History of the Non-Fiction Film,* Oxford, Oxford University Press.

Bennett, W. L., Gresett, L. A., & Haltom, W. (1985). Repairing the News: A Case Study of the News Paradigm *Journal of Communication* 35, 50–68.

Bird, S. E., & Dardenne, R. W. (1987). Myth, Chronicle, and Story: Exploring the Narrative Qualities of News. In Carey, J. W. (Ed.) *Media, Myths and Narratives: Television and the Press,* Newbury Park, CA, Sage Publications

Boczkowski, P. (2004). *Digitizing the News: Innovation in Online Newspapers.* (W. E. Bijker, W. B. Carlson, & T. Pinch, Eds.). Cambridge, MA: The MIT Press.

Bock, M. A. (2012). *Video Journalism: Beyond the One Man Band.* New York, NY: Peter Lang.

Carlebach, M. L. (1997). *American Photojournalism Comes of Age,* Washington, D.C., Smithsonian Institution

Carlson, M. (2006). War Journalism and the "KIA Journalist": The Cases of David Bloom and Michael Kelly *Critical Studies in Mass Communication,* 23, 91–111.

Carlson, M. (2007). Blogs and Journalistic Authority. *Journalism Studies,* 8, 264–279.

Chatman, S. (1978). *Story and Discourse: Narrative Structure in Fiction and Film,* Ithaca, NY, Cornell University Press

Chatman, S. (1990). *Coming to Terms: The Rhetoric of Narrative in Fiction and Film,* Ithaca, Cornell University Press

Coffman, E. (2009). Documentary and Collaboration: Placing the Camera in the Community. *Journal of Film and Video,* 61, 62–78.

Cook, T. E. (2005). *Governing the News: The News Media as a Political Institution,* Chicago, University of Chicago Press

Dahlgren, P. (1987). Tuning in the News: TV Journalism and the Process of Ideation In Vidal-Beneyto, J. & Dahlgren, P. (Eds.) *The Focused Screen.* Strasbourg, Amela, Council of Europe

Deuze, M. (2005). What Is Journalism? *Journalism,* 6, 442–464.

Ellis, J. (1992). *Seeing Things: Television in the Age of Uncertainty,* London, I.B. Tauris.

Flint, J., & James, M. (2009). Current TV to Shift from Video Format. *LAtimes.com.*

Gans, H. (1979). *Deciding What's News: A Study of CBS Evening News, NBC Nightly News, Newsweek and Time, 25th Anniversary Edition.* New York, NY: Pantheon Books (Random House).

Garcelon, M. (2006). The 'Indymedia' Experiment: The Internet as Movement Facilitator Against Institutional Control. *Convergence,* 12, 55–82.

Gillmor, D. (2004). *We the Media: Grassroots Journalism by the People, for the People* Sabastopol, CA, O'Reilly Media

Gitlin, T. (1980). *The Whole World Is Watching: Mass Media in the Making and Unmaking of the New Left,* Berkeley, CA, University of California Press.

Glasgow University Media Group (1976). *Bad News,* London, Routledge & Kegan Paul.

Hall, S. (1973). Encoding and Decoding in the Television Discourse. *Stencilled Paper 7,* University of Birmingham, CCCS.

Hanitzsh, T. F. H., Mellado, C., Anikina, M., Berganza, R., Cangoz, I., Coman, M., … Yuen, E. K. W. (2010). Mapping Journalism Cultures Across Nations. *Journalism Studies,* 12(3), 273–293.

Hartley, J. (1982). *Understanding News,* London, Metheun.

Hess, S. (1981). Washington Reporters. *Society* 18, 55–66.

Knobloch, S., Patzig, G., Mende, A.-M. & Hastall, M. (2004). Affective News: Effects of Discourse Structure in Narratives on Suspense, Curiousity and Enjoyment While Reading News and Novels. *Communication Research* 31, 259–287.

Kracauer, S. (1947). *From Caligari to Hitler: A Psychological History of the German Film,* Princeton, NJ, Princeton University Press.

Lang, T. (2004). The Longer View.

Massing, M. (2009). Out of Focus *Columbia Journalism Review.*

Mcmanus, J. (1994). *Market-Driven Journalism: Let The Citizen Beware?,* Thousand Oaks, CA, Sage Publications.

Mindich, D. T. Z. (1998). *Just the Facts: How Objectivity Came to Define American Journalism.* New York: New York University Press.

Molotch, H., & Lester, M. (1974). News as Purposive Behaviour: On the Strategic Use of Routine Events, Accidents and Scandals. *American Sociological Review,* 39, pp. 101–112.

Montgomery, M. (2006). Broadcast News, the Live "Two-Way" and the Case of Andrew Gilligan. *Media Culture & Society,* 28, 233–259.

Nichols, B. (1991). *Representing Reality: Issues and Concepts in Documentary,* Bloomington, IN, Indiana University Press.

Nichols, B. (2001). *Introduction to Documentary,* Bloomington, IN, Indiana University Press.

Parr, B. (2005). Things I Wish I'd Known Before I Became a Citizen Journalist. *Neiman Reports.*

Peters, J. D. (2001). Witnessing. *Media, Culture & Society* 23, 707–723.

Platon, S., & Deuze, M. (2003). Indymedia Journalism. *Journalism,* 4, 336–355.

Raymond, G. (2000). The voice of authority: The local accomplishment of authoritative discourse inlive news broadcasts. *Discourse Studies* 2, 354–379.

Rosen, J. (1999). *What Are Journalists For?,* New Haven, CT, Yale University Press.

Rosen, J. (2000). Questions and Answers about Public Journalism. *Journalism Studies* 1, 679–694.

Rouch, J. (2003). *Cine-Ethnography* Minneapolis, University of Minnesota Press.

Schwartz, D. (1992). To Tell the Truth: Codes of Objectivity in Journalism. *Communication,* 13, 95–109.

Schwartz, D. (1999). Objective Representation: Photographs as Facts. In Brennen, B. & Hardt, H. (Eds.) *Picturing the Past,* Urbana, University of Illinois Press.

Schudson, M. (2001). The Objectivity Norm in American Journalism. *Journalism,* 2(2), 149–170.

Sigal, L. (1973). *Reporters & Officials: The Organization and Politics of Newsmaking,* Lexington, MA, D.C. Heath & Co.

Singer, J. B. (2003). Who are these guys? *Journalism,* 4, 139–163.

Smith, J. (2012, May 14). Christopher Sharp v. Baltimore City Police Department et. al. US Department of Justice document 207-35-10.

Sontag, S. (2003). *Regarding the Pain of Others,* New York, Farrar, Staus and Giroux.

Sontag, S. (2004). Regarding the Torture of Others. *The New York Times Magazine.*

Stoller, P. (1992). *The Cinematic Griot,* Chicago, University of Chicago Press.

Taylor, J. (1998). *Body Horror,* Manchester, UK, Manchester University Press.

Tuchman, G. (1978). *Making News: A Study in the Construction of Reality,* London, The Free Press, A Division of Macmillan Publishing Co. Inc.

Tuchman, G. (1973). Making News by Doing Work: Routinizing the Unexpected. *The American Journal of Sociology,* 79(1), 110–131.

Van Dijk, T. (1985). Structures of News in the Press. In Van Dijk, T. (Ed.) *Discourse and Communication: New Approaches to the Analysis of Mass Media Discourse and Communication.* Berlin, deGruyter.

Vaughn, D. (1995). The Man with the Movie Camera. In Jacobs, L. (Ed.) *The Documentary Tradition: from Nanook to Woodstock* New York, Hopkinson and Blake.

Weber, M. (1947). *The Theory of Social and Economic Organization,* New York, Free Press.

Zelizer, B. (1990). Achieving Journalistic Authority Through Narrative. *Critical Studies in Mass Communication* 7, 366–376.

Zelizer, B. (2007). On "Having Been There": Eyewitnessing as a Journalistic Key Word. *Critical Studies in Media Communication,* 24, 408.

CHAPTER TWENTY-FIVE

Occupy Wall Street AND Social Media News Sharing AFTER THE Wake OF Institutional Journalism

KEVIN MICHAEL DELUCA & SEAN LAWSON

The internet has upended everything. The sudden and endlessly proliferating platforms of social media have only intensified the effects. One institution that has been upended is journalism. Although people still debate institutional journalism's future, it is already dead, part of the many institutions and practices rendered archaic by the internet. If institutional journalism is still around, it is on the cusp of huskness, a form whose force has dissipated, a zombie journalism kept alive only by attempts to cannibalize emerging forms of social media news sharing. In noting the death of institutional journalism, we are not celebrating it. We love newspapers and news. The death of institutional journalism is not the same thing as the death of news. The sharing of news will go on, indeed, it may be intensified by the sharing of news on multiple social media platforms. As we will be suggesting, the death of institutional journalism is not the death of a public service vital to democracy, but the death of a particular form of institutional knowledge commodification for the sake of power and profit. It has already been replaced in many ways by the multiple emergences of social media news sharing.

In this chapter we will start by deconstructing the unfortunate dynamic of the debate about zombie institutional journalism versus social media news sharing. We understand that the conventional terms used are "journalism" and "citizen journalism," but we think those terms are part of the problem. In deconstructing and displacing this debate, we hope to clear the ground for starting to imagine the different practices and politics that social media news sharing makes possible.

Part of this imagining requires acknowledging the transformative possibilities of new technologies. Although not advocating technological determinism, we are also not adopting technological idiocy. We will explain our position briefly, before exploring the possibilities of social media news sharing through the example of Occupy Wall Street (OWS).

SMASHING IDOLS

"Where to begin in philosophy has always—rightly—been regarded as a very delicate problem, for beginning means eliminating all presuppositions" (Deleuze, 1995:129). So begins "The Image of Thought," Chapter III of Deleuze's key book, *Difference and Repetition*. To repeat, for Deleuze, to begin to think "means eliminating all presuppositions." His work is instructive for us because if we want to think something new we must follow Deleuze's insistence that we must start with abandoning what "everybody knows" (1995:129–30).

The debate about the future of journalism and the internet too often posits an idealized image of institutional journalism so that in comparison social media news sharing is always found wanting. Institutional journalism, as "everybody knows," is about accuracy, objectivity, responsibility, and the necessary fourth estate watchdog. Finally, institutional journalism's trump card has always been its claim to be the lifeblood of democracy, the only business protected in the Constitution, and the 1st Amendment no less. In reality, these seeming truisms are mere fairytales for the naïve. Yet fairytales have power.[1]

Although social media is often criticized for its inaccuracies, lack of objectivity, and irresponsibility, it is no match for institutional journalism's reign of errors in terms of both pervasiveness and significance. The recent Boston Marathon bombings provide a dramatic example. Though the discussion board Reddit.com has been criticized after people posting on it wrongly identified a missing Brown University student as one of the bombing suspects, it is important to remember that institutional journalists also enthusiastically reported that very same "lead," even if some of them wished to blame Reddit after the fact (Greenfield, 2013). But institutional journalism made many more mistakes that could not be blamed on the supposedly corrupting influence of the internet. The *New York Post* reported that 12 people had been killed by the bombs. The paper also reported that a Saudi national had been taken into custody as a suspect. Although the *New York Post* may seem an easy target, it was not alone. NBC, CBS, and the *Los Angeles Times* all reported news of the Saudi suspect. *The Wall Street Journal* reported that five additional bombs had been found. *The New York Times* counted three unexploded bombs. It seems that nearly everyone reported that a bomb had gone off in the John F. Kennedy Library and that cell phone service had been cut off to the area. All of these "facts" reported by prominent news media organizations, some of them the very paragons of institutional journalism, were wrong.

After a couple days passed, perhaps the most embarrassing example of institutional journalism occurred. On Wednesday afternoon CNN reported an exclusive scoop that, in the words of CNN anchor John King, "a dark-skinned individual" had been arrested. Fox News and the Associated Press soon also reported the breaking news. Yet once again, institutional journalism got it wrong. As Jon Stewart memorably commented, "It's exclusive because it was completely fucking wrong!" Dean Edward Wasserman, of the Graduate School of Journalism at the University of California, Berkeley worries, "passing along what you hear but haven't bothered to check out or confirm or verify is now part of what the news media thinks they should be doing.... The standards of conversation have replaced the standards of publication" (Hohmann, 2013).

Wasserman's complaint seems to suggest that this is a recent problem, but such a claim requires amnesia. In 1996 Richard Jewell was crucified in a trial by media as the Atlanta Summer Olympics bomber. Lest we conclude that it is only bombings that institutional journalism bungles, we have recently experienced much graver examples. The calling of the 2000 U.S. presidential election for George Bush over Al Gore by institutional journalism has had truly damaging global repercussions with which we are still living. More recently, the numerous institutional journalism snafus with respect to the Iraq War should have put the final nails in the coffin of institutional journalistic credibility. The mistakes were so numerous and the cheerleading so egregious, both *The New York Times* and the *Washington Post* felt compelled to publish public apologies (Okrent, 2004; DSWright, 2013; Calderone, 2013).

The example of the Iraq War highlighted over and over again the tendency of institutional journalism to be inaccurate, lack objectivity, and be irresponsible. The Iraq War also exposed the lie that institutional journalism is a watchdog of the government. As the apologies from *The New York Times* and *Washington Post* admitted, institutional journalism tends to be an organ for orthodoxy and a puppet for power. We were reminded of such obsequious behavior just recently with the release of the report on torture from the independent, bipartisan Constitution Project Task Force on Detainee Treatment.[2] In a scathing indictment of the US government's torture practices, the task force confirmed that of course waterboarding and other forms of torture are indeed torture and that "the nation's highest officials bear some responsibility for allowing and contributing to the spread of torture." This indictment extends to the *The New York Times* and the *Washington Post* as exemplars of institutional journalism that stopped calling torture "torture" at the behest of the Bush Administration, as then executive editor of *The New York Times* Bill Keller admitted. A study by Harvard's Kennedy School of Government starkly revealed the results:

> we found a significant and sudden shift in how newspapers characterized waterboarding. From the early 1930s until the modern story broke in 2004, the newspapers that covered waterboarding almost uniformly called the practice torture or implied it was torture: *The New York Times* characterized it thus in 81.5% (44 of 54) of articles on the subject….By contrast, from 2002 – 2008, the studied newspapers almost never referred to waterboarding as torture. *The New York Times* called waterboarding torture or implied it was torture in just two of 143 articles (1.4%)…. In *The New York Times*, 85.8% of articles (28 of 33) that dealt with a country other than the United States using waterboarding called it torture or implied it was torture (Greenwald, 2010a).

After the task force report, *The New York Times* Public Editor was troubled: "language matters. When news organizations accept the government's way of speaking, they seem to accept the government's way of thinking" (Sullivan, 2013). Former *Salon* and *Guardian* blogger Glenn Greenwald has compiled a compendium of complicity between compliant institutional journalism and the government. As he concludes, "the NYT played an active and vital role in enabling the two greatest American crimes of the last decade: the attack on Iraq and the institutionalizing of a torture regime. As usual, those who pompously prance around as watchdogs over political elites are their most devoted and useful servants" (Greenwald, 2010b).

Though desperate defenders of institutional journalism at times admit to all of its flaws, they nonetheless cling to the notion that journalism is democracy's only hope. As the Newspaper Guild's slogan goes, "Democracy Depends on Journalism." It is a fine slogan, but all slogans merit suspicion. There are two weaknesses to this slogan. These weaknesses emerge out of the interpretation of the 1st Amendment put into practice. First, if we look at the 1st Amendment, it is not so much about protecting institutional journalism as it is about protecting the rights to think and share one's thoughts: "Congress shall make no law respecting an establishment of religion, or prohibiting the free exercise thereof; or abridging the freedom of speech, or of the press; or the right of the people peaceably to assemble, and to petition the Government for a redress of grievances." This is a beautiful theory of democracy: people need to think, talk to each other, and then talk to their government. In practice, it's not quite as beautiful. If "the press" is supposed to be the mechanism by which the people become informed citizens, then that mechanism is broken. As many surveys of the American people have revealed, they are woefully ill-informed about the basics of their government (PR Newswire 2009; Romano, 2011). Assuming that institutional journalism was effective, Americans would know basic information about their democracy. So if institutional journalism is the lifeblood of democracy, American democracy is anemic.

The second major problem is that the 1st Amendment tries to protect news sharing from government oppression, but the writers of the Constitution could not imagine the powers of modern-day corporations, from which there are no constitutional protections. As we mentioned earlier, America's institutional journalism is not interested in democracy, but instead is obsessed with profits and power befitting its role as a tool of large corporations. For decades now scholars have been noting the growing concentration of ownership of institutional journalism and the dangerous consequences of such concentration. Ben Bagdikian's (1983) classic, *Media Monopoly*, has been usefully updated over the years, most notably by Robert McChesney. All the updates tend to highlight ever-increasing concentration of ownership. In what he terms "the holy trinity of the global media system," McChesney highlights the extensive holdings of Time Warner, Disney, and Murdoch's News Corporation (2000:91–100).[3] In *Rich Media, Poor Democracy*, McChesney argues, "the media have become a significant *anti-democratic* force in the United States and, to varying degrees, worldwide. The wealthier and more powerful the corporate media giants have become, the poorer the prospects for participatory democracy.... if we value democracy, it is imperative that we restructure the media system so that it reconnects with the mass of citizens who in fact comprise 'democracy'" (2000:2–3). As we will argue, in an improvised manner, such a restructuring is now happening via internet platforms that enable social media news sharing. The OWS protest movement provides a window into these emergent practices.

The OWS protests provide a compelling example of the problems with a rich media.[4] Since OWS started as a fundamental challenge to the status quo corporatocracy that controls America, the dominant institutional journalism organizations predictably responded to OWS with malevolent neglect. OWS began September 17th in Zucotti Park in the Wall Street area of New York City. For its first 8 days, OWS was subject to a total major newspaper blackout. This blackout was for a protest of one of the major issues of our time in the heart of the world's financial district in the most important city in the world's only superpower. OWS finally made it onto the pages of *The New York Times* on Sunday, September 25th with "Gunning for Wall Street, With Faulty Aim" (Bellafante, 2011), but it was a modest appearance—a story from the Metropolitan Desk in Section MB. OWS finally made the privileged front page October 1st with the story "Wall Street Occupiers, Protesting Till Whenever" (Kleinfeld and Buckley, 2011). OWS fared worse in the other four major papers. *The Washington Post* did not cover OWS until October 3rd (Dobnik, 2011) and not until October 15th does OWS make the front page (Wallsten, 2011). The *Los Angeles Times* chimed in on September 30th in a front-page story that worried about the future of the movement (Susman, 2011). OWS suffered a belated October 11th arrival in the headlines of *USA Today* (Hampson, 2011). The *Wall Street Journal* was committed to ignoring the protests at its doorstep, printing no front-page stories in the first 25 days. TV news replicated this pattern.

What makes the details of this major newspaper blackout more revealing and damning is that from much earlier on the OWS protests were deemed worthy of international news coverage. On September 19th England's *Guardian* ran the story "The call to occupy Wall Street resonates around the world" (White and Lasn,

2011) and followed that up two days later with "Occupy Wall Street: the protesters speak" (Harris, 2011). The *Agence France-Presse* covered OWS's first action on September 17th: "Protesters blocked in bid to 'occupy' Wall Street" (Andrade, 2011). *China Daily* reprinted the *AFP* story on September 19th. As we will discuss later, this institutional journalism blackout in the US was more than countered by vibrant and multifaceted social media news sharing. As Castells documents in his book *Networks of Outrage and Hope*, by October 9th OWS had spread across the US, with over a thousand events occurring in all 50 states (2012:164–65).

We want to make one final point about institutional journalism and democracy. As institutional journalism writhes in its death throes, paywalls are often touted as a miracle cure. So, oddly, we have institutional journalism advocates championing journalism as the heart of democracy while simultaneously proposing paywalls as the savior of institutional journalism. The cognitive dissonance is high. Fences may make good neighbors, but paywalls do not promote participation and democracy.

A MOMENT OF SILENCE

With *The State of the News Media 2013,* The Pew Research Center's Project for Excellence in Journalism compiled the longest epitaph in the history of death, documenting the demise of institutional journalism, a digital archive of death by a thousand cuts. There were a number of striking cuts.[5] The first line reads, "In 2012, a continued erosion of news reporting resources converged with growing opportunities for those in politics, government agencies, companies and others to take their messages directly to the public." Once again, we are witnessing the decline of institutional journalism specifically, not news sharing generally. As Pew lamentably notes, "newsmakers and others with information they want to put into the public arena have become more adept at using digital technology and social media to do so on their own, without any filter by the traditional media." Print newspapers have seen advertising revenue drop over 40% and employment drop by at least 30% since 2000. Many American cities, headlined by New Orleans, no longer even have a daily print newspaper. Digital advertising goes overwhelming (72%) to six internet titans, effectively shutting out institutional journalism corporations, which are left grasping at paywalls. "*The New York Times* reports that its circulation revenue now exceeds its advertising revenue, a sea change from the traditional revenue split of as much as 80% advertising dollars to 20% circulation dollars." *Newsweek* shuttered its print edition, leaving only a wobbly *TIME* as "the only major print news weekly left standing." On TV, "Local TV audiences were down across every key time slot and across all networks in 2012.... Average revenue for news-producing stations declined by more than a third (36%) from

2006–2011." Overall, "Nearly a third of U.S, adults, 31%, have stopped turning to a news outlet." In perhaps the unkindest cut of all, the death of institutional journalism has gone largely unnoticed: "60% of the American public have heard little or nothing about the news industry's financial struggles."

NEW PRACTICES OF SOCIAL MEDIA NEWS SHARING

Technologies transform conditions of possibilities. We have no interest in getting mired in the agonizingly inane debate over technological determinism and so settle for the simple claim that media change the capacities of societies and individuals. An oral culture differs from a print culture and both differ from our mobile panmediated culture of myriad media (DeLuca, Sun, & Peeples, 2011; Angus, 2000). The notorious Marshall McLuhan, following Harold Innis (1951/2008) put it elegantly, "For the 'message' of any medium or technology is the change of scale or pace or pattern that it introduces into human affairs" (1964:24). The key point that we want to emphasize here is that a medium/technology works on scale, pattern, and pace, thus transforming structures of space, time, and speed in which we are immersed. Quite simply, reality is transformed, for far from being static, reality is always in processes of becoming (Whitehead, 1929/1979). The examples are endless. Wall Street trading now operates through the internet at beyond-human levels of speed and quantity (Adler, 2012) and US soldiers rain death via drones on villagers in Pakistan from an air-conditioned trailer in Nevada.

As the internet becomes the central organizing principle of post-human societies, social theorists have reimagined society literally and metaphorically as a network (Benkler, 2006; Castells, 2012, 2000; Kadushin, 2012; DeLanda, 2006). Yocahi Benkler compellingly demonstrates how the internet, "the move to a communications environment built on cheap processors with high computation capabilities, interconnected in a pervasive network" (2006:3), has enabled the transformation of the industrial economy into a "networked information economy" that "has made human creativity and the economics of information itself the core structuring facts" (2006:4). As Benkler and others summarize, traditional mass media tend to be centralized, one-to-many in form, commercial, professional-produced, and proprietary. Social media tend to be decentralized, many-to-many, nonmarket, peer-produced, non-proprietary, open-source platforms, commons based, and free or inexpensive in access and distribution (2006:1–32, 59–67, 209–219). Technology is central. "I place at the core of this shift the technical and economic characteristics of computer networks and information. These provide the pivot for the shift toward radical decentralization of production" (2006:18). Individuals with mobile panmediation smartphones become decentered knots of world-making that displace traditional mass media institutions as internet platforms enable individuals to reach millions of others and to engage in "effective, large-scale cooperative efforts—peer production of information, knowledge, and culture" (2006:5). Prominent examples of crowd-driven internet platforms include Twitter, Facebook, YouTube, Reddit, Vine, Tumblr, Instagram, Vimeo, Eventbrite, Wikipedia, LinkedIn, the GNU Project, Linux, Pinterest, and Yelp.

As Benkler, Castells, and others note, this radical decentralization puts individuals outside of easy institutional control, enabling efforts that bypass political and corporate authorities. Throughout *The Wealth of Networks*, Benkler celebrates how internet platforms make it possible for individuals to operate outside of commercial markets and proprietary models: "Ubiquitous low-cost processors, storage media, and networked conductivity have made it practically feasible for individuals, alone and in cooperation with others, to create and exchange information, knowledge, and culture in patterns of social reciprocity, redistribution, and sharing, rather than proprietary, market-based production. The basic material capital requirements of information production are now in the hands of one billion people around the globe who are connected to each other more or less seamlessly. These material conditions have given individuals a new practical freedom of action" (2006:462). Almost accidentally, the complex of technologies we call the internet have transformed the structure of society and created possibilities for new forms of social relations, economic practices, knowledge production, and sharing that threaten the assumptions and practices of capitalism. In describing the Arab Spring, Spain's *Indignadas*, and Occupy Wall Street, Manuel Castells champions the role of "Internet social networks, as these are spaces of autonomy, largely beyond the control of governments and corporations that had monopolized the channels of communication as the foundation of their power, throughout history" (2012:2). Benkler and Castells are not simplistic optimists. They, along with pessimist Evgeny Morozov (2011), provide somber reminders that governments and corporations will not yield control willingly, will not go gently into the good night. We have seen an alarming rise around the world in government surveillance and censorship of the internet. But this is precisely because, as Benkler, Castells, and others describe, the decentralizing technologies of the internet create opportunities for politics and activism that exceed the control of any centralized government. The OWS protests are a dramatic example.

In the first month of OWS protests, in arguably the United States' five most important newspapers, a total of 104 stories appeared (*The New York Times* 46, *Washington Post* 23, *LA Times* 17, *Wall Street Journal* 10, *USA Today* 8). On social media, OWS created a torrent of activity. According to the social analytics company PeopleBrowsr, mentions on Twitter of Occupy Wall Street were already at 4,300 on the first day of protests, 9,466 on September 25th, 25,148 on October 2nd, and 47,856 on October 14th nearly a month into the occupation protests. After three weeks, *The New York Times* acknowledged the social media activity. They cited the founder of Trendrr.com noting that on October 7th "the Twitter conversation was producing an average of 10,000 to 15,000 posts an hour on Friday about Occupy Wall Street" (Preston, 2011). The same article estimated at least 10,000 OWS videos on YouTube after three weeks. After three months that number had reached about 91,400. These numbers suggest an initial consequence of OWS' social media news sharing: institutional journalism was rendered surprisingly irrelevant. With a smartphone in her pocket, an Occupy activist camping in Zucotti Park or Salt Lake City's Pioneer Park could become a panmedia outlet, a decentered knot of video, photographs, and blogging that documents and creates and circulates the Occupy events. At the speed of the internet, events move from one person's tent to millions of people throughout the "twitterverse" and Facebook and YouTube and the world. After a photo of UC Davis Police Lieutenant John Pike pepper spraying student protesters is posted to Reddit, Pike is transformed into the global meme "Casually Pepper Spray Everything Cop."[6] Panmediation, which suggests that with the emergence of smartphones we live in and access mobile spaces of multiple media immersed in the wi-fi cloud, is a crucial process in the new social media worlds of the network society.

OCCUPY PANMEDIATION

While Benkler, Castells, and others have called attention to various new forms of participation and production enabled by the rise of internet and social media, in this final section we wish to highlight some of the particular technologies and practices of social media news sharing that were amplified by or emerged from the OWS protest movement. We note the fact that several of these technologies and practices have become increasingly common among social media news sharers and institutional journalists alike, even as they have evolved to meet the needs of new situations. Finally, we note that OWS participants not only made extensive use of existing social media technologies for recruiting, organizing, and sharing news of OWS events, but also have worked to create new social media technologies, some of which are already available for use by the next wave of protests and social media news sharing, wherever and whenever they occur.

There is, of course, a growing panoply of internet and social media technologies being used by social media news sharers, but in OWS it became clear that one technological artifact was at the center of the panmediation universe, the smartphone and its proliferating number of "apps." The smartphone emerged as a hub in a distributed network for sharing and acting on video, audio, textual, imagery, and geographic information in real time.

Though existing smartphone apps provide access to any number of social media platforms, three social media platforms have emerged in recent protest activities as particularly important. The first is the micro-blogging platform, Twitter. Twitter came to prominence as a tool for protesters to organize and spread word of their protest in the summer of 2009 with widespread and often violent protests in Iran's capitol. Some went so far as to call this event the "Twitter Revolution" (Grossman, 2009; Keller, 2010). Though this was almost certainly an overstatement of Twitter's role in that particular protest (Esfandiari, 2010), it nonetheless provided an early glimpse of the role that Twitter would come to play in later protests, especially the OWS protests. Whether Twitter and other social media really were a driving force behind the Iranian protests is less important than the widespread belief, including among some activists, that these technologies were at the heart of the protest movement. This belief provided inspiration to later protestors, especially OWS protestors, who really did make social media central to carrying out and sharing news of their activities. In OWS, Twitter, often accessed via smartphone apps, became a central platform for coordinating protest activities, for lateral communication among protest participants, but also for sharing news in various formats (text, images, videos) about the protest with a wider online audience. Of these, video emerged as one of the most important types of news content shared by protestors via the use of smartphone apps.

Thus, the second type of platform that emerged as central to OWS social media news sharing was "livestreaming," which was most often enabled by smartphone apps provided by sites such as livestream.com and ustream.com. Livestream platforms take advantage of a smartphone's video camera and connection to increasingly ubiquitous hi-speed data networks. Of course, a smartphone user can take a video with his or her camera and then later upload that video to a site like YouTube. But the user of a livestream app points the phone's video camera at the object of interest and streams *live* video to his or her profile page on the internet. The livestream emerged as one of the dominant forms of news sharing during OWS. This was especially the case when authorities took action to ground news helicopters and to prevent on-the-ground journalists access to the protests. While the major institutional journalism outlets could easily have their large news vans and video crews blocked by authorities, each protestor with a smartphone in his or her pocket could broadcast the live video that was once the exclusive preserve of corporate media outlets. In the end, far more of the OWS protests, especially incidents of police brutality, were documented via the use of smartphone-enabled livestreams, the links to which were shared with the world via Twitter, than by the cumbersome capabilities of institutional journalism (Captain, 2012; Martin, 2012; Preston, 2011).

Finally, one of the biggest disadvantages of thousands of protestors providing live updates of unfolding events with a worldwide audience is the exceedingly fragmentary nature of the accounts that they provide. "The story," if it can be said to exist at all, must be culled from millions of fragments of text, images, video, and audio. Blogging platforms, which were the oldest social media platform utilized by sharers of OWS news, provide a platform for taking these fragments and re-presenting them in narrativized form. But the move from thousands or even millions of fragments to a coherent narrative is exceedingly difficult. Thus, one of the newest platforms that emerged as crucial for social media sharers and institutional journalists covering OWS was a social media aggregation site called Storify.com (Morales, 2011). Storify allows its users to easily aggregate fragmentary bits of social media data, including tweets, Facebook status updates, video clips, images, quotes from blog posts and news stories, and more, into a single document that can then be shared via other social media, like Twitter, embedded into a blog post, or used as a dynamic notebook for writing a later news story or blog post. Storify was used extensively by OWS participants to aggregate, archive, and share their own story with the world (Storify, n.d; Stearns, n.d; Adbusters, n.d.), as well as by institutional journalists who were themselves struggling to keep up with and make sense of the flood of data produced by this fast-moving, panmediated event (Mitchell, 2011; Sonderman, 2011).

These technologies and practices are increasingly being used by social media news sharers and institutional journalists alike and are evolving to meet the needs of new situations as they emerge. In 2013, NPR journalist Andy Carvin published a book, *Distant Witness*, documenting his coverage of the Arab Spring by relying on social media posts, primarily Twitter, from on-the-ground participants across the Middle East (Carvin, 2013).

The OWS protestors, however, did not just make novel use of existing smartphone apps and web platforms. They worked to create new tools for protest and social media news sharing. Such projects have included the creation of apps to help protestors avoid police blockades called "kettling," for smoothing the flow of lateral communication among protestors, for enabling the "people's microphone" (a human voice-powered call-and-response technique used to circumvent bans on megaphones), for notifying friends and family in the event of one's arrest, and much, much more (Gere, 2011; Mack, 2011; Connor, 2011; Messieh, 2011). Hackers participating in the OWS movement also created new sites for aggregating multiple livestreams (Kingkade, 2011), new social media platforms specially attuned to the needs of protest movements (Captain, 2011), and "mesh networks" providing impromptu digital networks at protest sites (Kessler, 2011).

In the age of social media, protest movements like OWS do not just rely on new technologies and practices of news sharing, they are also hotbeds of innovation where the future of social media news sharing is being created. Although the corpse of institutional journalism continues to haunt discussions of the future of news, the emerging practices of social media news sharing are transforming the landscape of news beyond the imagination and control of the political and corporate authorities of a moribund modernism, inventing creative, cooperative, and decentralized news processes for a world of networks.

NOTES

1. Kaufhold et al. (2010). is a perfect illustration of this delusional dynamic.
2. See http://www.constitutionproject.org/task-force-of-detainee-treatment/ and specifically http://www.constitutionproject.org/pdf/Task_Force_on_Detainee_Treatment_Mission_and_Members.pdf, accessed 4/28/2013.
3. Columbia University's School of Journalism also provides extensive information on the concentration of media ownership. See http://www.cjr.org/resources/, accessed 4/28/2013.
4. This discussion is extended in DeLuca et al. 2012.
5. See http://stateofthemedia.org, accessed 4/28/2013.
6. See http://knowyourmeme.com/memes/casually-pepper-spray-everything-cop, accessed 4/28/2013.

REFERENCES

Adbusters (n.d.). May Day Updates on #OCCUPYWALLSTREET. *Storify*. Retrieved from http://storify.com/adbusters/may-day-updates-on-occupywallstreet

Adler, Jerry (2012). Raging Bulls: How Wall Street Got Addicted to Light-Speed Trading *Wired Business*. August 3. http://www.wired.com/business/2012/08/ff_wallstreet_trading/.

Agence France Presse (2004, 12 August). Leading US Daily Admits Underplaying Stories Critical of White House Push for Iraq War. *Agence France Presse*. Retrieved from http://www.commondreams.org/headlines04/0812-01.htm

Andrade, M. (2011). Protesters Blocked in Bid to 'Occupy' Wall Street. *Agence France Presse*, 17 September, Available at: http://www.google.com/hostednews/afp/article/ALeqM5jwB_zy0tcSXCUegKjGvlfkRmt7UAok

Angus, Ian (2000). *Primal Scenes of Communication: Communication, Consumerism, and Social Movements*. Albany: State University of New York Press.

Atherton, Kelsey (2013). "Cyber Attacks Are America's Top Security Threat. That's Better News Than It Sounds." *Popular Science*. March 14. http://www.popsci.com/technology/article/2013-03/cyber-attacks-were-named-top-security-threat-%E2%80%99s-better-news-it-sounds.

Badiou, Alain (2012). *The Rebirth of History: Times of Riots and Uprisings*. Translated by Gregory Elliott. 1st ed. New York: Verso.

Bagdikian, B. H. (1983). *The media monopoly*. Boston: Beacon Press.

Bellafante, G. (2011). Gunning for Wall Street, With Faulty Aim. *The New York Times*, 23 September, Available at: http://www.nytimes.com/2011/09/25/nyregion/protesters-are-gunning-for-wall-street-with-faulty-aim.html

Benkler, Y (2006). *The Wealth of Networks*. New Haven: Yale University Press.

Calderone, M. (2013, 24 March). Washington Post Defends Not Running Article on Iraq Media Failure. *Huffington Post*. Retrieved from http://www.huffingtonpost.com/2013/03/24/washington-post-iraq-media-failure_n_2944227.html

Captain, S. (2011, 27 December). Occupy Geeks Are Building a Facebook for the 99%. *Wired Threat Level*. Retrieved from http://www.wired.com/threatlevel/2011/12/occupy-facebook/all/1

Captain, S. (2012, 6 January). Livestreaming Journalists Want to Occupy the Skies with Cheap Drones. *Wired Threat Level*. Retrieved from http://www.wired.com/threatlevel/2012/01/occupy-drones/

Carvin, A. (2013). *Distant Witness*. New York: CUNY Journalism Press.

Castells, Manuel (2000). *The Rise of the Network Society: The Information Age: Economy, Society and Culture*. New York: Wiley.

Castells, M. (2000). Materials for an Exploratory Theory of the Network Society. *British Journal of Sociology* 51:1, pp 5–24.

Castells, M. (2012). *Networks of outrage and hope: Social movements in the internet age*. John Wiley & Sons.

Connor, T. (2011, 24 October). Occupy Wall Street Sympathizer Creates "I'm Getting Arrested" App to Help Protesters. *New York Daily News*. Retrieved from http://www.nydailynews.com/new-york/occupy-wall-street-sympathizer-creates-arrested-app-protesters-article-1.963970.

Deibert, Ronald, and Rafal Rohozinski (2010). *Access Controlled: The Shaping of Power, Rights, and Rule in Cyberspace*. Combridge, MA: MIT Press.

DeLanda, Manuel (2006). *A New Philosophy of Society: Assemblage Theory and Social Complexity*. 1st ed. New York: Continuum.

Deleuze, Gilles (1995). *Difference and Repetition*. New York: Columbia University Press.

DeLuca, K., Y. Sun, & J. Peeples (2011). Wild Public Screens and Image Events from Seattle to China: Using Social Media to Broadcast Activism Beyond the Confines of Democracy. In *Transnational Protests and the Media* Ed. by Cottle and Lester. New York: Peter Lang.

DeLuca, Kevin M., Sean Lawson, and Ye Sun (2012). Occupy Wall Street on the Public Screens of Social Media: The Many Framings of the Birth of a Protest Movement. *Communication, Culture & Critique* 5(4): 483–509. doi:10.1111/j.1753-9137.2012.01141.x.

Dobnik, V. (2011). Occupy Wall Street' Protests Grow. *The Washington Post*, 3 October, p. A3.

DSWright (2013, 25 March). Bill Keller: New York Times Published Bad, Unsourced, Unskeptical Stories in Run Up to Iraq War. *Firedog Lake*. Retrieved from http://news.firedoglake.com/2013/03/25/bill-keller-new-york-times-published-bad-unsourced-unskeptical-stories-in-run-up-to-iraq-war/

Esfandiari, G. (2010, 7 June). The Twitter Devolution. *Foreign Policy*. Retrieved from http://www.foreignpolicy.com/articles/2010/06/07/the_twitter_revolution_that_wasnt

Gere, D. (2011, 31 January). Sukey Apps Help Protesters Avoid Police Kettles. *Wired UK*. Retrieved from http://www.wired.co.uk/news/archive/2011-01/31/sukey-protest-app

Greenfield, R. (2013, 19 April). How Reddit Fueled the Scanner-Happy Media to Out Innocent Boston "Suspects." *The Atlantic Wire*. Retrieved from http://www.theatlanticwire.com/technology/2013/04/reddit-police-scanner-innocent-boston-suspects/64384/

Greenwald, G. (2010a, 30 June). New Study Documents Media's Servitude to Government. *Salon.com*. Retrieved from http://www.salon.com/2010/06/30/media_258/

Greenwald, G. (2010b, 3 July). Bill Keller's Self-Defense on "Torture." *Salon.com*. Retrieved from http://www.salon.com/2010/07/03/keller_2/

Grossman, L. (2009, 17 June). Iran Protests: Twitter, the Medium of the Movement. *Time*. Retrieved from http://www.time.com/time/world/article/0,8599,1905125,00.html

Hampson, R.I., (2011). Anti-Wall Street Protests Face Question: Now What? *USA Today*, 10 October, Available at: http://www.usatoday.com/NEWS/usaedition/2011-10-11-1Aprotests-CV_CV_U.htm

Harris P, (2011). Occupy Wall Street: The Protesters Speak. *The Guardian*, 21 September, Available at: http://www.guardian.co.uk/world/blog/2011/sep/21/occupy-wall-street-protests

Hohmann, J. (2013, 18 April). Media Shrug at Boston Blunders. *Politico*. Retrieved from http://www.politico.com/story/2013/04/boston-marathon-media-mistakes-90304.html

Homshaw, P. (2011, 17 November). Syrian Protesters Use iPhone App Souria Wa Bas to Fight Government. *appolicious.com*. Retrieved from http://www.appolicious.com/articles/10231-syrian-protesters-use-iphone-app-souria-wa-bas-to-fight-government

Innis, Harold A. (1951/2008). *The Bias of Communication*. 2nd Edition. Toronto: University of Toronto Press, Scholarly Publishing Division.

Jacques, Martin (2012). *When China Rules the World: The End of the Western World and the Birth of a New Global Order: Second Edition*. Revised. New York: Penguin Books.

Kadushin, Charles (2012). *Understanding Social Networks: Theories, Concepts, and Findings*. New York: Oxford University Press, USA.

Kaufhold, K., Valenzuela, S., & Gil de Zuniga, H. (2010). Citizen Journalism and Democracy. *J&MC Quarterly* V. 87, # ¾ 2010, pp 515–529. Accessed at: http://www.academia.edu/242384/Citizen_Journalism_and_Democracy_How_User-Generated_News_Use_Relates_to_Political_Knowledge_and_Participation

Keller, J. (2010, 18 June). Evaluating Iran's Twitter Revolution. *The Atlantic*. Retrieved from http://www.theatlantic.com/technology/archive/2010/06/evaluating-irans-twitter-revolution/58337/

Kessler, S. (2011, 14 November). How Occupy Wall Street Is Building Its Own Internet. *Mashable*. Retrieved from http://mashable.com/2011/11/14/how-occupy-wall-street-is-building-its-own-internet-video/

Kleinfeld, N.R., & Buckley, C. (2011). Wall Street Occupiers, Protesting Till Whenever. *The New York Times*, 30 September, p. A1.

Kingkade, T. (2011, 26 December). New Protest Apps Crowd-Sourced from Occupy Wall Street Hackers. *Huffington Post*. Retrieved from http://www.huffingtonpost.com/2011/10/26/occupy-wall-street-a-diy-tech-tools-protest_n_1032518.html

Mack, E. (2011, 26 October). Shouty App Helps Occupy Protesters Be Heard. *CNET News*. Retrieved from http://news.cnet.com/8301-17938_105-20126132-1/shouty-app-helps-occupy-protesters-be-heard/

Martin, A. (2012, 3 January). Occupy Wall Street's Livestream Operators Arrested. *The Atlantic Wire*. Retrieved from http://www.theatlanticwire.com/national/2012/01/occupy-wall-streets-livestream-operators-arrested/46921/

McLuhan, Marshall (1964). *Understanding Media: The Extensions of Man*. New York: Signet Books.

McChesney, R. (2000). *Rich Media, Poor Democracy*. New York: The New Press.

Messieh, N. (2011, 3 November). Protest4: A Mobile App for Connecting Activists. *The Next Web*. Retrievedfromhttp://thenextweb.com/apps/2011/11/03/protest4-a-mobile-app-for-connecting-activists/

Mitchell, J. (2011, 17 November). How Storifying Occupy Wall Street Saved the News. *ReadWrite Web*. Retrieved from http://readwrite.com/2011/11/17/how_storifying_occupy_wall_street_saved_the_news_o

Midler, Paul (2009). *Poorly Made in China: An Insider's Account of the Tactics Behind China's Production Game*. New York: John Wiley and Sons.

Morales, C. (2011, 22 November). Storify Seen as Innovative Tool to Thwart "Media Blackout" of Social Movement Coverage. *Journalism in the Americas Blog*. Retrieved from https://

knightcenter.utexas.edu/blog/storify-seen-innovative-tool-thwart-media-blackout-social-movement-coverage

Morozov, E. (2011). "Whither Internet Control?" *Journal of Democracy* 22(2): 62–74.

Morozov, Evgeny (2012). *The Net Delusion: The Dark Side of Internet Freedom*. Reprint. New York: PublicAffairs.

Okrent, D. (2004, 30 May). Weapons of Mass Destruction? Or Mass Distraction? *The New York Times*. Retrieved from http://www.nytimes.com/2004/05/30/weekinreview/the-public-editor-weapons-of-mass-destruction-or-mass-distraction.html?pagewanted=all&src=pm

Preston, J. (2011, 11 December). Occupy Video Showcases Live Streaming. *The New York Times*. Retrieved from http://www.nytimes.com/2011/12/12/business/media/occupy-movement-shows-potential-of-live-online-video.html?_r=1&

PR Newswire (2009, 2 December). 83 Percent of U.S. Adults Fail Test on Nation's Founding. *PR Newswire*. Retrieved from http://www.prnewswire.com/news-releases/83-percent-of-us-adults-fail-test-on-nations-founding-78325412.html

Romano, A. (2011, 20 March). How Dumb Are We? *The Daily Beast*. Retrieved from http://www.thedailybeast.com/newsweek/2011/03/20/how-dumb-are-we.html

Sonderman, J. (2011, 4 November). Journalists Exchange Tips on Covering Local Occupy Wall Street Protests. *Pynter.org*. Retrieved from http://www.poynter.org/latest-news/mediawire/152148/journalists-exchange-tips-on-covering-local-occupy-wall-street-protests/

Stearns, J. (n.d.). Tracking Journalist Arrests at Occupy Protests Around the Country. *Storify*. Retrieved from http://storify.com/jcstearns/tracking-journalist-arrests-during-the-occupy-prot

Storify (n.d.). Here's What Happened At Occupy The News: A Movement and the Media. *Storify*. Retrieved from http://storify.com/storify/tune-in-thursday-for-occupy-the-news-a-movement-and-the-media

Sullivan, M. (2013, 12 April). "Targeted Killing", "Detainee" and "Torture": Why Language Choice Matters. *The New York Times*. Retrieved from http://publiceditor.blogs.nytimes.com/2013/04/12/targeted-killing-detainee-and-torture-why-language-choice-matters/

Susman, T. (2011). Occupy Wall Street Protesters Driven By Varying Goals. *Los Angeles Times*, 29 September, Available at: http://www.latimes.com/news/nationworld/nation/la-na-wall-street-protest-20110930%2C0%2C6859500.story

The New York Times (2004, 26 May). The Times and Iraq. *The New York Times*. Retrieved from http://www.nytimes.com/2004/05/26/international/middleeast/26FTE_NOTE.html

Wallsten, K. (2011). Obama Looks to Harness Anti-Wall St. Angst. *The Washington Post*, 15 October, p. A1.

Whitehead, Alfred North (1979). *Process and Reality*. 2nd ed. New York: Free Press.

White, M., & Lasn, K. (2011). The Call to Occupy Wall Street Resonates Around the World. *The Guardian*, 19 September, Available at: http://www.guardian.co.uk/commentisfree/cifamerica/2011/sep/19/occupy-wall-street-financial-system

CHAPTER TWENTY-SIX

The Activist AS Citizen Journalist

SUE ROBINSON AND MITCHAEL L. SCHWARTZ

Making use of the digital technologies that allow them to bypass traditional journalism, communities' most active citizens such as nonprofit directors, civil rights advocates, volunteers, officials, and lobbyists tap into "citizen journalistic" techniques to argue, persuade and connect with constituents. Activists can blog, tweet, comment, and post on social-media sites like Facebook and YouTube, developing large followings and connecting with previously untapped audiences. This chapter explores exactly how citizen activists are turning to self-generated journalism to help control the information flow around their niche interests. Using a case study of a hot-button educational issue in the Midwestern capital city of Madison, Wisconsin, in the United States, we take as our thesis that activists are employing acts of "citizen journalism" to amplify causes, network and bridge power structures in this city. Understanding how that happened in a specific case study can help reveal the intricate ways in which today's new "citizen journalist" has matured.

To document this maturity, we chose the K-12 (primary and secondary education, namely from Kindergarten to Grade 12) minority achievement gap as a way to access a singular public discourse with active citizen bloggers and Facebook posters as well as a well-covered issue in the media during our study period. The minority achievement gap in the mid-sized, Progressive city was one of the worst in the United States; African American and Latino students woefully trailed their white counterparts at all education levels in areas such as reading test scores. Half the black males in Madison were failing to graduate high school in four years – compared to 88% of whites in the town. In September 2011, the Urban League of Greater Madison had won a planning grant to develop a publicly funded charter school called Madison Preparatory Academy that would educate African American and Latino boys in grades 6–12. The proposal immediately generated controversy, and a year of intense public debate ensued—borne out in the pages of the local news outlets as well as in education blogs, Facebook Group pages and other social media.

We wanted to know: What roles are activist citizen journalists adopting in this information exchange about Madison Prep and the minority achievement gap in this city, what techniques do they use, and what relationship do they have with journalism and journalists? To offer a response to these questions, we interviewed two-dozen "citizen journalists," primarily activists, bloggers and other public-domain writers on the K-12 minority achievement gap in addition to a dozen reporters who covered the issue during this time. We offer a small portion of that data in this chapter. Specifically, we examine the communicative practices of two opposing high-profile activists during the year that Madison Prep was proposed and the aftermath of its demise from Sept. 1, 2011, until Sept. 1, 2012. Identifying these two citizen journalists as "super-contributors," we approach their case-study contributions by calling on actor-network theory to explore how they are fitting into (and helping to reformulate) the overall media ecosystem around a specific issue.

CITIZEN JOURNALISM AND CITIZEN ACTIVISTS

In the first volume of *Citizen Journalism: Global Perspectives*, Allan and Thorsen (2009) and their contributing authors noted that regular people are working alongside the U.S.'s non-partisan journalists to produce newsworthy information via mobile content, photos, wikis, and other platforms. In doing so, citizens amplify alternative voices (Carpentier et al. 2009; Woo Young 2009), bypass media (Thorsen 2009), and offer a more authentic sense of reality (Zayyan and Carter 2009). At first scholars only considered those people who were writing blogs as true "citizen journalists," but by 2009, many scholars were categorizing just about anything in any blog, commenting space or tweet as such (Deuze et al. 2007; Lewis et al. 2010; Robinson and DeShano 2011; Rosenberry and St. John 2010). In 2010, Jan Schaffer urged people to rename them "new media makers" who commit "random and organized 'acts of journalism.'… We do a disservice to emerging players in the new media ecosystem and to our own understanding of what's evolving by lumping them all under one rubric" (Schaffer 2010, p. 177). In Allan's (2013) accounting of "citizen witnessing," he argued that citizens gain voice via interactive media to be alternative, protest, bear witness, break news, or comment on the world. All of these acts offer citizens new ways to negotiate their citizenship, he wrote. Many scholars have investigated activists' "citizen journalism" to resist oppressive regimes (Hamdy 2009; Tapas 2011; Yang 2009) or to fuel grassroots movements (Chadwick 2006; Karpf 2012; Kreiss 2012).

Burt (2000) thought of these citizen-journalist activists as "network entrepreneurs" who fill "structural holes" (Burt 1992) in networks, often by linking institutions and citizens. Bruns et al. (2009) called them "super-contributors" (2009: 204). Super-contributors are so successful because of the way they network communicatively such as circulating information from mainstream publications. Thus, Bruns et al. (2009, p. 204) argued that:

> successful [news] services will be those that take special care to establish good relations with the relatively small number of users who are regular or prolific contributors—in a very real sense, these super-contributors are even more vital to the health of citizen journalism communities than staff are.

ACTOR NETWORK THEORY AND "SUPER-CONTRIBUTORS"

The framing theory of exploration for this data is called Actor Network Theory (ANT), which was formalized in the late 80s by three sociologists—Latour, Callon and Law. Latour (2007) defined it as an alternative sociological theory about associations that eschewed macro social structures in favor of understanding the *relationships* through which action occurs and power is created. Scholars using the ANT approach must understand the *relational materiality* of one's words alongside their work and in context with their relationships to people, data, ideologies etc. Turner (2005) presented ANT as an ideal theoretical lens for analyzing the impact of digital media on journalistic practices; several media scholars have applied ANT to journalistic endeavors (Anderson 2013; Boczkowski 2004; Hemmingway 2008; Howard 2002; Weiss and Domingo 2010).

One way to explore an ANT approach is by slicing off a tiny piece of a network, one or two nodes, and investigate their interactions with other agents as well as with the material that contribute to a process of meaning-making (Callon 2005, 1986). Callon (1986) suggested that in any given network problematization (such as a minority achievement gap), particularly strong actors will opt into "a system of alliances and associations" with both people and entities to become an "obligatory passage point" (8). As an OPP—which operates via structural holes—an actor intentionally sets out to mobilize disparate groups toward some common goal.[1]

In this chapter we shall take two central nodes in our chosen network of the K-12 educational system and the minority achievement gap issue public in Madison, Wis. Essentially we want to investigate how these actors performed as citizen journalists to amplify their positions in the entire network—to become, in other words, obligatory passage points. Our chosen actors were opposing community activists:

1) Kaleem Caire, a civil-rights activist who is head of the Urban League of Greater Madison and the founder of the Madison Prep proposal and who rallied his social networks on Facebook

2) TJ Mertz, a Progressive blogger interested in improving the school system in general but working against the Madison Prep proposal.

What role are they playing as actors in the information network around the minority achievement gap? How are they using citizen journalism to negotiate their citizenship as Allan (2013) might describe? What are the characteristics of these "acts of journalism"—as Schaffer labels them? First, we document the way in which Caire and Mertz perform in this information network around the minority achievement gap issue.[2] Then we document how their "acts of journalism" work in the information-exchange network before concluding the chapter.

KALEEM CAIRE: CIVIL RIGHTS ACTIVIST AND FACEBOOK POSTER

Kaleem Caire has been the president and CEO of the Urban League of Greater Madison (ULGM), a nonprofit that serves minority interests, since 2010. A Madison native, Caire worked on these issues for 10 years in Washington, D.C., where he founded the Black Alliance for Educational Options and for which he commissioned the nation's first comprehensive study of high school graduation rates (Ginsberg-Schutz 2010). In 2010, Caire served as one of 45 reviewers for an education initiative by U.S. President Barack Obama called Race to the Top, before returning to Wisconsin with his wife and five kids. He became a prominent local figure with his proposal of Madison Prep, which he hoped would help close the growing academic chasm between white and minority students. But Madison Prep needed the approval of Madison Metropolitan School District's Board of Education (BOE), which was made up of eight elected citizens. At the time of the proposal, the BOE comprised one minority (an African American man), mostly Progressive lay people without an education background, and two active bloggers. The proposal did not need a general populace vote, but it did need broad public support for the cash-strapped BOE to approve.

A major source for local journalists,[3] Caire supplemented his message in three local listservs (all with a minority-community focus), Facebook (both his personal one, which he keeps public, and the public ULGM group page), YouTube and emails. During the period of our study, he posted 94 times on the equity gap on Facebook, had 3,760 Facebook friends and 1,600 likes on the ULGM page, and posted seven Madison Prep videos on the ULGM YouTube channel. The *School Information System* blog posted Caire's e-mails in their entirety; Caire or his writings were mentioned or sourced in nearly 400 articles, comments, blogs or posts during this time period.

Caire carefully considered each of his writings to make sure it maintained his work's empowering tones. "Am I conveying the message that I want people to hear? Am I helping them learn? Am I challenging them in a way that will get a result?" He sometimes reconsidered and deleted posts that did not serve these goals, such as one critical of Gov. Scott Walker, after publishing. His dozens of posts stimulated lively discussions: Of his 94 Facebook posts, 13 had 10–39 comments; 5 had 40–69 comments, and 4 had 70+ comments. "I put a question out there or a provocative statement and see what people say…I just use social media to take advantage of figuring out what people really think." His Facebook posts also promoted the cause such as this one on Sept. 28, 2011: "Click on the following link to sign our petition today and tell the Madison School Board you support Madison Prep." In this way, he used his platforms to spread information, increase support, build coalition and augment networks.

Yet, his methods did not always succeed in building the necessary relational linkages. For example, after publishing a post critical of BOE member Marj Passman (which he did to push the conversation, he said), she became alienated. Caire asked for a meeting to repair the relationship, but Passman refused. He used the anecdote to philosophize about his activism techniques, noting that social media can only be effective with offline interactions.

> So I wrote back to her, "I have feelings about things that have been said and done, and I know you do too. But I would like to talk with you." People hide behind their social media. People hide behind their friends. People hide behind newspaper articles. But won't talk to you.

Caire believed that connections between citizens and institutions are often neglected, which makes addressing community problems like achievement gaps problematic.

Despite Passman's alienation, Caire did become an obligatory passage point for this network: Every single one of those interviewed named him as a key influencer or an information source on this issue. Caire prioritized relationship-based action as the foundation to vibrant community:

> Leaders of Madison…they've been impatient about the pace of the problem, and they're like, "Where is this going? All they're doing is having conversations?" And I'm like, "You guys, you've gotta have relationships in order to get anything done."

As a key information source in mainstream news, blogs and forums, Caire created connections between his constituency and the broader Madison population, as well as between platforms. He was Facebook "friends" with most of the other prominent activists, school officials, major funders and others engaged in this achievement gaps issue in town. In turn, Caire hoped his work empowered constituents by bringing them into local institutional and communicative networks. He viewed his primary role as a community bridge. He used journalistic techniques to fill a "network structural node" (Burt 1992) on his way to becoming that obligatory passage point.

TJ MERTZ: PROGRESSIVE BLOGGER

Mertz became aware of the gap in the mid-1990s when he moved to town as a doctoral student in education with a specialty in racial history in school systems. With two kids in the Madison school system, Mertz started attending every school board meeting and participating in an early collaborative blog in the early 2000s named *School Information System* (SIS). In 2006 he was asked to be part of a citywide committee, called the Equity Task Force, to look at the gap and propose some solutions to the school district. After a couple years, few of the recommendations were adopted but the original members remained active and Mertz kept in touch with many of the members. Around the same time, Mertz stopped posting to SIS after a disagreement with the moderator over a headline. In February 2007, he began his own blog called *Advocating for Madison Public Schools* (AMPS). By 2013, AMPS had more than 200 followers with the average blog post on the gap attracting 500–700 hits and a handful of comments.

> What influence I have had comes from my reputation and my relationships, but there is no institutional power inherent in what I am doing, and I have no budget. Still, I can bring out 20 people to a BOE meeting. I can build coalition. That's what I aim to do. I believe in the power of information strongly.

During the year of the Madison Prep debate that included various budget proposals to cut programs, Mertz rallied activists he knew to speak in favor of programs like the Boys and Girls Club. His blog posts appeared regularly on several local news sites such as WisOpinion.com, national blogs such as the Big Education APE and SeattleEducation2010 as well as within a few popular Twitter hashtag conversations such as #sosmarch. In addition to the blog posts, Mertz was prolific on his own Facebook page and a public Facebook group that had formed in 2011 to stop a charter-school bill but was still active. He also regularly commented on other people's blogs, such as one run by BOE member Ed Hughes. By the end of the year of our study, Mertz had posted 17 blog entries and 249 Facebook posts about the achievement gaps, building a "curriculum" of content hammering at the single point about equity.

He viewed AMPS and his commenting in various citizen spaces throughout Madison's virtual community as an opportunity to be an activist but also to add "complexity" and to "offer alternative understandings." His blogging and commenting encompassed an obligation he felt to the Madison community; for Mertz, his activities in the public realm were part of a *process* of civic engagement:

> Complex issues go in one ear and out the other but when you have it in written form you can go back and respond. You can frame issues. You can influence discussion in quantity. You can keep hammering on your talking point and offering new evidence and maintaining the same themes. You can catch one person's attention with this and they share it, and it can go viral.

The materiality of the blog posts represented an opportunity to enhance his own role as a potential obligatory passage point in this particular network around the minority achievement gap. Every post represented not just Mertz' civic fulfillment, but also a chance for him to act as an agent in other people's engagement. Thus, Mertz also actively positioned himself via his blog as a structural node that aimed to bridge communities and enhance public debate about important educational issues.

Indeed, soon after he began AMPS, Mertz said he started getting emails from parents about their kids' mistreatment or their classroom's lack of resources. These exchanges—often made in confidence—led him to be measured in his blog: "I have to keep my relationships." Everything that Mertz wrote, he did so with his network in mind. He "worked" not only his networks but those he was not part of. As one example, during a controversial plan to open a 4-year-old kindergarten in an economically challenged part of town called Allied Drive, Mertz blogged about the proposal's chances and:

> then it started bothering me a lot. And so I reached out some to some friends of mine who were in Allied Drive...And they then reached out to people from the Allied Drive neighborhood associations and got them involved. And we fought and we won. And they ended up with two centers instead of one. But that was all because of these relationships.

Mertz described many such times when his production work online prompted offline civic action. His blog became a facilitating node between "regular people" like parents and the institutional school officials. He was often slipped obscure reports and hidden data that became points of agency in the issue during this time period.

But to act successfully as this structural node, Mertz adapted what he called an "academic style." His average blog post contained dozens of links to fact-based pieces of evidence that became agents in this information exchange as well. For example, one post in which he stated, "I'm not going do as many hyperlinks to sources as I usually do," still contained 17 links in 2,730 words.

> You want to give people the opportunity to sift through the evidence themselves. It also indicates an intellectual journey you took. It gives you some kind of authority, that "this is not just what TJ says; he has evidence behind it." I don't want them to not take my word for it. I want conversations around it to start.

The reports came from think tanks and academic journals, news articles and Progressive web sites. He took an advanced regression analysis class so he could analyze reports better. He said he strove for balance and is careful about language, talking about the "equity of educational opportunity" so as not to marginalize the issue, for example.

And yet while Mertz' blog—often cited as an information influencer in our interviews—occupied a structural hole in these education networks, his attempts to negotiate a position as an obligatory passage point did not fully succeed. Mertz' virtual communities consisted of mostly fellow Progressives who knew each other personally. Those he listed as influencers tended to be institutional officials such as BOE members, fellow academics or other liberal activists. He lamented that his style and known Progressive-centric readership could sometimes be intimidating. "Some people will contact me in private and say, 'I want to talk to you about that,' and I say, 'Well why don't you post a comment and we can have a public conversation.' They don't want to do that." Some of what he wrote about—race, class—generated reticence within public discussion. And sometimes, written online discussion engendered misunderstandings or defensiveness. In a September 8, 2011, post, he came out in his blog against Madison Prep, and he followed it up with half a dozen detailed posts. These moves lost him his relationship with Caire, who sent him a private email following the December 2011 vote asking him to "stop saying you know me. You don't know me," essentially severing him from Caire's network. These realities of disempowerment, history, politics, ideology, and also personalities combined with the posts and emails offer "a complex web of interrelations" (Callon 1986, p. 4) that ultimately inhibit Mertz and other activists from achieving the kind of mass acceptance for their writings they would like. Citizen-journalist activists must constantly confront and diffuse the tension that inevitably arises with agenda-driven content. This evidence demonstrates why the maintenance of communicative networks is so key for them to achieve their goals.

"ACTS OF JOURNALISM" AS ACTS OF NETWORKING

In our 278 mainstream news articles, Caire showed up as a source or mention in 122 of them, compared to 12 times for Mertz. Both Caire and Mertz reported close relationships with the reporters who covered the issue. Though both posted links to local newspaper stories, neither Mertz nor Caire mentioned reporters as influential sources. In general, both bypassed media—rarely commenting on news articles and preferring their own methods of communication.

In their content-production protocol, however, both Caire and Mertz performed like journalists: talking to multiple sources, fact-checking rumors, and finding documents. Caire noted that although ULGM has occasionally had interns to help with research, he did his own research by following newspapers, academic research, and listservs. Mertz described the reporting process of a blog post:

> "I know this person overreacts on these hot-button issues… but I honestly know that they were informed; that there's a greater truth here." So then you start asking around other people and find out what the greater truth is. Sometimes, I'll go to the documents if they're there. I'll often ask an administrator, "Can I get the documentation on what is program implementation here?" And I'll see if it matches. That doesn't mean that I necessarily trust the documents either. But then you start to get a fuller picture. I've got both parents and teachers and also community members, grandparents… I know a lot of people all over town. "Oh that's so-and-so's neighborhood. That's so-and-so's school." And just kind of reach out. "Hey what's going on with this? I heard some stuff."

This "reaching out" is constant network maintenance for both Mertz and Caire, trying to become indispensible within the information flow. Mertz responded to up to 25 emails a day on this issue typically within a few hours. He talked to 5–10 people a week on the phone—both institutional sources and "regular people" as well as activists and friends—during this period of time about this issue. Though he sometimes emailed or called a reporter, he preferred to post with a link.

Caire chose Facebook posting or "working" community events to "reach out" and stressed the importance of face-to-face relationships. He said that he has met 90% of his 3,760 Facebook friends, and had close personal or professional relationships with 60% of them. He regularly walked through local neighborhoods, going door-to-door to elicit citizen opinions—working his "beat" like a reporter. And then he commented on those communal interactions online, providing the factual material to make the points he heard in the streets, as in this July 12, 2012, post: "According to the US Justice Department, in their report 'Homicide Trends in the United States 1980–2008,' whites still commit the majority of crimes in the USA but Blacks lead in the rate of victims of homicides…" With each post, Caire hoped to provoke discussion between networks, make influential nodes in other networks not at the events or in the visited communities aware, and, ultimately, bring about change. Every like, post, link, document, source became used as an agent in determining the direction of the flow of information. The digital ability to report, frame and disseminate (and then argue, persuade and re-frame) allowed these citizen-journalist activists to alter the conversation.

CONCLUSION

One major thesis in the first volume of *Citizen Journalism: Global Perspectives* proffered that citizen journalists do not act in a vacuum, but are creating a new kind of "journalism as social networking" (Bruns et al. 2009). For example, Bruns et al. (2009) demonstrated a hyper interconnectedness between the citizen-journalist "super-contributors" and the peace march organizers they studied. This chapter's actor-network analysis has shown that this kind of communicatively networked embeddedness is so powerful, the journalists in this information network had become ancillary to the whole media ecosystem for this issue—still part of it but operating as a peripheral rather than a central node, at least for activists who are citizen-journalist "super-contributors." This represents a major shift from traditional information flows in which journalists occupied central nodes in information-flow networks while activists had to jockey for press attention (Shoemaker & Reese 1995).

This suggests that if we rely on citizens to conduct all of society's necessary information gathering (as opposed to professional journalists), we elevate the activist network while diminishing the power of journalistic dissemination and influence dispersed throughout our communities. This could be deleterious if we desire *collective* and *objectively informed* deliberation. Ultimately neither Mertz nor Caire with their activism could fully bridge disparate communicative networks in the way they needed to. Though both expressed the desire to communicate with those they disagreed with, both operated in agenda-driven social and professional circles governed by homophily. During the yearlong public debate, both Mertz and Caire lost key network relationships through their public writings (such as Caire's fallout with a BOE member and Mertz' fallout with Caire). And although he was a top influencer among our data, Caire failed to attain the broad public support he needed and the BOE ultimately voted down the Madison Prep proposal.

That said, both men used their private and public connections and platforms to offer alternative perspectives and illuminate important problems that their "journalism" uncovered. These two activist "super-contributors" hoped that through various strategies of public-content contribution they could push a victory on their issue by linking both inside and outside their networks. This could also be thought about in Actor Network Theory terms as "the work of mediation," which "must be done at every moment to restore or maintain the links between actors" (Harman, 2009, 116). For Caire, face-to-face networking prompted provocative posting; for Mertz, online-offline "reporting" compelled his link-laden posts. Both were careful in their public writings, always mindful of their relationship maintenance. And, perhaps most importantly, with each writing, the activist became more active, using the "citizen-journalism" contribution to civically engage and to propel others to be more involved. This is the key finding for this chapter and an important note about how information exchange is shifting empowerment towards civic engagement via online journalistic production by citizens.

Obviously this chapter has only touched the surface of actor-network transformation within the issue of the minority achievement gap. One limitation of using ANT in this way, besides lacking the space to do such a mapping justice, is the impossibility of truly capturing such a complex network with a descriptive technique. Through this kind of analysis, however, we were able to document the relationships around two activists attempting to be the filler in Madison's "structural holes" (Burt 2005) to become "obligatory passage points" (Callon 1986) through citizen journalism. By telling the story of Caire and Mertz, we have shown how local media ecologies are evolving to accommodate the online activist—leaving out the journalist in some information circulatory patterns. More research should delve into how the entire community of citizen online contributors is changing media ecosystems as social-media platforms—and techniques of citizen journalism—mature. It is important that we consider which communities various citizen journalists are contributing to, how that content is being circulated, whom they are influencing and what all that interaction means for knowledge outcomes.

NOTES

1. Special thanks to Chris Anderson of CUNY for the idea to apply Callon's "obligatory passage points" concept to this data.
2. The data in this chapter derived from a two-year-long study of Madison, WI's public discussion around the minority achievement gap. We collected and coded 278 news articles, 161 blog entries, and 847 Facebook posts that mentioned the gap in some way. We coded these for authors, sources (both primary and secondary for the author), and use of evidence such as URLs and reports. We then interviewed 32 of the people authoring this content. We asked questions about their motivations, standards, and general practices of public authorship and asked them about specific posts or incidents as well. In this chapter, we used frequency data but also employed ANT as an analytical framework. We read through the transcripts along with the articles and posts to make linkages and follow how content spurred more content or how offline incidents prompted online, public discussion.
3. As found in both the content and the interviews.

REFERENCES

Allan, S., 2013. *Citizen witnessing: Revisioning journalism in times of crisis (Key concepts in journalism).* Cambridge: Polity.

Allan, S., and Thorsen, E. (eds) 2009. *Citizen journalism: Global perspectives.* New York: Peter Lang.

Anderson, C. 2013. *Rebuilding the news: Metropolitan journalism in the digital age.* Philadelphia: Temple University Press.

Boczkowski, P. 2004. The process of adopting multimedia and interactivity in three online newsrooms. *Journal of Communication,* 54 (2), 197–213.

Bruns, A., Wilson, J., and Saunders, B. 2009. Citizen journalism as social networking: Reporting the 2007 Australian federal election. In: Allan, S., and Thorsen, E., eds. *Citizen journalism: Global perspectives.* New York: Peter Lang, 197–207.

Burt, R. 2000. The network entrepreneur. In: Swedberg, R., ed. *Entrepreneurship: The social science view.* Oxford: Oxford University Press, 281–307.

Burt, R. 1992. *Structural holes: The social structure of competition.* Cambridge, MA: Harvard University Press.

Callon, M. 1986. Some elements of a sociology of translation: Domestication of the scallops and the fishermen of St Brieuc Bay. In: Law, J., ed. *Power, action and belief: A new sociology of knowledge?* London: Routledge, 96–223.

Callon, M. 2005. Why virtualism paves the way to political impotence: A reply to Daniel Miller's critique of the laws of the markets. *Economic Sociology: European Electronic Newsletter,* 6 (2), 3–20.

Carpentier, N., De Brabander, L. and Cammaerts, B. 2009. Citizen journalism and the North Belgian Peace March. In: Allan, S. and Thorsen, E., eds. *Citizen journalism: Global perspectives.* New York: Peter Lang, 163–174.

Chadwick, A. 2006. *Internet politics: States, citizens, and new communication technologies.* New York: Oxford University Press.

Deuze, M., Bruns, A. and Neuberger, C. 2007. Preparing for an age of participatory news. *Journalism Practice,* 1 (3), 322–338.

Ginsberg-Schutz, M. 2010. Kaleem Caire: Change agent. *Isthmus,* 23 December 2010. Available from: http://www.thedailypage.com/isthmus/article.php?article=31675 [Accessed 21 June 2013].

Hamdy, N. 2009. Arab citizen journalism in action: Challenging mainstream media, authorities and media laws. *Westminster Papers in Communication and Culture,* 6 (1), 92–112.

Harman, G. 2009. *Prince of networks: Bruno Latour and metaphysics.* Melbourne: re.press books.

Hemmingway, E. 2008. *Into the newsroom: Exploring the digital production of regional television news.* New York: Routledge.

Howard, P., 2002. Network ethnography and the hypermedia organization: New media, new organizations, new methods. *New Media & Society,* 4 (4), 550–574.

Karpf, D. 2012. *The MoveOn effect: The unexpected transformation of American political advocacy.* New York: Oxford University Press.

Kreiss, D. 2012. *Taking our country back: The crafting of networked politics from Howard Dean to Barack Obama.* New York: Oxford University Press.

Latour, B. 2007. *Reassembling the social: An introduction to actor-network-theory.* New York: Oxford University Press.

Lewis, S. Kaufhold, K. and Lasorsa, D. 2010. Thinking about citizen journalism: The philosophical and practical challenges of user-generated content for community newspaper editors. *Journalism Practice,* 4 (2), 163–179.

Robinson, S., and DeShano, C. 2011. "Anyone can know": Citizen journalism and the interpretive community of the mainstream press. *Journalism,* 12 (8), 963–982.

Rosenberry, J., and St. John, B. 2010. *Public journalism 2.0: The promise and reality of a citizen engaged press.* New York: Routledge.

Schaffer, J. 2010. Open source interview: Civic and citizen journalism's distinctions. In: Rosenberry, J., and Burton, St. John, eds., *Public Journalism 2.0: The promise and reality of a citizen-engaged Press.* New York: Routledge, 176–182.

Shoemaker, P., and Reese, S. 1995. *Mediating the message.* New York: Longman Trade.

Tapas, R. 2011. The "story" of digital excess in revolutions of the Arab Spring. *Journal of Media Practice,* 12 (2), 189–196.

Thorsen, E. 2009. Blogging the climate change crisis from Antarctica. In: Allan, S., and Thorsen, E., eds. *Citizen journalism: Global perspectives.* New York: Peter Lang, 107–120.

Turner, F. 2005. Actor-networking the news. *Social Epistemology,* 19 (4), 321-324.

Weiss, A., and Domingo, D. 2010. Innovation process in online newsrooms as actor networks and communities of practice. *New Media & Society,* 12 (7), 1156–1171.

Woo Young, C. 2009. OhmyNews: Citizen journalism in South Korea. In: Allan, S., and Thorsen, E., eds. *Citizen journalism: Global perspectives.* New York: Peter Lang, 1143–152.

Yang, G. 2009. *The power of the Internet in China: Citizen activism online.* New York: Columbia University Press.

Zayyan, H. and Carter, C. 2009. Human rights and wrongs: Blogging news of everyday life in Palestine. In: Allan, S., and Thorsen, E., eds. *Citizen journalism: Global perspectives.* New York: Peter Lang, 85–94.

List of Contributors

Karina Alexanyan is a researcher and analyst with the Berkman Center for Internet and Society at Harvard University, USA, as well as with the Social Media Research Foundation. Karina's research focuses on global social media, with an emphasis on Russia. She received her PhD in communications from Columbia University, her MA in communication from NYU and BA in linguistics and modern languages (French and Russian) from the Claremont Colleges.

Stuart Allan is Professor of Journalism and Communication in the Cardiff School of Journalism, Media and Cultural Studies at Cardiff University, UK. He has authored several books, including *Citizen Witnessing: Revisioning Journalism in Times of Crisis* (Polity Press 2013). His recent edited books include *The Routledge Companion to News and Journalism* (Routledge 2012). Currently, he is conducting research examining the uses of digital imagery in news reporting, while also writing an alternative historiography of war photography.

Mary Angela Bock is Assistant Professor of Journalism at the University of Texas at Austin, USA. A former television journalist, her research interests include visual rhetoric and photojournalistic practice. She authored *Video Journalism: Beyond the One-Man Band* (Peter Lang 2012), and co-edited *The*

Content Analysis Reader (Sage 2008) with Klaus Krippendorff. Her work has also appeared in *Visual Communication Quarterly*, *International Journal of Press/Politics*, and the *Encyclopedia of Journalism*.

Lilie Chouliaraki is Professor of Media and Communications at the London School of Economics, UK. She has published extensively on the mediation of suffering, humanitarian communication and war and conflict reporting, as well as on Discourse Theory and Analysis. She is the author/editor of seven books, including *The Spectatorship of Suffering* (Sage 2006), *Self-Mediation: New Media, Citizenship and Civic Selves* (Routledge 2012) and *The Ironic Spectator: Solidarity in the Age of Post-humanitarianism* (Polity Press 2013).

Kayt Davies is Senior Lecturer in Journalism at Edith Cowan University in Perth, Western Australia. Trained in journalism via cadetship, she has worked in international news in London, as a business journalist in Perth, and she has edited community newspapers and national magazines. She has a psychology degree and her PhD was an ethnographic study of editors. Her recent academic research interests include independent media in West Papua, the interface of journalism and research ethics, journalism as a research methodology, and International Humanitarian Law.

Kevin Michael DeLuca is Professor in the Department of Communication, University of Utah, USA. Kevin's scholarship explores how media enable and constrain forms of activism and how certain events interrupt the world as it is and offer possibilities for alternative futures. He is the author of dozens of essays and the book *Image Politics: The New Rhetoric of Environmental Activism* (Routledge 2005). In a world confronting catastrophic ecological and social problems, he treasures the space to think and engage the issues of our times.

Nik Gowing has been a main presenter for the BBC's international 24-hour news channel BBC World News, since 1996. He has presented The Hub with Nik Gowing, BBC World Debates, Dateline London, plus location coverage of major global stories. Previously he worked at ITN, where he was bureau chief in Rome and Warsaw, and Diplomatic Editor for Channel Four News (1988–1996). In 1994 he was a fellow at the Joan Shorenstein Barone Center in the J. F. Kennedy School of Government, Harvard University. In 2008–9 he was a visiting fellow at the Reuters Institute at Oxford University, for whom he wrote *Skyful of Lies and Black Swans*.

Chris Greer is Professor of Sociology and Co-Director of the Centre for Law, Justice and Journalism at City University London, UK. His current research develops the concepts of 'scandal' and 'trial by media' to analyse the transforming relations between the news media and institutional power in a context of digital proliferation and market volatility, and a culture of political mistrust. He is founding and current co-editor of *Crime, Media, Culture: An International Journal* (Sage).

Lei Guo is a PhD candidate in the School of Journalism, University of Texas at Austin, USA. Her research focuses mainly on alternative and citizen media, the development of agenda setting theory, and international communication. Her work has been published or is forthcoming in a number of leading peer-reviewed journals such as *Journal of Broadcasting & Electronic Media*, *International Communication Gazette* and *Journal of Computer-Mediated Communication*.

Yasmin Ibrahim is Reader in International Business and Communications at Queen Mary, University of London, UK. Her ongoing research on new media technologies explores the cultural dimensions and social implications of the appropriation of ICTs in different contexts. Beyond new media and digital technologies, she writes on political communication and political mobilisation from cultural perspectives. Her other research interests include globalization, Islam, visual culture and memory studies.

Yomna Kamel is a media researcher pursuing her PhD at the Media School of Bournemouth University, UK. She began her media career as a journalist with UAE's daily Gulf News. Her work includes writing for Arabic and English publications, in addition to teaching media studies at various universities in Egypt and the UAE. She holds a BA in Mass Communication and Political Science from the American University in Cairo and an MA in Journalism Studies from Cardiff University. Her research interests focus on comparative news coverage, media and activism and contemporary Arab media.

Trevor Knoblich is a project director, consultant and author with Frontline SMS in the USA, an award-winning software company that builds tools for managing text messages and data (winner of a 2011 Knight News Challenge award). Trevor serves as Director for Media Projects, advising organizations on issues of journalism, mobile technology, and citizen participation in news. He has worked as a federal policy reporter in Washington, DC, and has served as a global humanitarian response coordinator. He writes regularly about the nexus of international journalism, mobile technology, and the use of platforms for data collection.

Sean Lawson is Assistant Professor in the Department of Communication, University of Utah, USA. He writes about the relationships among science, technology, and security with an emphasis on new media, information, and communication technologies. Topics of interest include cybersecurity policy, surveillance, network-centric warfare, and military use of social media. He is author of *Non-linear Science and Warfare: Chaos, Complexity and the US Military in the Information Age* (Routledge 2013).

Lisa Lynch is Associate Professor of Journalism at Concordia University in Montreal, Quebec, Canada. Her research is broadly situated at the intersection of technology, culture, and political change. Since 2008, she has been researching and writing about Wikileaks, and has published articles on the site in publications ranging from *Digital Journalism* to *Radical History Review*. She is currently at work on a monograph that explores how journalism negotiates the ever-increasing boundary skirmishes between traditional, institutional sites of facticity and newer, contingent sites of authority.

Donald Matheson is Senior Lecturer in Media and Communication at the University of Canterbury, New Zealand. Donald is the author of *Media Discourses* (Open University Press 2005) and co-author with Stuart Allan of *Digital War Reporting* (Polity Press 2009). He co-edits the journal *Ethical Space: The International Journal of Communication Ethics*. He writes on news discourse, communication ethics and journalistic practices, with a particular emphasis on emerging digital textual forms.

Eugene McLaughlin is Professor of Criminology at City University London, UK, and a member of its Centre for Law, Justice and Journalism. His current research and publications concentrate on: trial by media and media justice; the criminology of scandal; and the policing of risk societies. His most recent books are *Criminological Perspectives* (Sage, 3rd edition, 2013), *The Sage Handbook of Criminological Theory* (Sage 2013) and *The Sage Dictionary of Criminology* (Sage, 3rd edition, 2012).

Graham Meikle is Professor of Social Media at the University of Westminster, UK. His authored books include *Social Media: Cultures and Politics of Sharing* (Routledge 2013), *Media Convergence: Networked Digital Media in Everyday Life*, co-authored with Sherman Young (Palgrave Macmillan 2012), *Future Active: Media Activism and the Internet* (Routledge 2002) and *Interpreting News* (Palgrave Macmillan 2009). He is also co-editor, with Guy Redden, of *News Online: Transformations and Continuities* (Palgrave Macmillan 2011).

Mette Mortensen is Associate Professor in the Department of Media, Cognition and Communication, The University of Copenhagen, Denmark. She has written the monograph *Kampen om Ansigtet: Fotografi og Identifikation*

(*Facial Politics: Photography and Identification*, 2012) and co-edited several books and special issues of journals. She has published articles in *The International Journal of Cultural Studies*, *Global Media and Communication*, and *Digital Journalism*. She is currently writing *Eyewitness Images and Journalism: Digital Media, Participation, and Conflict* to be published by Routledge.

Last Moyo is a Senior Visiting Fellow at the Midlands State University in Zimbabwe. He previously lectured in Media Studies departments at the University of Witwatersrand, UN-Mandated University of Peace Programme (East Asia), and National University of Science and Technology, Zimbabwe. His research interests are in digital media cultures and practices, critical political economy, and comparative media systems.

Lindsay Palmer is a PhD candidate in Film and Media studies at the University of California, Santa Barbara, USA. Before attending UCSB, Lindsay worked as a television and online news producer. Her current research focuses on media industries, digital technologies and practices, citizen journalism, and feminist theory. She has published in several peer-reviewed journals, including *Genders*, *Feminist Review*, *Television and New Media*, and *Continuum*.

Kristina Riegert is Professor in Media and Communication Studies at the Department of Media Studies, Stockholm University, Sweden. Most of her research has been comparatively oriented, focusing on factors important for how media represent the world such as globalisation, propaganda, war and crisis situations, national and regional identities and journalist cultures. Her abiding interest in Lebanon and the MENA-region developed into a project with Arabist Gail Ramsay on the nature and impact of top bloggers in three Arab mediascapes, looking at issues pertinent to cultural citizenship and digital media use.

Sue Robinson is Associate Professor in the School of Journalism and Mass Communication, University of Wisconsin-Madison, USA. Sue researches online journalism, collective memory and the news, information flow and authority, and civic engagement. She has served as the editor of a special issue for *Journalism Practice* on 'Community Journalism Midst Media Revolution,' authored a monograph called *Journalism as Process* in *Journalism and Communication Monographs*, and has published in journals such as *Journal of Communication*, *New Media & Society* and *Mass Communication & Society*.

James Rodgers is Lecturer in Journalism at City University London, UK. He is the author of *Reporting Conflict* (Palgrave Macmillan 2012) and *No Road Home: Fighting for Land and Faith in Gaza* (Abramis 2013). Before becoming an academic and author, he spent twenty years as a journalist: five for Reuters

Television, and fifteen for the BBC where he worked as a reporter, editor, producer, and presenter. He spent most of his BBC career (1995–2010) as a foreign correspondent, completing postings in Moscow, Gaza, and Brussels.

Clemencia Rodríguez is a Professor in the Department of Communication at the University of Oklahoma, USA. Her book *Fissures in the Mediascape: An International Study of Citizens' Media* (Hampton 2001) developed her 'citizens' media theory' to better understand the role of community/alternative media in our societies. She continues to explore how people living in the shadow of armed groups use community radio, television, video, digital photography, and the internet to shield their communities from armed violence, including in her book *Citizens' Media Against Armed Conflict: Disrupting Violence in Colombia* (University of Minnesota Press 2011).

Mitchael L. Schwartz is a PhD student in the School of Journalism and Mass Communication, University of Wisconsin-Madison, USA. His dissertation research investigates interactions between digital technologies and community activism in the communication ecologies of nonprofit groups and members. His work with Madison Commons explores the position of civic communication within contemporary network society.

Einar Thorsen is Senior Lecturer in Journalism and Communication at the Media School, Bournemouth University, UK, where he is also Convenor of the Journalism Research Group. His research focuses on online journalism, particularly during crisis and conflicts, and in response to political and environmental change. He co-edited the first volume of *Citizen Journalism: Global Perspectives* (Peter Lang 2009), and has published research on BBC News Online, Wikinews and WikiLeaks.

Neil Thurman is Senior Lecturer at City University London, UK. He has worked with interactive media since the early 1990s, including working for Thomson Corporation and Granada Plc. His experience ranges across technologies from interactive video discs, through CD-Roms and the web, to mobile apps. His research has been covered by publications including *The Wall Street Journal*, *Le Figaro*, and *The Guardian*.

Firuzeh Shokooh Valle is a PhD student in Sociology at Northeastern University, USA, and the Spanish Language Editor of the citizen media publication Global Voices Online. Her primary research interests focus on the intersection between technology and society in Latin America and the Caribbean. She is also a recognized journalist who has reported for numerous media outlets in her native Puerto Rico and abroad.

Kush Wadhwa is Senior Partner at Trilateral Research & Consulting in the UK. He provides independent, non-partisan advisory services with respect to emerging technologies in security, surveillance, ICT and health issues. He focuses, in particular, on issues of strategic policy development related to the future direction of privacy, data protection and surveillance, as well as other emerging applications.

Silvio Waisbord is Professor in the School of Media and Public Affairs at George Washington University, USA. His recent books include *Reinventing Professionalism: Journalism and News in Global Perspective* (2012), and the edited volume *Media Sociology: A Reappraisal* (2013), both published by Polity Press, and *Vox Populista* published by Gedisa in 2013. He is editor-in-chief of the *International Journal of Press/Politics*. He earned a PhD in sociology from the University of California, San Diego.

Hayley Watson is an Associate Partner at Trilateral Research & Consulting LLP, a niche research and advisory consultancy bringing together strategy, technology and policy in the UK. She is author of a number of articles relating to the development of citizen journalism in society as well as the impact of citizen journalism on the publicity of terror. Her main area of expertise includes: the role of technology in relation to security, the development of citizen journalism, and the role of social media in crisis management.

Index

Simon Cottle, *General Editor*

From climate change to the war on terror, financial meltdowns to forced migrations, pandemics to world poverty, and humanitarian disasters to the denial of human rights, these and other crises represent the dark side of our globalized planet. They are endemic to the contemporary global world and so too are they highly dependent on the world's media.

Each of the specially commissioned books in the *Global Crises and the Media* series examines the media's role, representation, and responsibility in covering major global crises. They show how the media can enter into their constitution, enacting them on the public stage and thereby helping to shape their future trajectory around the world. Each book provides a sophisticated and empirically engaged understanding of the topic in order to invigorate the wider academic study and public debate about the most pressing and historically unprecedented global crises of our time.

For further information about the series and submitting manuscripts, please contact:

Dr. Simon Cottle
Cardiff School of Journalism
Cardiff University, Room 1.28
The Bute Building, King Edward VII Ave.
Cardiff CF10 3NB
United Kingdom
CottleS@cardiff.ac.uk

To order other books in this series, please contact our Customer Service Department at:

(800) 770-LANG (within the U.S.)
(212) 647-7706 (outside the U.S.)
(212) 647-7707 FAX

Or browse online by series at:

www.peterlang.com